Advances in Experimental Medicine and Biology

Advances in Experimental Medicine and Biology provides a platform for scientific contributions in the main disciplines of the biomedicine and the life sciences. This series publishes thematic volumes on contemporary research in the areas of microbiology, immunology, neurosciences, biochemistry, biomedical engineering, genetics, physiology, and cancer research. Covering emerging topics and techniques in basic and clinical science, it brings together clinicians and researchers from various fields.

Advances in Experimental Medicine and Biology has been publishing exceptional works in the field for over 40 years, and is indexed in SCOPUS, Medline (PubMed), Journal Citation Reports/Science Edition, Science Citation Index Expanded (SciSearch, Web of Science), EMBASE, BIOSIS, Reaxys, EMBiology, the Chemical Abstracts Service (CAS), and Pathway Studio.

2019 Impact Factor: 2.450 5 Year Impact Factor: 2.324

More information about this series at http://www.springer.com/series/5584

Heide Schatten
Editor

Ovarian Cancer: Molecular & Diagnostic Imaging and Treatment Strategies

Springer

Editor
Heide Schatten
Department of Veterinary Pathobiology
University of Missouri-Columbia
COLUMBIA, MO, USA

ISSN 0065-2598 ISSN 2214-8019 (electronic)
Advances in Experimental Medicine and Biology
ISBN 978-3-030-73358-2 ISBN 978-3-030-73359-9 (eBook)
https://doi.org/10.1007/978-3-030-73359-9

This Springer imprint is published by the registered company Springer Nature Switzerland AG
The registered company address is: Gewerbestrasse 11, 6330 Cham, Switzerland

Preface

Ovarian cancer is the second most common gynecological cancer in women, and it still has the highest mortality rate being the fifth-leading cause of cancer death among women in the United States. This is in part related to the fact that ovarian cancer is still the most difficult to diagnose in early stages of the disease, as early tumor biomarkers are not yet available and this cancer often has no symptoms at the early stages. However, because of the urgency to achieve progress in early diagnosis and more effective ovarian cancer treatment, encouraging new research has been initiated on several levels to combat the disease.

While general treatment of ovarian cancer includes cytoreductive surgery and chemotherapy that typically employs platinum-based drugs and taxanes, personalized medicine plays an increasingly important role to combat this heterogeneous disease which is based on new detailed insights on cell and molecular levels. Basic research, improved imaging modalities as well as new clinical trials have opened up new avenues to treat the heterogeneous disease with new possibilities of patient-specific approaches. While only modest progress has been made in early detection of the disease, treatment of advanced stages of ovarian cancer has resulted in steady progress, and intense efforts are underway to understand cell migration, epithelial–mesenchymal transition points, and metastasis on genetic, cell, and molecular levels that have become possible with newly developed research methods.

The advent of molecular and genomic technologies has significantly improved our understanding of the biological processes underlying ovarian cancer which has been enhanced by the Cancer Genome Atlas that has identified mutations in human ovarian cancer genomes that may play a role in tumor progression and modifications of cellular metabolism. Targeted therapies are now available to inhibit specific signaling pathways that are aberrant in ovarian cancer cell populations, and we are now able to image signaling molecules with specific markers in live cells in culture. Progress has also been made in designing nanoparticles that offer the potential for imaging and targeted ovarian cancer treatment. The joint initiatives and efforts of advocate patients, ovarian cancer survivors, basic researchers, statisticians, epidemiologists, and clinicians with various and specific expertise have allowed close communication for more specific and targeted treatment. Major forces supporting these efforts are the Department of Defense, the American Cancer Society, and several other Foundations that recognized the need for intensified advocacy to find treatments for the disease that represents an under-studied area of research. Multi-modal approaches are oftentimes required to manage ovarian cancer and achieve positive outcomes which require patient-specific evaluation and analysis for specific management.

This book on molecular and diagnostic imaging and treatment strategies of ovarian cancer is one of two companion books with the second one being focused on cell and molecular biology of ovarian cancer. Both books include new and exciting aspects of ovarian cancer research with chapters written by experts in their respective fields who contributed their unique expertise in specific ovarian cancer research areas and include cell and molecular details that are important for the specific subtopics. Comprehensive and concise reviews are included of key topics in the field. Cutting edge new information is balanced with background information that is readily understandable for the newcomer, ovarian cancer patients, and for the experienced ovarian cancer researcher alike. Chapters include current and futuristic roadmap of ovarian cancer management; epithelial ovarian cancer and cancer stem cells; therapeutic strategies to overcome immune suppression; pharmacological effects of natural components against ovarian cancer and mechanisms; modeling the early steps of ovarian cancer dissemination in an organotypic culture of the human peritoneal cavity; PAX8, an emerging player in ovarian cancer; single cell RNA sequencing of ovarian cancer: promises and challenges; enforced expression of METCAM/MUC18 decreases in vitro motility and invasiveness and tumorigenesis and in vivo tumorigenesis of human ovarian cancer BG-1 cells; 3D models for ovarian cancer; and ovarian cancer stem cells: characterization and role in tumorigenesis.

It is a great pleasure and timely to edit this book on molecular and diagnostic imaging and treatment strategies of ovarian cancer depicting areas in ovarian cancer that have impacted new treatment strategies. I am most grateful to the outstanding contributors for sharing their unique and specific expertise with the scientific community. My sincere thanks to all for their most valuable contributions.

COLUMBIA, MO, USA Heide Schatten

Contents

1 Current and Futuristic Roadmap of Ovarian Cancer Management: An Overview

Orlandric Miree, Sanjeev Kumar Srivastava, Santanu Dasgupta, Seema Singh, Rodney Rocconi, and Ajay Pratap Singh

1.1 Introduction

Ovarian cancer (OC) is an ominous diagnosis for women and, irrespective of slightly decreasing trends in incidence and mortality, holds its rank as the deadliest gynecological malignancy worldwide [1]. Globally, it accounted for 295,414 new diagnoses and 184,799 deaths in 2018, of which an estimated 22,240 diagnoses and 14,070 patient deaths occurred in the United States alone as per an estimate of the American Cancer Society [2, 3]. Various potential risk factors associated with OC development include age at menarche and menopause, parity and infertility, lactation, use of contraception, pre-existing benign gynecologic conditions, prior gynecologic surgery, hormone replacement therapy, obesity, diet, nutrition, exercise and physical activity, and environmental factors [4]. Due to the lack of effective methods for early diagnosis and a largely asymptomatic progression, OC is often diagnosed at an advanced stage limiting the options for the treatment. OC has a complex biology driven through multiple genetic alterations as well as overlapping and compensatory pathways that makes its clinical management extremely challenging. Further, OCs develop therapeutic resistance against existing therapies contributing to the poor survival of patients [5]. Over the years, we have improved our understanding of OC genetics and biology, which has provided hope for an advancement in detection procedures and therapeutic options. In the following sections, we summarize these aspects and discuss the pertinent literature. We also discuss about existing barriers hindering progress and provide our perspective on the future of OC clinical management.

O. Miree · S. K. Srivastava
Department of Pathology, College of Medicine, University of South Alabama, Mobile, AL, USA

Department of Oncologic Sciences, Mitchell Cancer Institute, University of South Alabama, Mobile, AL, USA

S. Dasgupta
Department of Medicine, The University of Texas Health Science Center at Tyler, Tyler, TX, USA

S. Singh · A. P. Singh (✉)
Department of Pathology, College of Medicine, University of South Alabama, Mobile, AL, USA

Department of Oncologic Sciences, Mitchell Cancer Institute, University of South Alabama, Mobile, AL, USA

Department of Biochemistry and Molecular Biology, College of Medicine, University of South Alabama, Mobile, AL, USA
e-mail: asingh@health.southalabama.edu

R. Rocconi
Division of Gynecologic Oncology, Mitchell Cancer Institute, University of South Alabama, Mobile, AL, USA

H. Schatten (ed.), *Ovarian Cancer: Molecular & Diagnostic Imaging and Treatment Strategies*, Advances in Experimental Medicine and Biology 1330, https://doi.org/10.1007/978-3-030-73359-9_1

1.2 Ovarian Cancer: Origin and Histological Subtypes

Ovarian cancer is a heterogeneous disease characterized by phenotypic and molecular variations. The vast majority of ovarian tumors (benign and malignant) develop from one of the three ovarian cell types: epithelial cells, sex cord-stromal cells, and germ cells. Tumors originated from epithelial cells account for around 90% of the malignant cases followed by 5–6% from sex cord-stromal cells and less than 3% from germ cells [6, 7]. As discussed below, we further divide these tumors into different subtypes based on differences in their development, presentation, and behaviors (Fig. 1.1).

1.2.1 Epithelial Ovarian Cancers

Epithelial OCs (EOCs) are the most common and aggressive form of OC that is further classified into five major histological subtypes: high-grade serous, low-grade serous, endometrioid, clear cell, and mucinous. These histological subtypes differ in their prevalence and molecular features. Advances in the understanding of molecular characteristics of the histological subtypes have led to further grouping of the subtypes into a dualistic system—Type I or Type II (Fig. 1.2).

1.2.1.1 Type I Epithelial Ovarian Cancer

Type I EOCs have defined precursor lesions, infrequently harbor TP53 mutation, and are generally indolent and slow-growing. They also usually present at a younger age. Type I carcinomas include endometrioid, clear cell, mucinous, and low-grade serous carcinomas [8]. Endometrioid carcinomas arise from cells that resemble cells of the endometrium. They represent about 15% of all OCs and most are graded as either grade I or II with as low as 5% being grade III [9]. Malignant transformation of endometriosis has been associated with endometrioid carcinomas [10, 11]. Clear cell carcinomas account for about 10% of ovarian tumors and, much like endometrioid carcinomas, they are also associated with endometriosis. Mucinous carcinomas account for less

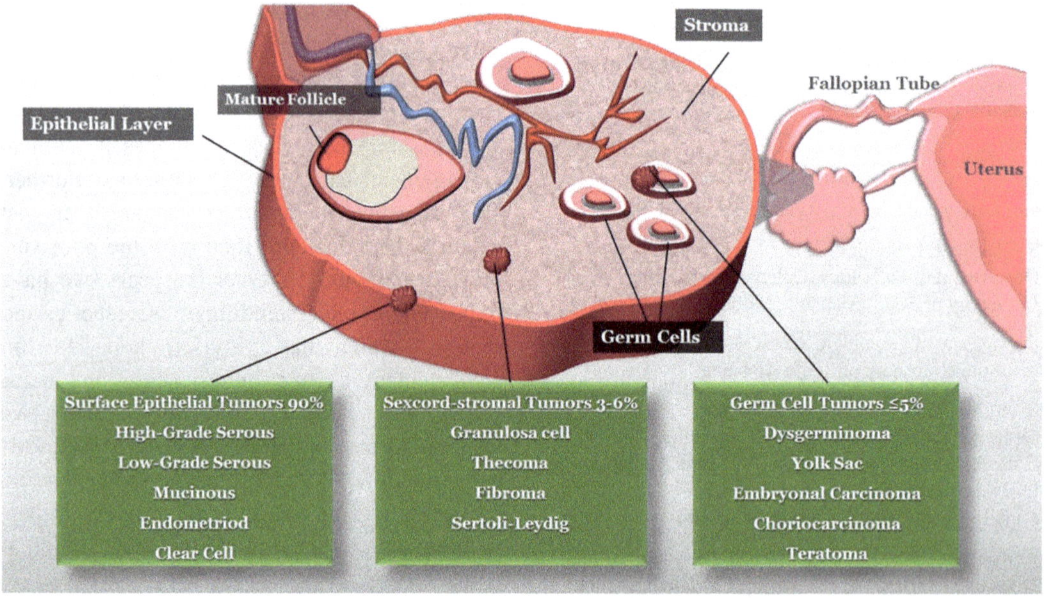

Fig. 1.1 Cellular origin of ovarian tumors. Ovarian cancers originate from three cell types and are named accordingly. Most common subtype is epithelial ovarian cancer, which originates from the surface epithelial cells. Other two less common subtypes, sex cord-stromal and germ cell tumors, originate from stromal and germ cells, respectively

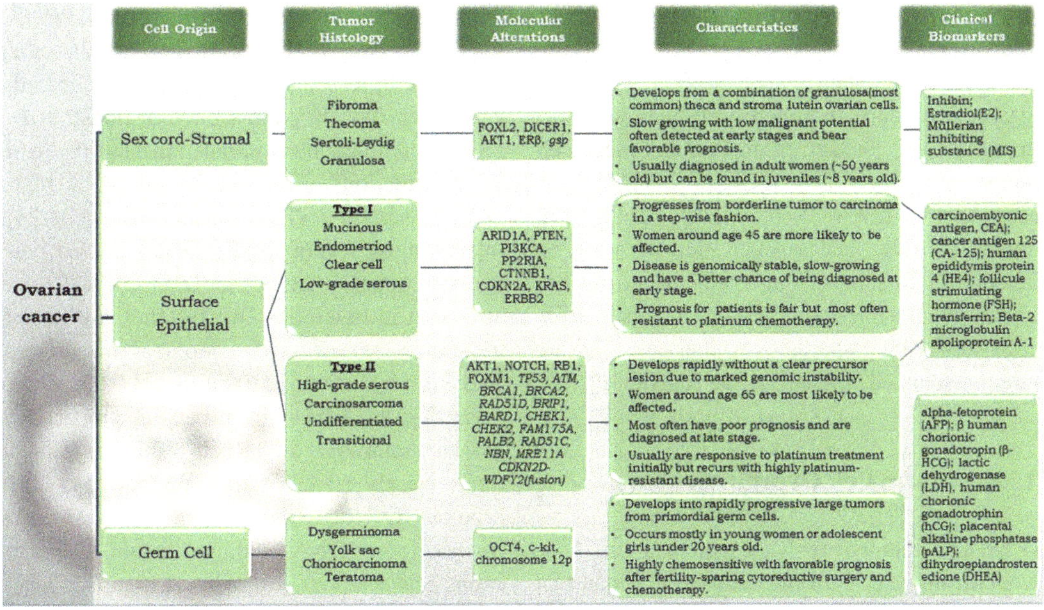

Fig. 1.2 Histological and molecular heterogeneity of ovarian cancers. Ovarian cancers vary significantly in their histological features and the molecular alterations that they carry. This also leads to their differential clinical presentation and prognosis

than 5% of all ovarian carcinomas and can be either intestinal or Mullerian type depending on the type of cells present in the tumor [9]. Intestinal mucinous carcinomas are the most common type and many of them contain goblet and/or Paneth cells. They are also usually presented with a thick mucus. Low-grade serous carcinomas account for less than 5% of ovarian carcinomas and are slow-growing that are thought to arise from benign cystadenoma [9]. These tumors, unlike high-grade serous carcinomas, are characteristic for the absence of nuclear atypia and harbor few genetic mutations (including a wild-type p53) and are described as genetically stable [11, 12]. Both low-grade serous and mucinous carcinomas develop in a stepwise manner from benign to borderline tumors [11, 12].

1.2.1.2 Type II Epithelial Ovarian Cancer

Type II carcinomas are clinically aggressive with unknown precursor lesions. In addition, they usually present at an older age and almost ubiquitously possess mutant TP53 [8]. Type II carcinomas include high-grade serous as well as undifferentiated carcinoma and carcinosarcoma. The spectrum of aberrations in type II carcinomas differs significantly from the histotypes grouped in Type I Carcinomas. High-grade serous carcinomas have the worst prognosis due to their aggressive nature and are usually diagnosed at an advanced stage. Unlike low-grade serous carcinomas, this histotype is almost always tubal in origin with no clear precursor lesion and thought to develop de novo instead of a stepwise function. These carcinomas are markedly more sensitive to platinum drugs than those in Type I group [8].

1.2.2 Sex Cord-Stromal Ovarian Cancers

Ovarian tumors of sex cord-stromal cells are believed to derive from stromal (fibroblast, theca, and Leydig cells) and primitive ovarian sex cord cells (granulosa and Sertoli cells) [13]. They can either be composed of a singular cancerous cell type or a combination of different cell types [14]. The vast majority of sex cord-stromal tumors are characterized by their endocrine functions since their cells produce androgens, estrogens, and corticoids [9, 15]. In fact, granulosa and theca cell tumors are often described as being hyper-estrogenic, whereas Sertoli and Leydig cell tumors are described as hyper-androgenic [15]. In contrast, the most commonly occurring sex cord-stromal tumors, fibromas, arise from collagen-forming spindled stromal cells and are rarely associated with endocrine function [9, 15].

1.2.3 Germ Cell Ovarian Cancers

Germ cell tumors are heterogeneous and arise from primitive cells of the embryonic gonad that are destined to form eggs. The most common types of germ cell tumors are dysgerminomas and yolk sac tumors, and both tumor types occur most frequently at younger ages (premenopause). They have nearly 100% 5-year survival rates if detected early [7, 13, 16]. Other forms of ovarian germ cell tumors include embryonal carcinoma, choriocarcinoma, polyembroma, mixed germ cell, and teratomas [16]. Interestingly, germ cell tumors are known to harbor abnormalities in chromosme12p (isochromosome 12p and/or 12p overrepresentation) and express either all or a combination of three tumor markers, alpha-fetoprotein, β human chorionic gonadotropin, and lactic dehydrogenase, which are considered the diagnostic markers for these tumor types [7, 17].

1.3 Molecular Alterations Associated with Ovarian Cancer Pathogenesis

1.3.1 Nuclear Genetic Alterations

Over the past years, numerous studies have identified molecular alteration in various genes in different OC subtypes (Fig. 1.2). Most of the indolent type I ovarian tumors (except clear cell carcinoma) are often characterized with mutations in *KRAS*, *BRAF*, *PIK3CA*, *PTEN*, *ARID1A*, *CTNNB1*, and *CDKN2A* [18, 19]. On the other hand, multiple sequencing studies in clinically aggressive type II tumors have demonstrated a high genomic/chromosomal instability with *TP53* mutations occurring in ≥96% of high-grade serous OCs examined [19–21]. Interestingly, several studies also revealed relatively frequent mutations in *TP53* (16–52%) are also seen in mucinous OC, but they are rare in low-grade serous and endometrioid tumors. There is an excess of 2300 different TP53 mutations that have been identified in human OCs with a majority (~70%) being missense mutations. Nearly 80% of the TP53 mutations that promote OC development and aggressiveness occur in the DNA-binding domain (between exons 5 and 8) [20, 22]. Also, a high frequency (46% of 119 cases and 57% of 43 cases analyzed) of inactivating mutations in the tumor suppressor *ARID1A* has been reported in clear cell ovarian carcinomas [19].

Data from DNA sequencing of 258 EOCs have demonstrated that 31% of the patients had harmful deleterious germline and/or somatic mutations in homologous recombination genes including *ATM*, *BRCA1*, *BRCA2*, *BRIP1*, *BARD1*, *CHEK1*, *CHEK2*, *FAM175A*, *PALB2*, *RAD51C*, *NBN*, *MRE11A*, and *RAD51D*. Notably, similar rates of homologous recombination gene mutations are also observed among non-serous and serous adenocarcinoma cases [23]. One study demonstrated a more frequent accumulation of p53 in OCs harboring *BRCA1* mutation than those with *BRCA2* mutation [24]. Other studies have also shown a positive correlation between *P53* and *BRCA1* mutations with advanced stage (stage III) OCs [25, 26]. On the other hand, *ERBB2* (gene encoding for HER2) has been found to be amplified in about 19% of mucinous ovarian tumors and in 14% clear cell carcinoma [27, 28].

Whole genome (next-generation) sequencing identified various mutations that were frequent in type II ovarian tumors involving genes of NOTCH, RB1, FOXM1, and homologous recombinant repair pathways [8]. An integrated genomic analysis of high-grade serous ovarian carcinomas reported a higher rate of *CCNE1* amplification in OCs bearing wild type-*BRCA* (26%, 212 cases) compared to altered-BRCA molecule (8%, 103 cases) [21]. Amplification of *AKT1* gene at chromosomal region 14q32.33 was also suggested to serve as a predictive marker for platinum therapy response as it was associated with a poor overall and progression-free survival of OC patients treated with platinum-based therapy [29]. Interchromosomal fusion gene *CDKN2D-WDFY2* has been detected in about 20% (12/60) of high-grade serous adenocarcinoma cases and shown to activate PI3K/AKT pathway, making it potentially useful for the classification of sub-lineages [30]. Nonhistone DNA-binding factor *HMGA2* is a high-mobility-group AT-hook (HMGA) protein, which binds to AT-rich sequences in the minor groove of the DNA helix [31]. *HMGA2* is a critical regulator of cell growth, differentiation, apoptosis, and neoplastic transformation and is frequently overexpressed in various cancers including OCs. Notably, Aberrant *HMGA2* expression was found to be an early event in driving OC progression in vivo. Mechanistically, HMGA2 is repressed by *let-7 miRs* in vivo and in vitro [31]. Concomitant overexpression of HMGA2 (64%, 18/30 cases) and reduced expression of *let-7s* (*≥1-fold*) has been reported in high-grade papillary serous carcinomas [31]. Another study examining 117 OC cases has reported overexpression of HMGA2 in 39% of cases [32]. An earlier study examined 100 primary OCs and observed that high HMGA2 expression predicted poor prognosis in terms of both progression-free and overall survival [29]. This study also demonstrated that loss of let-7 induces HMGA2 expression in OC model. Collectively, these studies have suggested loss of let-7miRs and concomitant overexpression of HMGA2 as a potential biomarker for OC prognostication. In a mutational profiling study of Brenner tumors, sporadic (nonrecurrent), mutations were reported in genes involved in cell cycle control and DNA repair (*CDK12*, *BRIP1*, *FANCC*, *NUMA1*, *ERCC5*) as well as genes important for epigenetic regulation of histone and DNA modification (*SETD2*, *KDM6A*, *KAT6B*, and *TET2*). Moreover, Brenner tumors were shown to be negative for *TP53* and *TERT* promoter mutations and exhibited *MDM2* and *CCND1* amplifications in 75% and 25% of the four malignant cases studied [33].

1.3.2 Mitochondrial Genetic Alterations

Mitochondria are unique organelles that regulate the oxidative phosphorylation system (OXPHOS), generate cellular energy, possess their own DNA (mtDNA), and follow maternal inheritance [34, 35]. The human mtDNA is a double-stranded 16.5-kb closed circular molecule encoding 12S and 16S rRNAs, 22 tRNAs, and 13 proteins essential for the mitochondrial respiratory complex assembly and functions [34, 35]. Due to the lack of protective histones, mutation rate of mtDNA is approximately ten times higher than that of nuclear DNA (nDNA) in cancer cells, and these mutations can be easily detected due to their high copy number. Numerous studies have detected somatic mtDNA mutations encompassing the coding and noncoding regions of the mtDNA in malignant tumors of various anatomic origins [34, 35]. Studies have also established, although few in number, a role of human patient-derived mtDNA mutations in cancer progression [34–37].

In an earlier study, mtDNA mutations in noncoding D-loop, 12S rRNA, 16SrRNA, and coding cytochrome B gene (*CYTB*) were detected in 36% (9/25) of OC subjects [38]. Another study identified 86 overall polymorphic mtDNA sequence variants and 9 somatic mtDNA-D-loop mutations in 26% (9/35 cases) of the OC tumors [39]. Notably, in the same group, different D-loop mtDNA mutations were also detected in 24% (4/17 cases) of patients with bilateral OCs. Interestingly, metastatic cancers from different patients with bilateral OCs demonstrated the presence of identical mtDNA mutation [39]. A comprehensive analysis of 102 OC cases revealed frequent mtDNA mutations in the 12S rRNA (nucleotide position 772, 773, and 780) in stage IIIC endometrioid tumors [40]. Mutations (np-1657) were observed exclusively in stage-IV serous tumors, benign cystadenomas, and borderline tumors (np-8221delA). Moreover, a high frequency (81%) of TC insertion (np 310) was detected only in early-stage serous subtypes including benign cystadenomas, borderline, and stage I tumors [40]. In more recent studies, analysis of D-loop mtDNA region in 93 OCs identified an association between 73A/G, 207G/A, and 523C/del polymorphism with increased risk of OC. On the other hand, a decreased risk of OC with 249A/del and 263A/G mtDNA sequence variants were observed [41]. Interestingly, in another study, clonal mtDNA mutations in borderline ovarian tumors and peritoneal implants were also identified in nearly 40% of the women studied [42]. Altogether, these studies demonstrate the presence of somatic mtDNA mutations in various progressive stages of OCs, suggesting their functional role in OC development and progression. Specific mtDNA sequence variants could also potentially be associated with an increased risk of OC development.

1.4 Molecular Pathways Associated with Ovarian Cancer Pathogenesis

Many of the genes discussed in the previous section are integral parts of important molecular pathways associated with the pathogenesis of ovarian and other malignancies (Fig. 1.3). Below, we discuss some of the pathways that are well explored in OC and considered important targets for therapy.

1.4.1 PI3K/PTEN/AKT Pathway

Phosphoinositide 3-kinase (PI3K) and Phosphatase and tensin homology (PTEN) proteins regulate multiple biological functions and phenotypes by regulating the activation status of its substrate protein kinase B (PKB), also referred to as AKT. *PTEN* is an established tumor suppressor gene, and its mutations, deletions, or loss of functions are frequently reported in OCs [43]. PI3K is a heterodimeric protein with a regulatory subunit (p85) and a catalytic subunit (p110). The p85 subunit of PI3K protects the p110 subunit from degradation and inhibits its enzymatic activity until it is bound and activated by receptor tyrosine kinases (RTKs), after this activation, a conformational change allows the p110 subunit to translocate and phosphorylate membrane lipid phosphatidylinositol-4,5 biphosphate $(PIP)_2$ to phosphatidylinositol-3,4,5-triphosphate $(PIP)_3$ [44]. AKT(PKB) has a high affinity for PIP_3 and is thus recruited to the membrane from the cytosol. This interaction causes conformational changes in AKT opening up its phosphorylation sites (Thr308 and Ser473) for kinases including phosphoinositide-dependent kinases (PDK1 and PDK2), integrin-linked kinase (ILK), etc. [45]. PTEN serves as a negative regulator of AKT as it dephosphorylates PIP3 to PIP2, preventing the downstream effects of its activation [43]. *Activation* of AKT induces multiple downstream signaling pathways such as NF-κB, mTOR, and p53 to affect tumor cell proliferation, cell division, and survival [46]. Studies have demonstrated that the inhibition of PI3K/AKT decreased the activation of p70S6K1, induced the expression of p21 and p27, and blocked the activation of Rb, ultimately leading to the suppression of proliferation and promoting phenotypic aggressiveness of OC cells [47]. PI3K/PTEN/AKT signaling has also been shown to be involved in the regulation of cancer stemness and chemoresistance [48]. 3D spheroids developed from OC cell lines, SKOV3 and HO8910, possessed high phosphorylated-AKT1 and low PTEN expression [49]. Further, it was shown that the treatment of these spheroids with a PI3K/AKT pathway inhibitor decreased the expression of markers associated with mesenchymal phenotype and stemness, and sensitized them to paclitaxel toxicity [49]. In cisplatin-resistant A2780 cells, it was shown that activation and overexpression of AKT stabilizes nuclear localization of PPM1D protein and prevented its degradation, and that inhibition of AKT caused PPM1D degradation and sensitized the cells to cisplatin [50]. Mammalian target of rapamycin (mTOR), comprised of two complexes (mTORC1 and mTORC2), is an impor-

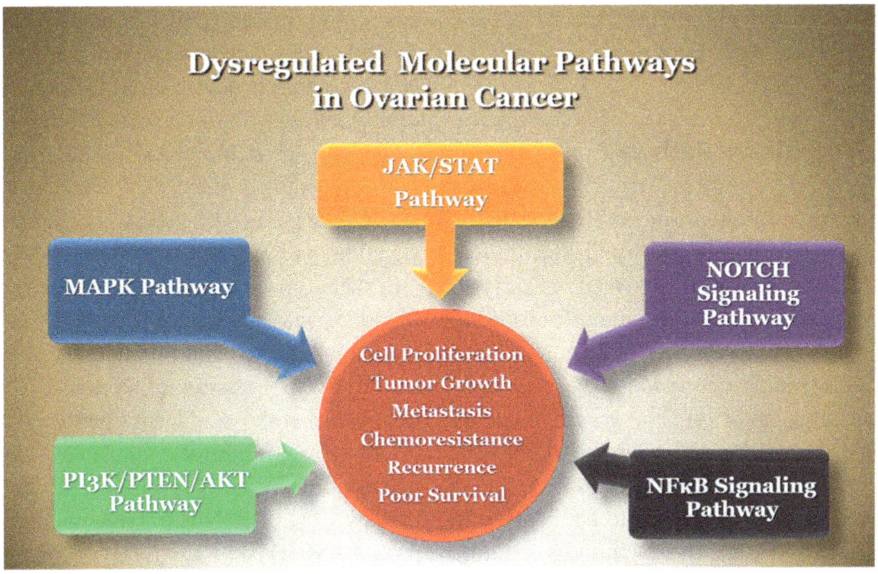

Fig. 1.3 Dysregulated molecular pathways in ovarian cancer. Several molecular pathways are altered in ovarian cancer of which the most common are presented here. Dysregulation of these molecular pathways promotes ovarian cancer pathogenesis by inducing growth and malignant tumor phenotypes (chemoresistant, metastatic, proliferative, etc.) that ultimately contribute to poor clinical outcomes

tant downstream target of PI3K/AKT pathway that also phosphorylates AKT in a feedback mechanism at the serine residue 473 through its mTORC2 complex [51]. Studies have demonstrated that PI3K/AKT/mTOR pathway was frequently activated in OC through genetic aberrations suggestive of its significant role in ovarian tumor pathogenesis in both type I and type II cancers [52].

1.4.2 JAK/STAT Pathway

The JAK/STAT pathway is shown to be constitutively active in OC and plays important role in pathogenesis [53]. Inhibition of this pathway suppresses ascites production and peritoneal metastasis of the ovarian tumor cells [53]. Pharmacological inhibition of Jak2 is also shown to suppress the nuclear translocation of STAT3 as well as the growth of ovarian tumor xenografts [54]. In a more recent study, it was demonstrated that the activation of STAT3 via phosphorylation at Tyr705 and loss of protein inhibitor of activated STAT3 (*PIAS3*) were the tumor-initiating events in the distal fallopian tube, potentially driving high-grade serous ovarian carcinoma (HGSC). Also, the overexpression of *STAT3* in secretory epithelial cells of the fallopian tube resulted in tumor promotion and metastasis in a preclinical mouse model [55]. The role of constitutively active STAT3 in ovarian tumorigenesis and prognosis has also been suggested in a study that revealed expression of pSTAT3 in 86% of OC cases across different histotypes. Constitutive activation of STAT3 was observed in 63% of the HGSC cases and tumors with intense nuclear pSTAT3 staining were associated with poorer survival rates [56]. The JAK/STAT pathway is also shown to be involved in OC stemness. In a murine model, the conditional deletion of *Apc*, *Pten*, and *Trp53* in the ovarian surface epithelium resulted in the development of high-grade metastatic ovarian carcinoma. Further, cell lines derived from this tumor model expressed several stemness-associated markers [57]. Importantly, $CD24^+$ cells had high phosphorylation of Stat3 as compared to $CD24^-$ cells and the inhibition of Jak2 decreased the Stat3 phosphorylation and survival of $CD24^+$ cells. In addition, Jak2 inhibition in this model decreased metastases and prolonged the survival of mice [57].

1.4.3 MAP Kinase Pathway

Mitogen-activated protein kinase (MAPK), also known as extracellular signal-regulated protein kinase (ERK), is aberrantly activated in nearly all cancer types and promotes tumor cell proliferation, survival, metastasis, and chemoresistance [58]. RAS activation via gene mutation or aberrant upstream receptor tyrosine kinase signaling has been associated with induction of MAPK pathway in various cancers [43, 59–62]. *KRAS* mutations have been suggested to be involved in the genesis of low-grade ovarian tumors resulting in a hyper-activated proliferative phenotype [63, 64]. It has been shown that treatment of OC cells with Endothelin-1, an autocrine regulatory factor, causes the phosphorylation-mediated activation of MAPK-1. Moreover, it is suggested that the Endothelin-1-induced activation of MAPK pathway occurs, in part, via EGFR/RAS-dependent signaling [65]. Treatment with interleukin-8 (IL-8) is also shown to activate MAPK-1/-3 pathway through its binding to CXCR-1/2 receptors. A role of EGFR in mediating the IL-8-induced activation of MAPK-1/-3 in SKOV-3 cells is also reported [66]. MAPK pathway activation has also been suggested to promote a gonadotropin (FSH and LH)-stimulated pathogenesis in OC [67].

1.4.4 Notch Signaling Pathway

The Notch signaling pathway plays an important role in cell-fate determination and embryogenesis. The deregulation of this pathway has been associated with angiogenesis and progression and metastasis of many cancers [68, 69]. Genomic profiling has demonstrated that the alterations in the Notch pathway are among the most prevalent genomic changes in serous ovarian carcinoma [70]. Hairy and enhancer of split-1 (HES1), a repressive transcriptional factor overexpressed in OC, has been shown to be a target of Notch signaling and could serve as a potential therapeutic target [71]. In one study, about 22% of HGSOC tumors were found to have aberrant activation of Notch signaling due to amplification of ligands of this pathway, Jag1, Jag2, and Notch 3 [21, 72]. The hyper-activation of Jag-1/-2 and Notch3 has also been shown to be likely mediated through p53 family genes [73]. The significance of Notch signaling pathway in the promotion of malignant OC phenotype has been reported in a study that found expansion of cancer stem cells (CSCs) and induction of platinum chemoresistance upon forced overexpression of Notch3. Conversely, treatment with a Notch pathway inhibitor (γ-secretase inhibitor) depleted CSCs and enhanced the sensitivity of OC cells to platinum therapy [74]. The Notch 1 intercellular domain protein (NICD) acts as a receptor for membrane-bound ligands Jagged1, Jagged2, and Delta1 to govern cell-fate determination. Upon ligand activation through the released NICD, it forms a transcriptional activator complex with RBPJ/RBPSUH and affects differentiation, proliferation, and apoptosis. One earlier study demonstrated NICD protein overexpression in 76% (16/21) OCs and 100% (3/3) established human OC cell lines [75]. Moreover, blockade of NICD expression in all three (CaOV3, OVCAR3, and SKOV3) OC cell lines markedly inhibited their growth in vitro [76]. Another study showed that stable transfection of the NICD1 increased the proliferation and clonogenic survival of A2780 OC cells [77]. Other than Notch1, more than two-fold higher expression of another Notch family member *Notch3* was detected in 63% OCs [78]. Abundant Notch3 expression was associated with OC aggressiveness, recurrence, and poor overall survival. Jag1 serves as a primary Notch ligand in OC and peritoneal mesothelial cells. Impairing Jag1 in mesothelial cells remarkably inhibited their proliferation and adhesion, implicating a Jag-Notch3-mediated promotion of OC dissemination in the peritoneal cavity.

1.4.5 NF-κB Signaling Pathway

NF-κB is a complex of proteins (RelA/p65, c-Rel, RelB, p50, p52) that functions as a transcription factor. An overexpression of complex subunits, p65 and p50, has been reported in OC and associated with its progression, aggressiveness, stemness, and chemoresistance as well as the poor survival of patients [75, 79]. NF-κB signaling is described as either canonical (NEMO-dependent) or non-canonical (NEMO-independent) based on its dependence on IκB kinase adaptor molecule NEMO for activation [79]. NF-κB signaling has been reported to be critical for the survival of ovarian tumor spheroids [80]. Moreover, multiple studies have shown that NF-κB pathway is activated in OC cells exhibiting resistance to apoptosis and chemotherapy drugs (cisplatin, carboplatin, paclitaxel, and erlotinib), and its pharmacological inhibition leads to chemo-sensitization [75, 81]. In advanced OC, NF-κB signaling is shown to regulate the proliferative cell populations of tumors by directly regulating the enzyme, aldehyde dehydrogenase (ALDH), which is associated with cancer stemness. Furthermore, it is shown that the loss of RelB drastically reduces the spheroid-forming ability, ALDH expression and activity, and tumor growth in mouse models of human OC [82]. Moreover, inhibition of NF-κB activity in tumor-bearing mice promotes antitumor immune responses [83]. Interestingly, it is also shown that NF κB signaling increases the expression levels of immunosuppressive cytokines, IL6 and IL-8, in OC cells and potentiates tumor growth [75].

1.5 Diagnostic and Prognostic Markers in Ovarian Cancer

Ovarian cancer suffers significantly from the lack of reliable early diagnostic and prognostic biomarkers [84]. Though the concept of detecting the presence or absence of a common marker for OC via simple assay may seem fairly straightforward, the daunting task has been to find a bona fide marker or a group of biomarkers. The quest to find a serum biomarker for EOC began in 1965 with CEA (carcinoembryonic antigen). Originally, CEA was described as a serum biomarker for mucinous colon cancer but nearly a decade later it became a biomarker for OC and was being detected via blood testing [85]. CEA levels were found to be significantly elevated in 77% of patients with serous cystadenocarcinoma and associated with poorly differentiated tumors [86]. It was also seen that elevated serum CEA levels correlated with less favorable tumor response to chemotherapy. Unfortunately, CEA has proven to be unreliable as a standalone biomarker for OC. Studies have shown that more than 30% of OC have no detectable CEA levels in serum and individual serum levels of CEA are variable and seemed to fall during terminal illness [86]. Shortly after the emergence of CEA as a serum biomarker, cancer antigen 125 (CA125 a.k.a. MUC16) was proposed as a specific serum biomarker for OC [85]. Even though CA125 tests are ordered in the United States to evaluate ovarian tumors prior to surgery, it is approved by the Food and Drug Administration (FDA) for cancer surveillance only in women with a known diagnosis of OC. It has not been approved for preoperative usage as there has been no survival benefit associated at any time point pre/post diagnosis. Despite CA125 being neither extraordinarily sensitive nor specific, it is still the only best-known serum biomarker for OC. There are several other promising OC serum biomarkers including LDH, β-hCG, inhibin B, and α-fetoprotein. The usefulness of these markers however is OC subtype specific [85]. One promising marker, human epididymis protein 4 (HE4), much like CA125, was cleared by the FDA for use in monitoring the previously diagnosed OC patients but not as a preoperative diagnostic. There is 2–3 month advantage of HE4 over CA125 in the ability to detect recurrence but there is still no clear evidence associating HE4 screening to improved patient survival [85].

The hunt for a serum OC biomarker has unveiled promising evidence that nucleic acids (free DNA, mRNA, microRNAs, and circulating tumor DNA (ctDNA)) could be just as useful, if not more useful, as protein serum markers [87]. Both normal and cancer cells can release cell-free DNA (cfDNA), mRNA, and microRNA that circulate in blood via release/secretion from proliferating or dying cells [88]. The idea of investigating cfDNA/ctDNA as cancer biomarkers resulted from earlier findings of high cfDNA concentrations in the blood of cancer patients that had detectable mutations and epigenetic alterations associated with cancer development, progression, and resistance to therapy [88]. There is also a correlation of cfDNA levels with metastasis, stage, and tumor size. However, genomic and epigenomic profiles of cfDNA in metastatic and primary tumors can be dynamic and vary even in a single patient if sampled at different time points. Fortunately, though, this type of biomarker monitoring is minimally invasive and therefore multiple blood analyses of cell-free nucleic acids can be performed allowing real-time monitoring of disease status, which is very useful in clinical settings. Notably, cell-free DNA monitoring cut the percentage of false positives determined through CA125 to nearly half (12.9% versus 23.4%) [89]. It is also important to note that high levels of cfDNA can also be reported in non-cancer-related conditions and thus quantitatively cfDNA levels are not specific enough to be used as a standalone diagnostic tool. Thus, a combinatory approach looking at a comprehensive panel of cfDNA and protein (i.e., HE4, CA125 levels, etc.) serum biomarker levels along with their genetic and epigenetic abnormalities could be a better approach for early diagnosis of OC [85, 89, 90]. Therefore, harvesting circulating tumor cells (CTCs), cfDNA, and extracellular vehicles (EVs) from bodily fluids (blood, cerebrospinal fluid, saliva, or urine) and looking for a more sensitive and specific set of biomarkers should be optimal for OC diagnosis [91].

1.6 Therapeutic Approaches for the Treatment of Ovarian Cancer

Currently, general protocol for treating women diagnosed with OC usually includes surgical removal of all visible mass (and some normal/non-cancerous surrounding tissue) and some type of drug intervention [92]. In some cases, OC tumors are also subjected to radiation therapy, but it is not a common primary treatment and thus will not be discussed in the following sections.

1.6.1 Surgical Strategies

Surgical procedures are utilized to determine the extent of the disease, remove as much of the mass as possible, and proactively deter possible recurrence. One of these procedures, salpingo-oophorectomy, completely removes one (unilateral) or both (bilateral) of the ovaries and fallopian tubes. For women, who wish to bear children, a unilateral procedure is the best way to preserve that ability. Hysterectomies are also a common procedure for treatment, which removes the uterus and, if necessary, the cervix. The omentum, a thin tissue covering the stomach and intestines, is a common secondary site for metastatic OC. Therefore, in some cases, the surgeons also perform omentectomy as well. Spread to the lymphatic system is also a common occurrence in OC. Therefore, in some cases, surgeons may also perform lymphadenectomy [93]. For women suffering from OC with distant metastasis, cytoreductive/debulking surgery is used that safely removes the optimal amount of tumor mass and may also remove metastatic tissues from other organs (i.e., spleen, liver, and colon) as well. These surgical procedures can be done alone or in conjunction with each other depending on the extent of the disease. Although surgical intervention has been typically associated with a survival benefit in OC patients, it is also associated with an increased rate of postoperative complications [94, 95].

1.6.2 Chemotherapy

Chemotherapy is an important facet of therapeutic intervention for OC. In fact, it works hand-in-hand with the surgical intervention to either reduce the mass before debulking (neoadjuvant) or to control the disease that remains after the procedure (adjuvant). Currently, 6 cycles of 21 day-increment treatment with a platinum-based drug (cisplatin/carboplatin) and a taxane (paclitaxel/docetaxel), intravenously, is standard for invasive epithelial OC [96]. Platinum-based drugs work by crosslinking DNA, blocking cell division, and resulting in apoptotic cell death

[97]. Taxanes prevent cell division by stabilizing microtubules in the cytoskeleton and inducing apoptosis [98]. The combination became established after significantly improved median overall survival was observed in patients receiving paclitaxel/cisplatin (37.5 months) as opposed to the previous standard combination of cyclophosphamide/cisplatin (24.4 months) [99].

1.6.3 Mechanism-Based Targeted Therapies

Though most OCs initially respond well to the standard platinum/taxane drug combination, the presence or development of resistance creates a major barrier in clinical management [100, 101]. Therefore, efforts have been or are being made to better understand the biology behind the inherent or acquired resistance that could help in identifying ways to compensate or revert it. Mechanism-based therapy can be an effective improvement upon standard chemotherapy due to their targeted nature towards the genes and cellular pathways underlying the sustenance of cancer.

1.6.3.1 Tumor Cell-Targeted Therapies

One tumor cell-based approach focuses on a well-studied tumor suppressor, p53, that maintains normal cell cycle progression, growth, response to DNA damage, and apoptosis signaling. Dysfunctional p53 is common in OC (frequency > 90% in high-grade serous OC) and associated with poor drug response and patient prognosis. This supports the idea that targeting p53 and restoring its normal tumor suppressive function could be beneficial for treating OC patients [102, 103]. Targeting p53, in fact, has shown some success both in vitro and in vivo marking decreased proliferation, increased apoptosis, and heightened sensitivity to cisplatin in OC tumors with dysfunctional p53 [104, 105]. Based on the known roles of p53 and Wee1 in the G1 and G2 DNA damage checkpoints, respectively, clinical trials using a Wee1 inhibitor, AZD1775, on patients with TP53-mutated OC revealed that OC patients with TP53 mutations were more sensitive to standard treatments when treated in combination. Refractory patients who experienced progression during first-line therapy and patients with chemoresistant OC exhibited improved progression-free survival with this drug combination [20, 106].

Aside from frequent mutations in p53 genes, nearly one-half of all high-grade serous OC (the most commonly diagnosed type) have defective DNA repair and nearly one-fourth of those cases carry deleterious germline mutations in the tumor suppressors *BRCA1* and *BRCA2*. OC patients that have tumors with BRCA mutations have better drug response and prognosis than non-*BRCA* ovarian cancers [107, 108]. Since homologous recombination is impaired when BRCA function is lost in OC cells and PARP has the ability to circumvent the damage repair process and promote the survival of mutant cells, BRCA-1/-2 mutant OC cells are targeted with poly-ADP-ribose polymerase (PARP) inhibitors (e.g., niraparib, olaparib, rucaparib) [108]. It has been shown that cells deficient in homologous recombination (HR) are more sensitive to damage by platinum drugs; therefore, the combination of drug-induced DNA damage, PARP inhibition, and impaired HR resulting from BRCA mutations show promise as a good therapeutic strategy for OC treatment [109]. In fact, the United States Food and Drug Administration (FDA) approved rucaparib, a PARP inhibitor, as a third line treatment for OC patients with germline mutations in 2016. Clinical trials revealed that rucaparib enhanced progression-free survival in platinum-sensitive OC patients [110]. Another PARP inhibitor, niraparib, has also been approved by FDA for OC maintenance therapy. In a clinical trial, niraparib-treated patients with germline BRCA mutation had 21 months of progression-free survival as compared to 5.5 months for the same patient population taking a placebo. Patients that did not have germline mutations also benefited from an improved progression-free survival of 9.3 months compared to 3.9 months for placebo-treated patients [107].

Another tumor cell-based approach, suicide gene therapy, has been utilized and shows promise for future improvements in treating OC. In suicide gene therapy, genes are delivered to tumor cells that either activate a prodrug or release a toxin that induces cell suicide [111]. This type of intervention has the potential to be extremely beneficial once a bonafide OC biomarker is discovered, in fact, studies have been done where known alterations in OC (i.e., expression of HE4 and p53 mutations) are used to discriminately deliver toxins and induce suicide expression that, in turn, sensitizes the cells to standard chemotherapy drugs [112–114].

1.6.3.2 Tumor Microenvironment-Targeted Therapies

Targeting Tumor Vasculature

Cancer cells require a continuous supply of nutrients and oxygen to survive, which is fulfilled by the development of the new vasculature through angiogenesis within the TME, a critical event necessary for OC progression. This process is supported by different angiogenic regulators/promoters including vascular endothelial growth factors (VEGFs) and their receptors (VEGFR/Ftl), angiopoietins (ANGPT) and ANGPT receptors (Tie1/2), etc. [115]. Studies have shown VEGF overexpression is associated with advanced malignant OC and promotes ascites formation [116]. Both single agent and combination chemotherapy with anti-angiogenic drugs resulted in the inhibition of tumor growth, metastasis, and ascites formation with an observed improvement in survival [117–119]. Bevacizumab, a monoclonal antibody that directly targets VEGF was approved by FDA for the treatment of recurrent platinum-resistant OC after clinical trials using this agent in combination with chemotherapy revealed significant improvement in median progression-free survival over those treated with standard chemotherapy alone (8.1 months vs. 3.9 months) [120]. However, no significant effect in overall survival was observed when OC patients were treated with bevacizumab in combination with chemotherapy there was no significant improvement in overall survival compared to the group treated with the standard therapy only (33.6 months vs. 32.9 months) [121]. VEGF independent strategies targeting angiogenesis in OC were also studied in the clinical trial with receptor tyrosine kinase inhibitors (i.e., cediranib, pazopanib, and sunitinib). Reports following clinical trials where recurrent OC patients were treated with cediranib, 30% of patients achieved a partial response to therapeutic intervention or had stable disease [122]. Four hundred fifty-six recurrent platinum-sensitive OC patients were evaluated for progression-free survival following three conditions: (1) placebo, (2) carboplatin with cediranib and placebo maintenance, and (3) carboplatin with cediranib and cediranib maintenance. The result of the study showed improved median progression-free survival in comparison to the placebo group when cediranib was combined with carboplatin and an even greater improvement when maintenance therapy was added (8.7, 9.9, 11.0 months, respectively). In all, these studies show that angiogenesis is a useful process to drug target for combatting recurrence in OC. This approach, unfortunately, still misses the mark on improving the long-term goal of alleviating the highly fatal nature of this malignancy.

Targeting Exosomes

Exosomes are small (50–200 nm) vesicles of endocytic origin. The exosomes derived from various cell types harbor a diverse array of nucleic acids and proteins and play a pivotal role in intercellular and intracellular communication [123–125]. Based on their functional role in cancer development and progression, exosomes are emerging as attractive tools for cancer treatment and detection. Similar to other cancers, OC cells release exosomes into the plasma, serum, and ascites where they communicate or are recognized and taken up by other cells to carry out pathogenic progression [126, 127]. Studies have shown that dysregulated microRNAs (miR-21, miR-141, miR-184, miR-193b, miR-200a, miR-105, let-7 miR, etc.) and proteins associated with malignant progression (CD63, CD44, annexin A3, HSP90, etc.) are found to be in or on exosomes derived from OC cells [126, 128, 129]. One study has shown overexpression of miR-21 in exosomes with the depletion of PDCD4 resulting in tumor spread while another study showed that the presence of miR-200 in exosomes had tumor suppressive functions [130, 131]. One recent study observed increased miR-205 expression in sera exosomes of OCs ($N = 333$) and demonstrated that exosomal miR-205 facilitates OC metastasis in vitro and in vivo by inducing angiogenesis and PTEN-AKT signaling pathway [132]. Another interesting study uncovered that soluble E-cadherin (sE-cad), an angiogenesis promoter, is not only highly expressed in metastatic OCs but also OC-derived exosomes [133]. Further in vivo and clinical analysis implicated the possible role of sE-cad-positive exosomes for metastatic dissemination. Together these data reiterate the need for a context-specific approach to utilizing OC exosomes in the clinic. Exosomes harboring specific oncogenic promoters (such as sE-cad) could be useful biomarkers for OC diagnosis and prognosis. On the other hand, targeting exosomes containing those oncogenic promoters could also be advantageous by attempting to attenuate the production of exosomes with the use of inhibitors (GW4869, Dimethyl amiloride) or by capitalizing on the expression of proteins that inhibit exosomal secretion resulted in the induction of an antitumorigenic niche [134–137]. Exosomes have been shown to have the ability to improve compound stability, bioavailability, solubility, and reduce compound toxicity which makes them not only attractive drug targets but also promising for usage as a potential drug delivery system [128, 138–140]. In fact, clinical trials have illustrated the feasibility of an exosome-mediated drug delivery system and determined that patients treated with exosome-delivered drugs responded with disease stability and antitumor immune responses [141–143].

Immunotherapy

The recent success with PD-1/PD-LI directed immunotherapy in melanoma led by the Noble Laureate Dr. James P Allison opens up a new avenue for cancer treatment [144, 145]. Treatment with PD-1/PD-L1 directed Nivolumab and Pembrolizumab (FDA approved) remarkably harnessed the immune system to prevent cancer development as well as progression to metastatic disease. Of note, immunotherapeutic treatment regimens can be employed for the treatment of various other malignancies including OC. Following this path, a phase II clinical trial with Nivolumab in 20 platinum-resistant recurrent OCs demonstrated promising clinical response and tolerance leading to enhanced overall survival and long-term disease control in around 15% of patients [146]. A more recent study evaluated the therapeutic efficacy of Niraparib in combination with Pembrolizumab in a pool of 62 recurrent OCs through a phase I and phase II clinical trial [147]. In concert, both the drugs were well tolerated and of the 62 cases, 3 patients achieved complete and 8 patients received partial treatment response. Twenty-eight patients were reported to have stable disease and 20 subjects progressed further. Although the overall success rate is limited with PD1/PD-L1 regimens, which is attributable to considerable variation in PD1/PD-L1 expression in OCs, its promise for treating subsets of aggressive OC patients alone or in combination with other immunotherapeutic targets remains feasible.

In addition to PD1-PD-LI-based treatment modalities, chimeric antigen receptor (CAR)-based T cell therapy is also gaining much attention in this era for their therapeutic benefits in various cancers [148]. For CAR-T-cell treatment in OCs, MUC16, mesothelin, HER2 and FRα are the most frequently targeted antigens. Among these antigens, MUC16-CAR-T-cells have been demonstrated to have killing effects on MUC16^{+} OC cells in vitro and in vivo [149]. Tumor microenvironment enriched in diverse array of oncogenic stimulators and neighboring cell populations including cancer-associated fibroblast (CAF) play an important role in cancer progression [150]. Utilizing this knowledge, therapeutic efficacy of a recently developed monoclonal antibody targeting microfibril-associated protein 5 (*MFAP5*), secreted predominately by CAFs has been tested in OC model [150]. Remarkably, MFAP5 blockade inhibited fibrosis, normalized tumor vascularization, and augmented chemotherapeutic sensitivity of OC cells in vivo. Other than the MFAP5-mAb, a recent study also examined the therapeutic potential of a human bispecific antibody (MUC16^{+}/

CD3$^+$, REGN4018) [151]. Interestingly, REGN4018 not only triggered T cell activation but also induced T cell-mediated killing of MUC16$^+$ OC cells in vitro and in vivo. Moreover, in a humanized immuno-competent mouse model (expressing human CD3 and MUC16), REGN4018 inhibited tumor growth and exhibited superior efficacy when given in combination with an anti-PD-1 antibody. These innovative approaches appear to have tremendous potential to effectively manage recurrent and drug-resistant OCs in the future. Natural killer (NK) cells could be an attractive immunotherapeutic mediator for treating OC. Immunopotention manipulates the immune system's ability to recognize antigens and elicit an antitumor immune response by enhancing cytokine expression that can promote NK cytotoxicity and illicit cytotoxic T-lymphocyte response in OC tumors [152, 153]. OC encounters frequent alteration in human leukocyte antigens (HLAs), a pivotal pathway for eliciting T cell-based antitumor response. NK cells can be activated in tumor cells lacking HLA expression and can uniquely elicit antitumor activity in an antigen-independent manner [154]. NK cell-based immunotherapeutic studies in human OC patients (clinical trials) are currently in progress, particularly for the treatment refractory OCs [154]. Collectively, in concert with conventional treatment, a new era with immunotherapy bears significant promise for better management of OC.

1.7 Conclusion and Future Perspectives

Tremendous efforts made in the past few decades undoubtedly enhanced our understanding of OC biology and aggressiveness, aiding in disease monitoring and guiding treatment. However, enormous complexity in histologic subtypes and associated enigmatic molecular architecture driving rapid progression and treatment failure still pose significant challenges in efficient management of OC. These challenges pose dire needs to develop novel treatments based on a deeper understanding of the distinct molecular characteristics and specific cellular/extracellular mechanisms. Hopefully in this decade, ongoing efforts with immunotherapy, targeted therapy, rapid development of cfDNA, ctDNA, and exosome-based noninvasive/liquid biomarkers in OC could open up future doors for better OC management. Furthermore, the development of cost-effective early diagnostic tools to improve upon the conventional diagnostic approaches is of paramount importance as detection of OC at the earliest time point can prevent its further progression to aggressive/lethal disease [155]. In addition, patient-specific next-generation sequencing strategies identifying novel altered molecular pathways will not only address the heterogeneity context of OC aggressiveness but also open up novel avenues for personalized treatment and care. With the advent of cutting-edge technologies in concert with deeper knowledge in OC biology, the future roadmap of efficient management for this dreaded malignancy with new and improved treatment and diagnostic strategies is promising.

Author Contributions O.M., S.K.S., and S.D. wrote the manuscript; S.D., S.S., R.P.R., and A.P.S. proofread and corrected it; O.M. prepared figures; A.P.S. conceptualized the idea, provided resources, and supervised the project. All authors read and approved the final manuscript.

Funding: Authors would like to acknowledge the funding from NIH/NCI [R01CA175772, R01CA224306, U01CA185490 (to A.P.S.); R01CA204801, R01CA231925 (to S.S.)] and USAMCI.

Conflicts of InterestThe authors declare no conflict of interest.

References

1. Siegel, R. L., Miller, K. D., & Jemal, A. (2019). Cancer statistics, 2019. *CA: A Cancer Journal for Clinicians, 69*(1), 7–34.
2. Momenimovahed, Z., et al. (2019). Ovarian cancer in the world: Epidemiology and risk factors. *International Journal of Women's Health, 11*, 287–299.
3. Torre, L. A., et al. (2018). Ovarian cancer statistics, 2018. *CA: A Cancer Journal for Clinicians, 68*(4), 284–296.
4. Reid, B. M., Permuth, J. B., & Sellers, T. A. (2017). Epidemiology of ovarian cancer: A review. *Cancer Biology & Medicine, 14*(1), 9–32.

5. Pokhriyal, R., et al. (2019). Chemotherapy resistance in advanced ovarian cancer patients. *Biomarkers in Cancer, 11*, 1179299X19860815.
6. Edson, M. A., Nagaraja, A. K., & Matzuk, M. M. (2009). The mammalian ovary from genesis to revelation. *Endocrine Reviews, 30*(6), 624–712.
7. Richards, J. S., & Pangas, S. A. (2010). The ovary: Basic biology and clinical implications. *The Journal of Clinical Investigation, 120*(4), 963–972.
8. Kurman, R. J., & Shih Ie, M. (2016). The dualistic model of ovarian carcinogenesis: Revisited, revised, and expanded. *The American Journal of Pathology, 186*(4), 733–747.
9. Chen, V. W., et al. (2003). Pathology and classification of ovarian tumors. *Cancer, 97*(10 Suppl), 2631–2642.
10. Shaaban, A. M., et al. (2014). Ovarian malignant germ cell tumors: Cellular classification and clinical and imaging features. *Radiographics, 34*(3), 777–801.
11. Nezhat, F. R., et al. (2015). New insights in the pathophysiology of ovarian cancer and implications for screening and prevention. *American Journal of Obstetrics and Gynecology, 213*(3), 262–267.
12. Mabuchi, S., Sugiyama, T., & Kimura, T. (2016). Clear cell carcinoma of the ovary: Molecular insights and future therapeutic perspectives. *Journal of Gynecologic Oncology, 27*(3), e31.
13. Koulouris, C. R., & Penson, R. T. (2009). Ovarian stromal and germ cell tumors. *Seminars in Oncology, 36*(2), 126–136.
14. Fuller, P. J., Leung, D., & Chu, S. (2017). Genetics and genomics of ovarian sex cord-stromal tumors. *Clinical Genetics, 91*(2), 285–291.
15. Horta, M., & Cunha, T. M. (2015). Sex cord-stromal tumors of the ovary: A comprehensive review and update for radiologists. *Diagnostic and Interventional Radiology, 21*(4), 277–286.
16. Pectasides, D., Pectasides, E., & Kassanos, D. (2008). Germ cell tumors of the ovary. *Cancer Treatment Reviews, 34*(5), 427–441.
17. Cossu-Rocca, P., et al. (2006). Chromosome 12p abnormalities in dysgerminoma of the ovary: A FISH analysis. *Modern Pathology, 19*(4), 611–615.
18. Hunter, S. M., et al. (2015). Molecular profiling of low grade serous ovarian tumours identifies novel candidate driver genes. *Oncotarget, 6*(35), 37663–37677.
19. Rojas, V., et al. (2016). Molecular characterization of epithelial ovarian cancer: Implications for diagnosis and treatment. *International Journal of Molecular Sciences, 17*(12), 2113.
20. Zhang, Y., et al. (2016). TP53 mutations in epithelial ovarian cancer. *Translational Cancer Research, 5*(6), 650–663.
21. Cancer Genome Atlas Research Network. (2011). Integrated genomic analyses of ovarian carcinoma. *Nature, 474*(7353), 609–615.
22. Ricci, F., et al. (2018). Recent insights into mucinous ovarian carcinoma. *International Journal of Molecular Sciences, 19*(6), 1569.
23. Pennington, K. P., et al. (2014). Germline and somatic mutations in homologous recombination genes predict platinum response and survival in ovarian, fallopian tube, and peritoneal carcinomas. *Clinical Cancer Research, 20*(3), 764–775.
24. Zweemer, R. P., et al. (1999). Accumulation of p53 protein is frequent in ovarian cancers associated with BRCA1 and BRCA2 germline mutations. *Journal of Clinical Pathology, 52*(5), 372–375.
25. Tirkkonen, M., et al. (1997). Distinct somatic genetic changes associated with tumor progression in carriers of BRCA1 and BRCA2 germ-line mutations. *Cancer Research, 57*(7), 1222–1227.
26. Wysham, W. Z., et al. (2012). BRCAness profile of sporadic ovarian cancer predicts disease recurrence. *PLoS One, 7*(1), e30042.
27. Anglesio, M. S., et al. (2013). Molecular characterization of mucinous ovarian tumours supports a stratified treatment approach with HER2 targeting in 19% of carcinomas. *The Journal of Pathology, 229*(1), 111–120.
28. Tan, D. S., et al. (2011). Genomic analysis reveals the molecular heterogeneity of ovarian clear cell carcinomas. *Clinical Cancer Research, 17*(6), 1521–1534.
29. Despierre, E., et al. (2014). Somatic copy number alterations predict response to platinum therapy in epithelial ovarian cancer. *Gynecologic Oncology, 135*(3), 415–422.
30. Kannan, K., et al. (2014). CDKN2D-WDFY2 is a cancer-specific fusion gene recurrent in high-grade serous ovarian carcinoma. *PLoS Genetics, 10*(3), e1004216.
31. Mahajan, A., et al. (2010). HMGA2: A biomarker significantly overexpressed in high-grade ovarian serous carcinoma. *Modern Pathology, 23*(5), 673–681.
32. Califano, D., et al. (2014). High HMGA2 expression and high body mass index negatively affect the prognosis of patients with ovarian cancer. *Journal of Cellular Physiology, 229*(1), 53–59.
33. Pfarr, N., et al. (2017). Mutational profiles of Brenner tumors show distinctive features uncoupling urothelial carcinomas and ovarian carcinoma with transitional cell histology. *Genes, Chromosomes & Cancer, 56*(10), 758–766.
34. Chatterjee, A., Dasgupta, S., & Sidransky, D. (2011). Mitochondrial subversion in cancer. *Cancer Prevention Research (Philadelphia, Pa.), 4*(5), 638–654.
35. Hertweck, K. L., & Dasgupta, S. (2017). The landscape of mtDNA modifications in cancer: A tale of two cities. *Frontiers in Oncology, 7*, 262.
36. Petros, J. A., et al. (2005). mtDNA mutations increase tumorigenicity in prostate cancer. *Proceedings of the*

National Academy of Sciences of the United States of America, 102(3), 719–724.
37. Dasgupta, S., et al. (2012). Mitochondrial DNA mutations in respiratory complex-I in never-smoker lung cancer patients contribute to lung cancer progression and associated with EGFR gene mutation. *Journal of Cellular Physiology, 227*(6), 2451–2460.
38. Liu, V. W., et al. (2001). High incidence of somatic mitochondrial DNA mutations in human ovarian carcinomas. *Cancer Research, 61*(16), 5998–6001.
39. Van Trappen, P. O., et al. (2007). Somatic mitochondrial DNA mutations in primary and metastatic ovarian cancer. *Gynecologic Oncology, 104*(1), 129–133.
40. Aikhionbare, F. O., et al. (2007). Mitochondrial DNA sequence variants in epithelial ovarian tumor subtypes and stages. *Journal of Carcinogenesis, 6*, 1.
41. Liu, S., et al. (2016). Identification of sequence nucleotide polymorphisms in the D-loop region of mitochondrial DNA as a risk factor for epithelial ovarian cancer. *Mitochondrial DNA. Part A, DNA Mapping, Seqencing, and Analysis, 27*(1), 9–11.
42. Girolimetti, G., et al. (2017). Mitochondrial DNA sequencing demonstrates clonality of peritoneal implants of borderline ovarian tumors. *Molecular Cancer, 16*(1), 47.
43. Saldanha, S. N., & Tollefsbol, T. O. (2014). Pathway modulations and epigenetic alterations in ovarian tumorbiogenesis. *Journal of Cellular Physiology, 229*(4), 393–406.
44. Jimenez, C., et al. (2002). The p85 regulatory subunit controls sequential activation of phosphoinositide 3-kinase by Tyr kinases and Ras. *The Journal of Biological Chemistry, 277*(44), 41556–41562.
45. Carnero, A. (2010). The PKB/AKT pathway in cancer. *Current Pharmaceutical Design, 16*(1), 34–44.
46. Blanco-Aparicio, C., et al. (2007). PTEN, more than the AKT pathway. *Carcinogenesis, 28*(7), 1379–1386.
47. Meng, Q., et al. (2006). Role of PI3K and AKT specific isoforms in ovarian cancer cell migration, invasion and proliferation through the p70S6K1 pathway. *Cellular Signalling, 18*(12), 2262–2271.
48. Roy, L., & Cowden Dahl, K. D. (2018). Can stemness and chemoresistance be therapeutically targeted via signaling pathways in ovarian cancer? *Cancers (Basel), 10*(8), 241.
49. Luo, X., et al. (2013). Enrichment of ovarian cancer stem-like cells is associated with epithelial to mesenchymal transition through an miRNA-activated AKT pathway. *Cell Proliferation, 46*(4), 436–446.
50. Ali, A. Y., et al. (2015). Akt confers cisplatin chemoresistance in human gynecological carcinoma cells by modulating PPM1D stability. *Molecular Carcinogenesis, 54*(11), 1301–1314.
51. Mabuchi, S., Hisamatsu, T., & Kimura, T. (2011). Targeting mTOR signaling pathway in ovarian cancer. *Current Medicinal Chemistry, 18*(19), 2960–2968.
52. Dobbin, Z. C., & Landen, C. N. (2013). The importance of the PI3K/AKT/MTOR pathway in the progression of ovarian cancer. *International Journal of Molecular Sciences, 14*(4), 8213–8227.
53. Wen, W., et al. (2014). Targeting JAK1/STAT3 signaling suppresses tumor progression and metastasis in a peritoneal model of human ovarian cancer. *Molecular Cancer Therapeutics, 13*(12), 3037–3048.
54. Hedvat, M., et al. (2009). The JAK2 inhibitor AZD1480 potently blocks Stat3 signaling and oncogenesis in solid tumors. *Cancer Cell, 16*(6), 487–497.
55. Saini, U., et al. (2018). STAT3/PIAS3 levels serve as "early signature" genes in the development of high-grade serous carcinoma from the fallopian tube. *Cancer Research, 78*(7), 1739–1750.
56. Rosen, D. G., et al. (2006). The role of constitutively active signal transducer and activator of transcription 3 in ovarian tumorigenesis and prognosis. *Cancer, 107*(11), 2730–2740.
57. Burgos-Ojeda, D., et al. (2015). CD24+ ovarian cancer cells are enriched for cancer-initiating cells and dependent on JAK2 signaling for growth and metastasis. *Molecular Cancer Therapeutics, 14*(7), 1717–1727.
58. Koul, H. K., Pal, M., & Koul, S. (2013). Role of p38 MAP kinase signal transduction in solid tumors. *Genes & Cancer, 4*(9–10), 342–359.
59. Mane, S. M., et al. (1990). RAS gene activation in acute myelogenous leukemia: Analysis by in vitro amplification and DNA base sequence determination. *Genes, Chromosomes & Cancer, 2*(1), 71–77.
60. Edkins, S., et al. (2006). Recurrent KRAS codon 146 mutations in human colorectal cancer. *Cancer Biology & Therapy, 5*(8), 928–932.
61. Harris, T. J., & McCormick, F. (2010). The molecular pathology of cancer. *Nature Reviews. Clinical Oncology, 7*(5), 251–265.
62. Li, H. T., et al. (2011). KRAS, BRAF and PIK3CA mutations in human colorectal cancer: Relationship with metastatic colorectal cancer. *Oncology Reports, 25*(6), 1691–1697.
63. Vereczkey, I., et al. (2011). Molecular characterization of 103 ovarian serous and mucinous tumors. *Pathology Oncology Research, 17*(3), 551–559.
64. Stewart, C. J., et al. (2012). KRAS mutations in ovarian low-grade endometrioid adenocarcinoma: Association with concurrent endometriosis. *Human Pathology, 43*(8), 1177–1183.
65. Vacca, F., et al. (2000). Transactivation of the epidermal growth factor receptor in endothelin-1-induced mitogenic signaling in human ovarian carcinoma cells. *Cancer Research, 60*(18), 5310–5317.
66. Venkatakrishnan, G., Salgia, R., & Groopman, J. E. (2000). Chemokine receptors CXCR-1/2 activate mitogen-activated protein kinase via the epidermal growth factor receptor in ovarian cancer cells. *The Journal of Biological Chemistry, 275*(10), 6868–6875.
67. Choi, K. C., Auersperg, N., & Leung, P. C. (2003). Mitogen-activated protein kinases in normal

and (pre)neoplastic ovarian surface epithelium. *Reproductive Biology and Endocrinology, 1*, 71.
68. Aster, J. C., Pear, W. S., & Blacklow, S. C. (2017). The varied roles of Notch in cancer. *Annual Review of Pathology, 12*, 245–275.
69. Tan, D. S., Agarwal, R., & Kaye, S. B. (2006). Mechanisms of transcoelomic metastasis in ovarian cancer. *The Lancet Oncology, 7*(11), 925–934.
70. Groeneweg, J. W., et al. (2014). Notch signaling in serous ovarian cancer. *Journal of Ovarian Research, 7*, 95.
71. Schreck, K. C., et al. (2010). The Notch target Hes1 directly modulates Gli1 expression and Hedgehog signaling: A potential mechanism of therapeutic resistance. *Clinical Cancer Research, 16*(24), 6060–6070.
72. Yamaguchi, E., et al. (2002). Expression of Notch ligands, Jagged1, 2 and Delta1 in antigen presenting cells in mice. *Immunology Letters, 81*(1), 59–64.
73. Sasaki, Y., et al. (2002). The p53 family member genes are involved in the Notch signal pathway. *The Journal of Biological Chemistry, 277*(1), 719–724.
74. McAuliffe, S. M., et al. (2012). Targeting Notch, a key pathway for ovarian cancer stem cells, sensitizes tumors to platinum therapy. *Proceedings of the National Academy of Sciences of the United States of America, 109*(43), E2939–E2948.
75. Harrington, B. S., & Annunziata, C. M. (2019). NF-kappaB signaling in ovarian cancer. *Cancers (Basel), 11*(8), 1182.
76. Rose, S. L., et al. (2010). Notch 1 signaling is active in ovarian cancer. *Gynecologic Oncology, 117*(1), 130–133.
77. Hopfer, O., et al. (2005). The Notch pathway in ovarian carcinomas and adenomas. *British Journal of Cancer, 93*(6), 709–718.
78. Aburjania, Z., et al. (2018). The role of Notch3 in cancer. *The Oncologist, 23*(8), 900–911.
79. Shih, V. F., et al. (2011). A single NFkappaB system for both canonical and non-canonical signaling. *Cell Research, 21*(1), 86–102.
80. Kleinschmidt, E. G., et al. (2019). Rgnef promotes ovarian tumor progression and confers protection from oxidative stress. *Oncogene, 38*(36), 6323–6337.
81. Momeny, M., et al. (2018). Blockade of nuclear factor-kappaB (NF-kappaB) pathway inhibits growth and induces apoptosis in chemoresistant ovarian carcinoma cells. *The International Journal of Biochemistry & Cell Biology, 99*, 1–9.
82. House, C. D., et al. (2017). NFkappaB promotes ovarian tumorigenesis via classical pathways that support proliferative cancer cells and alternative pathways that support ALDH(+) cancer stem-like cells. *Cancer Research, 77*(24), 6927–6940.
83. Wilson, A. J., et al. (2013). Tracking NF-kappaB activity in tumor cells during ovarian cancer progression in a syngeneic mouse model. *Journal of Ovarian Research, 6*(1), 63.
84. Coticchia, C. M., Yang, J., & Moses, M. A. (2008). Ovarian cancer biomarkers: Current options and future promise. *Journal of the National Comprehensive Cancer Network, 6*(8), 795–802.
85. Ueland, F. R. (2017). A perspective on ovarian cancer biomarkers: Past, present and yet-to-come. *Diagnostics (Basel), 7*(1), 14.
86. Khoo, S. K., & MacKay, E. V. (1976). Carcinoembryonic antigen (CEA) in ovarian cancer: Factors influencing its incidence and changes which occur in response to cytotoxic drugs. *British Journal of Obstetrics and Gynaecology, 83*(10), 753–759.
87. Bettegowda, C., et al. (2014). Detection of circulating tumor DNA in early- and late-stage human malignancies. *Science Translational Medicine, 6*(224), 224ra24.
88. Schwarzenbach, H., Hoon, D. S., & Pantel, K. (2011). Cell-free nucleic acids as biomarkers in cancer patients. *Nature Reviews. Cancer, 11*(6), 426–437.
89. Kamat, A. A., et al. (2010). Plasma cell-free DNA in ovarian cancer: An independent prognostic biomarker. *Cancer, 116*(8), 1918–1925.
90. Wimberger, P., et al. (2011). Impact of platinum-based chemotherapy on circulating nucleic acid levels, protease activities in blood and disseminated tumor cells in bone marrow of ovarian cancer patients. *International Journal of Cancer, 128*(11), 2572–2580.
91. Campos, C. D. M., et al. (2018). Molecular profiling of liquid biopsy samples for precision medicine. *Cancer Journal, 24*(2), 93–103.
92. du Bois, A., et al. (2005). 2004 Consensus statements on the management of ovarian cancer: Final document of the 3rd International Gynecologic Cancer Intergroup Ovarian Cancer Consensus Conference (GCIG OCCC 2004). *Annals of Oncology, 16*(Suppl 8), viii7–viii12.
93. Harter, P., et al. (2007). Pattern and clinical predictors of lymph node metastases in epithelial ovarian cancer. *International Journal of Gynecological Cancer, 17*(6), 1238–1244.
94. du Bois, A., et al. (2010). Potential role of lymphadenectomy in advanced ovarian cancer: A combined exploratory analysis of three prospectively randomized phase III multicenter trials. *Journal of Clinical Oncology, 28*(10), 1733–1739.
95. Harter, P., et al. (2019). A randomized trial of lymphadenectomy in patients with advanced ovarian neoplasms. *The New England Journal of Medicine, 380*(9), 822–832.
96. Cortez, A. J., et al. (2018). Advances in ovarian cancer therapy. *Cancer Chemotherapy and Pharmacology, 81*(1), 17–38.
97. Dasari, S., & Tchounwou, P. B. (2014). Cisplatin in cancer therapy: Molecular mechanisms of action. *European Journal of Pharmacology, 740*, 364–378.
98. Gligorov, J., & Lotz, J. P. (2004). Preclinical pharmacology of the taxanes: Implications of the differences. *The Oncologist, 9*(Suppl 2), 3–8.
99. Helm, C. W., & States, J. C. (2009). Enhancing the efficacy of cisplatin in ovarian cancer treatment—

Could arsenic have a role. *Journal of Ovarian Research, 2*, 2.
100. Lou, J. Y., et al. (2006). [Reversal of multi-drug resistance in ovarian cancer cell by RNA interference]. *Zhonghua Fu Chan Ke Za Zhi, 41*(6): 413–416.
101. Goto, T., et al. (2006). Gene expression profiles with cDNA microarray reveal RhoGDI as a predictive marker for paclitaxel resistance in ovarian cancers. *Oncology Reports, 15*(5), 1265–1271.
102. de Queiroz, R. M., et al. (2016). Changes in O-linked N-acetylglucosamine (O-GlcNAc) homeostasis activate the p53 pathway in ovarian cancer cells. *The Journal of Biological Chemistry, 291*(36), 18897–18914.
103. Collinet, P., et al. (2006). In vivo expression and antitumor activity of p53 gene transfer with naked plasmid DNA in an ovarian cancer xenograft model in nude mice. *The Journal of Obstetrics and Gynaecology Research, 32*(5), 449–453.
104. Kigawa, J., et al. (2002). Effect of p53 gene transfer and cisplatin in a peritonitis carcinomatosa model with p53-deficient ovarian cancer cells. *Gynecologic Oncology, 84*(2), 210–215.
105. Miettinen, S., & Ylikomi, T. (2009). Concomitant exposure of ovarian cancer cells to docetaxel, CPT-11 or SN-38 and adenovirus-mediated p53 gene therapy. *Anti-Cancer Drugs, 20*(7), 589–600.
106. Leijen, S., et al. (2016). Phase II study of WEE1 inhibitor AZD1775 plus carboplatin in patients with TP53-mutated ovarian cancer refractory or resistant to first-line therapy within 3 months. *Journal of Clinical Oncology, 34*(36), 4354–4361.
107. Parkes, E. E., & Kennedy, R. D. (2016). Clinical application of poly(ADP-ribose) polymerase inhibitors in high-grade serous ovarian cancer. *The Oncologist, 21*(5), 586–593.
108. Lord, C. J., & Ashworth, A. (2017). PARP inhibitors: Synthetic lethality in the clinic. *Science, 355*(6330), 1152–1158.
109. Rocha, C. R. R., et al. (2018). DNA repair pathways and cisplatin resistance: An intimate relationship. *Clinics (Sao Paulo), 73*(Suppl 1), e478s.
110. Swisher, E. M., et al. (2017). Rucaparib in relapsed, platinum-sensitive high-grade ovarian carcinoma (ARIEL2 Part 1): An international, multicentre, open-label, phase 2 trial. *The Lancet Oncology, 18*(1), 75–87.
111. Navarro, S. A., et al. (2016). Cancer suicide gene therapy: A patent review. *Expert Opinion on Therapeutic Patents, 26*(9), 1095–1104.
112. Rawlinson, J. W., et al. (2013). Adenoviral-delivered HE4-HSV-tk sensitizes ovarian cancer cells to ganciclovir. *Gene Therapy and Molecular Biology, 15*, 120–130.
113. Sher, Y. P., et al. (2013). Targeted endostatin-cytosine deaminase fusion gene therapy plus 5-fluorocytosine suppresses ovarian tumor growth. *Oncogene, 32*(9), 1082–1090.
114. Singh, P. P., et al. (2011). Purine nucleoside phosphorylase mediated molecular chemotherapy and conventional chemotherapy: A tangible union against chemoresistant cancer. *BMC Cancer, 11*, 368.
115. Nishida, N., et al. (2006). Angiogenesis in cancer. *Vascular Health and Risk Management, 2*(3), 213–219.
116. Bekes, I., et al. (2016). Does VEGF facilitate local tumor growth and spread into the abdominal cavity by suppressing endothelial cell adhesion, thus increasing vascular peritoneal permeability followed by ascites production in ovarian cancer? *Molecular Cancer, 15*, 13.
117. Welch, S. A., et al. (2010). Sorafenib in combination with gemcitabine in recurrent epithelial ovarian cancer: A study of the Princess Margaret Hospital Phase II Consortium. *International Journal of Gynecological Cancer, 20*(5), 787–793.
118. Xu, L., et al. (2000). Inhibition of malignant ascites and growth of human ovarian carcinoma by oral administration of a potent inhibitor of the vascular endothelial growth factor receptor tyrosine kinases. *International Journal of Oncology, 16*(3), 445–454.
119. Byrne, A. T., et al. (2003). Vascular endothelial growth factor-trap decreases tumor burden, inhibits ascites, and causes dramatic vascular remodeling in an ovarian cancer model. *Clinical Cancer Research, 9*(15), 5721–5728.
120. Husain, A., et al. (2016). Independent radiologic review of AURELIA, a phase 3 trial of bevacizumab plus chemotherapy for platinum-resistant recurrent ovarian cancer. *Gynecologic Oncology, 142*(3), 465–470.
121. Aghajanian, C., et al. (2015). Final overall survival and safety analysis of OCEANS, a phase 3 trial of chemotherapy with or without bevacizumab in patients with platinum-sensitive recurrent ovarian cancer. *Gynecologic Oncology, 139*(1), 10–16.
122. Matulonis, U. A., et al. (2009). Cediranib, an oral inhibitor of vascular endothelial growth factor receptor kinases, is an active drug in recurrent epithelial ovarian, fallopian tube, and peritoneal cancer. *Journal of Clinical Oncology, 27*(33), 5601–5606.
123. Patel, G. K., et al. (2019). Comparative analysis of exosome isolation methods using culture supernatant for optimum yield, purity and downstream applications. *Scientific Reports, 9*(1), 5335.
124. Kannan, A., et al. (2017). Genetic mutation and exosome signature of human papilloma virus associated oropharyngeal cancer. *Scientific Reports, 7*, 46102.
125. Philley, J. V., et al. (2017). Exosome secretome and mediated signaling in breast cancer patients with nontuberculous mycobacterial disease. *Oncotarget, 8*(11), 18070–18081.
126. Nakamura, K., et al. (2017). Exosomes promote ovarian cancer cell invasion through transfer of CD44 to peritoneal mesothelial cells. *Molecular Cancer Research, 15*(1), 78–92.
127. Shender, V. O., et al. (2014). Proteome-metabolome profiling of ovarian cancer ascites reveals novel components involved in intercellular communica-

tion. *Molecular & Cellular Proteomics, 13*(12), 3558–3571.
128. Dorayappan, K. D. P., et al. (2016). The biological significance and clinical applications of exosomes in ovarian cancer. *Gynecologic Oncology, 142*(1), 199–205.
129. Srivastava, S. K., et al. (2017). MicroRNAs in gynecological cancers: Small molecules with big implications. *Cancer Letters, 407*, 123–138.
130. Kobayashi, M., et al. (2014). Ovarian cancer cell invasiveness is associated with discordant exosomal sequestration of Let-7 miRNA and miR-200. *Journal of Translational Medicine, 12*, 4.
131. Vaksman, O., et al. (2014). Exosome-derived miRNAs and ovarian carcinoma progression. *Carcinogenesis, 35*(9), 2113–2120.
132. He, L., et al. (2019). Ovarian cancer cell-secreted exosomal miR-205 promotes metastasis by inducing angiogenesis. *Theranostics, 9*(26), 8206–8220.
133. Tang, M. K. S., et al. (2018). Soluble E-cadherin promotes tumor angiogenesis and localizes to exosome surface. *Nature Communications, 9*(1), 2270.
134. Trajkovic, K., et al. (2008). Ceramide triggers budding of exosome vesicles into multivesicular endosomes. *Science, 319*(5867), 1244–1247.
135. Pelosi, G., et al. (2012). DeltaNp63 (p40) and thyroid transcription factor-1 immunoreactivity on small biopsies or cellblocks for typing non-small cell lung cancer: A novel two-hit, sparing-material approach. *Journal of Thoracic Oncology, 7*(2), 281–290.
136. Chalmin, F., et al. (2010). Membrane-associated Hsp72 from tumor-derived exosomes mediates STAT3-dependent immunosuppressive function of mouse and human myeloid-derived suppressor cells. *The Journal of Clinical Investigation, 120*(2), 457–471.
137. Ostrowski, M., et al. (2010). Rab27a and Rab27b control different steps of the exosome secretion pathway. *Nature Cell Biology, 12*(1), 19–30.
138. Sun, D., et al. (2010). A novel nanoparticle drug delivery system: The anti-inflammatory activity of curcumin is enhanced when encapsulated in exosomes. *Molecular Therapy, 18*(9), 1606–1614.
139. Tian, Y., et al. (2014). Doxorubicin-loaded magnetic silk fibroin nanoparticles for targeted therapy of multidrug-resistant cancer. *Advanced Materials, 26*(43), 7393–7398.
140. Johnsen, K. B., et al. (2014). A comprehensive overview of exosomes as drug delivery vehicles – Endogenous nanocarriers for targeted cancer therapy. *Biochimica et Biophysica Acta, 1846*(1), 75–87.
141. Escudier, B., et al. (2005). Vaccination of metastatic melanoma patients with autologous dendritic cell (DC) derived-exosomes: Results of the first phase I clinical trial. *Journal of Translational Medicine, 3*(1), 10.
142. Dai, S., et al. (2008). Phase I clinical trial of autologous ascites-derived exosomes combined with GM-CSF for colorectal cancer. *Molecular Therapy, 16*(4), 782–790.
143. Morse, M. A., et al. (2005). A phase I study of dexosome immunotherapy in patients with advanced non-small cell lung cancer. *Journal of Translational Medicine, 3*(1), 9.
144. Sharma, P., & Allison, J. P. (2015). Immune checkpoint targeting in cancer therapy: Toward combination strategies with curative potential. *Cell, 161*(2), 205–214.
145. Sharma, P., et al. (2011). Novel cancer immunotherapy agents with survival benefit: Recent successes and next steps. *Nature Reviews. Cancer, 11*(11), 805–812.
146. Hamanishi, J., et al. (2015). Safety and antitumor activity of anti-PD-1 antibody, nivolumab, in patients with platinum-resistant ovarian cancer. *Journal of Clinical Oncology, 33*(34), 4015–4022.
147. Konstantinopoulos, P. A., et al. (2019). Single-arm phases 1 and 2 trial of niraparib in combination with pembrolizumab in patients with recurrent platinum-resistant ovarian carcinoma. *JAMA Oncology, 5*(8), 1141–1149.
148. Yan, W., Hu, H., & Tang, B. (2019). Advances of chimeric antigen receptor T cell therapy in ovarian cancer. *Oncotargets and Therapy, 12*, 8015–8022.
149. Chekmasova, A. A., et al. (2010). Successful eradication of established peritoneal ovarian tumors in SCID-Beige mice following adoptive transfer of T cells genetically targeted to the MUC16 antigen. *Clinical Cancer Research, 16*(14), 3594–3606.
150. Yeung, T. L., et al. (2019). Anticancer immunotherapy by MFAP5 blockade inhibits fibrosis and enhances chemosensitivity in ovarian and pancreatic cancer. *Clinical Cancer Research, 25*(21), 6417–6428.
151. Crawford, A., et al. (2019). A Mucin 16 bispecific T cell-engaging antibody for the treatment of ovarian cancer. *Science Translational Medicine, 11*(497), eaau7534.
152. Dou, J., et al. (2009). Antitumor efficacy induced by human ovarian cancer cells secreting IL-21 alone or combination with GM-CSF cytokines in nude mice model. *Immunobiology, 214*(6), 483–492.
153. Yu, Y., et al. (2008). rAAV/Her-2/neu loading of dendritic cells for a potent cellular-mediated MHC class I restricted immune response against ovarian cancer. *Viral Immunology, 21*(4), 435–442.
154. Nersesian, S., et al. (2019). Naturally killing the silent killer: NK cell-based immunotherapy for ovarian cancer. *Frontiers in Immunology, 10*, 1782.
155. Olusola, P., et al. (2019). Human papilloma virus-associated cervical cancer and health disparities. *Cell, 8*(6), 622.

2 Epithelial Ovarian Cancer and Cancer Stem Cells

Amr A. Soliman, Alaa A. Elzarkaa, and Eduard Malik

2.1 Ovarian Cancer, Treatment, and Recurrence

Epithelial ovarian cancer (EOC) accounts for approximately 238,000 new cases annually worldwide and is responsible for at least 150,000 deaths every year [1]. More than 70% of patients present with advanced stages III or IV [2]. The estimated 5-year overall survival (OS) of stages III and IV are 35% and 22%, respectively [3]. The standard treatment of EOC is primary debulking, aiming at complete cytoreduction followed by six cycles of combined carboplatin and paclitaxel and, eventually, bevacizumab [4, 5]. This treatment strategy is based on many studies showing a remarkable benefit for patients who receive a complete macroscopic resection through aggressive cytoreductive surgery [6–8]. Thus, patients with tumor residues <1 cm after primary debulking have relatively worse prognoses, while patients with tumor resides >1 cm have prognoses compared with those who do not undergo debulking [9, 10]. Carboplatin and paclitaxel significantly influence OS compared with other regimens [11–15]. Bevacizumab confers survival benefits upon patients with advanced EOC, particularly those with tumor residues after debulking surgery [16, 17].

Accumulating evidence shows that neoadjuvant chemotherapy followed by interval debulking leads to increases in complete tumor resection rates, with OS rates comparable with those of primary debulking followed by adjuvant chemotherapy [18, 19]. Moreover, this regimen achieves lower morbidity, mortality, and better quality of life [20]. These findings led oncologists to consider neoadjuvant chemotherapy followed by interval debulking as a possible therapeutic option in certain clinical situations, where primary debulking surgery is difficult to perform because of patients' unfavorable general conditions or a very advanced nonresectable tumor stage [4, 21]. Unfortunately, as many as 70% of patients with advanced EOC will experience recurrence after standard treatment [12, 15, 22]. Recurrences are so frequent in such patients that only 10–30% survive long term [23].

EOC recurrences are classified as "platinum-sensitive" and "platinum-resistant," according to the response to initial platinum-based therapy (see below). This classification determines platinum resistance according to recurrence based on clinical symptoms, clinically detectable disease or radiological evidence of disease recurrence, or

A. A. Soliman (✉) · E. Malik
Universitätsklinik für Gynäkologie und Geburtshilfe, Klinikum Oldenburg AöR, Carl von Ossietzky Universität, Oldenburg, Oldenburg, Germany
e-mail: amr.soliman@uni-oldenburg.de

A. A. Elzarkaa
El-Shatby University Maternity Hospital, University of Alexandria, Alexandria, Egypt

H. Schatten (ed.), *Ovarian Cancer: Molecular & Diagnostic Imaging and Treatment Strategies*, Advances in Experimental Medicine and Biology 1330, https://doi.org/10.1007/978-3-030-73359-9_2

both [24–26]. A patient is designated "platinum-sensitive" if she initially responds to platinum-based chemotherapy and does not experience a relapse for ≥6 months after initial treatment. Approximately 30–90% of these patients will respond to further platinum-based chemotherapy with a median survival of 2 years, although survival can range from a few months to more than a decade [27–29]. Many patients will receive multiple lines of treatment over time, but with few exceptions, will ultimately develop platinum-resistant disease. Patients who relapse within 6 months of completing first-line therapy are classified as "platinum-resistant" and typically have response rates <15% to subsequent chemotherapy, progression-free survival of 3–4 months, and median survival <1 year [30].

2.2 Cytotoxic Effects of Platinum and Platinum Resistance

The cytotoxicity of platinum therapy is mainly caused by DNA damage [31]. Platinum-based drugs react with guanine nucleotides to form platinum–DNA mono-adducts, which often react with a second purine nucleotide to form interstrand and intrastrand crosslinks, leading to increased cytotoxicity. Carboplatin and cisplatin form the same platinum–DNA crosslinks in vivo because of their identical cis-diamine ligands [31]. The formation of intrastrand and interstrand crosslinks leads to cell death through apoptosis or necrosis. These processes are irreversible unless the crosslinks are repaired. Apoptosis is executed by a series of cysteine proteases termed caspases [32]. Caspase activation leads to mitochondrial dysfunction [33] and DNA fragmentation [34].

Carboplatin and cisplatin share similar in vitro chemoresistance spectra and clinical indications, although cisplatin is possibly more effective for certain cancers [31]. Factors associated with resistance to platinum include those that limit the formation of cytotoxic platinum–DNA adducts and those that prevent cell death after platinum-adduct formation [35]. The former may result from reduced uptake of cisplatin into cells, increased efflux via alterations to transport proteins, or through inactivation of intracellular cisplatin by its conversion to cisplatin-thiol conjugates. The latter form of resistance may occur through increased DNA repair after adduct formation. The five major DNA repair mechanisms are as follows: nucleotide excision repair, mismatch repair, homologous recombination, base-excision repair, and translesion synthesis [35].

The cancer stem cell (CSC) model and the environment-mediated drug resistance model (EMDR) were proposed to explain the origin of drug-resistant cells [36]. The CSC model proposes that genetic or epigenetic alterations, or both, which occur in multipotent, tissue-specific adult stem cells, may induce malignant transformation to generate CSCs. CSCs possess stem cell-like properties, including self-renewal and cell division to form tumors that acquire further genetic or epigenetic alterations. Such alterations may contribute to the development of invasive properties that allow the tumor to metastasize to distant sites [30, 37, 38]. CSCs may be intrinsically resistant to chemotherapy through different mechanisms and may represent a major source of chemoresistant cells within tumors [39, 40].

In the EMDR model, resistance emerges as the cancer cells interact with their surrounding microenvironment and enter a quiescent state caused by the complex interplay between the tumor and its microenvironment. Tumors that develop a prominent desmoplastic reaction are associated with poor prognosis as well as with platinum resistance [30, 40].

2.3 CSCs

Stem cells are defined as cells that perpetuate through self-renewal and differentiation to mature cells of a particular tissue [41]. Stem cells must therefore be prospectively identified and carefully purified to study their properties. Unfortunately, isolation of tissue-specific stem cells could not be universally achieved, as somatic stem cells identification and isolation have been achieved only in a few instances. For example, human hematopoietic stem cells [42] generate and reconstitute the hematopoietic and immune (hematolymphoid) systems [41, 42]. Although CSCs were originally described in hematological cancers, they have been isolated from solid tumors [43, 44].

An important issue in stem cell biology is understanding the mechanisms that regulate self-renewal. Self-renewal, which is crucial to stem cell function, is a property of diverse stem cells required for their lifelong persistence. Moreover, whereas stem cells from different organs may vary in their developmental potential, all stem cells must self-renew and regulate the relative balance between self-renewal and differentiation. Understanding the regulation of the self-renewal of normal stem cells is fundamental to the understanding of the regulation of cancer cell proliferation because cancer can be considered a disease of unregulated self-renewal.

Another fundamental attribute of stem cells is their transient or long-term quiescence (also termed dormancy) [45], which is a component of the mechanism of regulated self-renewal. Accordingly, stem cells are often identified by their propensity to retain labeled DNA much longer than their rapidly proliferating offspring. Moreover, dormancy may serve as a crucial mechanism for the resistance of CSCs to chemotherapy. The dormancy of CSCs may explain the appearance of local recurrence or distant metastasis after long delays [45]. Figure 2.1 shows a hypothetical model of the CSC concept and its evolution.

2.4 Identification of CSCs

Many markers define CSC populations, and the most commonly reported for solid tumors are CD24, CD44, and CD133. CSC populations are commonly defined by the presence or absence of various combinations of cell surface proteins. Reacting the cells with antibodies against these markers readily identifies cell populations of interest, which are isolated using fluorescence-activated cell sorting [43]. Figure 2.2 shows the currently available markers used to identify different subsets of CSCs in different tumors.

2.5 Ovarian CSCs

CSCs isolated from ovarian cancer are associated with worse prognosis and recurrence [48]. The use of markers of ovarian CSCs, such as CD44, CD24, and CD133, is proposed by recent studies.

2.5.1 CD44

CD44 is a glycoprotein that is widely presented on the outer surface of many mammalian cells such as endothelial cells, epithelial cells, fibroblasts, and leukocytes [46]. CD44 is a surface marker of CSCs in many tissues such as breast, pancreas, gastric, prostate, head, neck, ovarian, and colon [47, 49]. CD44 is associated with diseases such as cancer, arthritis, interstitial lung disease, vascular disease as well as in wound healing and infections. Several studies focus on CD44–HA signaling and its implications in malignancies of solid organs such as breast and ovarian cancer [47, 49, 50].

A single gene encodes CD44, which is located on chromosome 11 in humans and chromosome 2 in mice. There are approximately 20 CD44 isoforms ranging from 80 to 200 kDa. The heterogeneity of this group is generated by post-transcriptional regulation, including alternative splicing and protein modifications. All isoforms are encoded by exons 1–5 and 16–18, whereas exons 6–15 and 19–20 are present in isoforms generated by alternative splicing [50, 51]. Specific to the tissue and isoform, CD44 plays roles in adhesion, motility, proliferation, and cell survival [52]. CD44 contains four major domains, including the conserved extracellular hyaluronan-binding domain and variably spliced regions, the transmembrane sequence, and the intracellular cytoskeletal/signaling domain. Figure 2.3 shows the structure and genomic organization of CD44 [51].

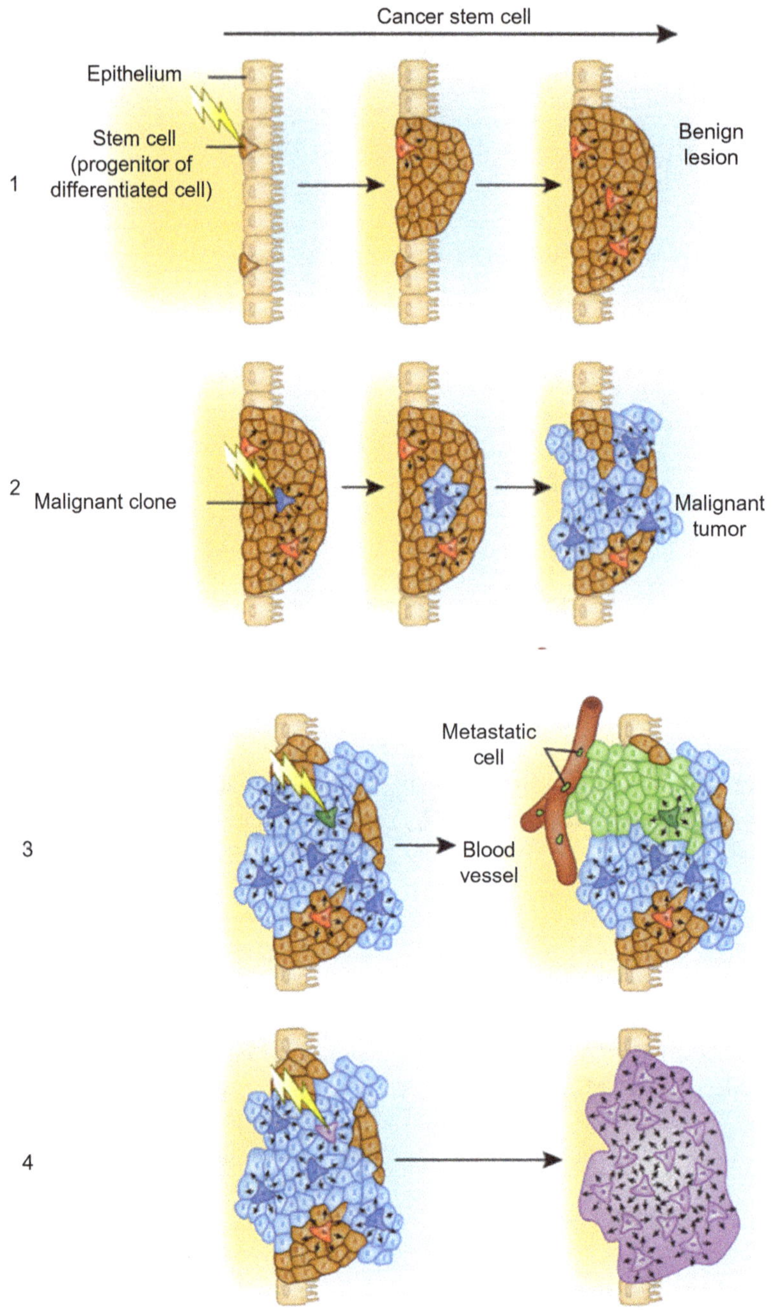

Fig. 2.1 A theoretical synthesis of the clonal Cancer stem cell evolution and CSC concepts. Top to bottom: clonal evolution drives tumor progression [46]. (1) The first oncogenic mutation (lightning arrow) occurs in a stem cell (or, alternatively, benign Stem cell in a progenitor or even a differentiated cell) of lesion (progenitor or a healthy epithelium), resulting in the growth of a genetically homogeneous benign lesion. (2) The second hit targets one of the cells in the benign lesion, which leads to the growth of a more malignant and invasive clone within the primary tumor. (3) A third hit in a cell within the malignant sub clone causes further transformation, visualized as entry into a blood vessel for distant metastasis. Genetically independent sub clones can coexist within the tumor. (4) Final mutational hit leads to tumor being entirely taken over by cells that behave as malignant cancer stem cells. Shown, left to right: at each tumor stage of this clonal evolution process, tumors and sub clones within tumors contain some cells that behave as CSCs

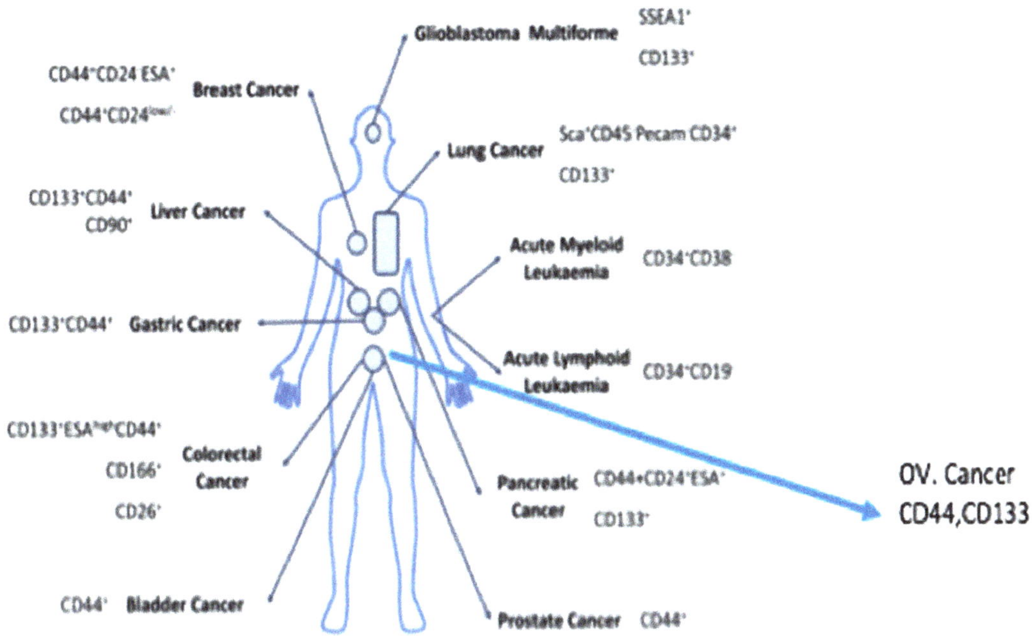

Fig. 2.2 Currently identified surface markers for CSC [47]

Interactions between CD44 and the extracellular matrix glycosaminoglycan hyaluronan (HA) are currently under investigation. HA is enriched in the stem cell niche and likely plays an integral role in the function of CD44 in CSCs [53, 54]. CD44 guides the epithelial stromal reaction with the extracellular microenvironment (ECM) to direct intracellular signaling and modifies the ECM. The extracellular domain of CD44 binds ECM components such as collagen, laminin, fibronectin, and HA [55]. CD44 contains binding sites for glycosaminoglycans other than HA, for example, osteopontin [56]. HA is the best-characterized CD44 ligand and possesses an immense repertoire of biological functions. HA, which is a cell-surface-associated glycosaminoglycan that is ubiquitous in extracellular and pericellular matrices, is synthesized and simultaneously secreted by transmembrane HA synthases as a 106–107 kDa polymer [56, 57].

CD44 modulates many signaling activities through interactions with its cytoplasmic tail. Treatment with soluble low or high molecular weight HA induces cell invasion and migration through CD44-mediated activation of Rho family GTPases. Hyaluronan–CD44 interactions initiate recruitment of signaling molecules such as Tiam1, p115, Rac1, Rho Gefs, Rho-associated protein kinase, and cSrc. Interactions with signaling molecules lead to activation of the PI3K signaling pathway and a number of cellular functions such as survival and invasion [57]. Figure 2.4 illustrates CD44-mediated signal transduction [57] and Fig. 2.5 shows an example of immunohistochemical detection of CD44.

2.5.2 CD24

Mouse CD24 was first identified as a heat-stable antigen 30 years ago, and the CD24 gene was molecularly cloned and found to encode a small protein whose mature form comprises 27 amino acid residues [58, 59]. Human CD24 is located on chromosome 6q21, as determined by in situ hybridization [60]. The CD24 isoforms isolated from different tissues or cell types have different molecular masses, ranging from 20 to 70 kDa, depending on cell or tissue type, demonstrating that the glycosylation of CD24 is highly variable and cell-type dependent [61]. CD24 is expressed by hematopoietic cells such as B cells and T cells

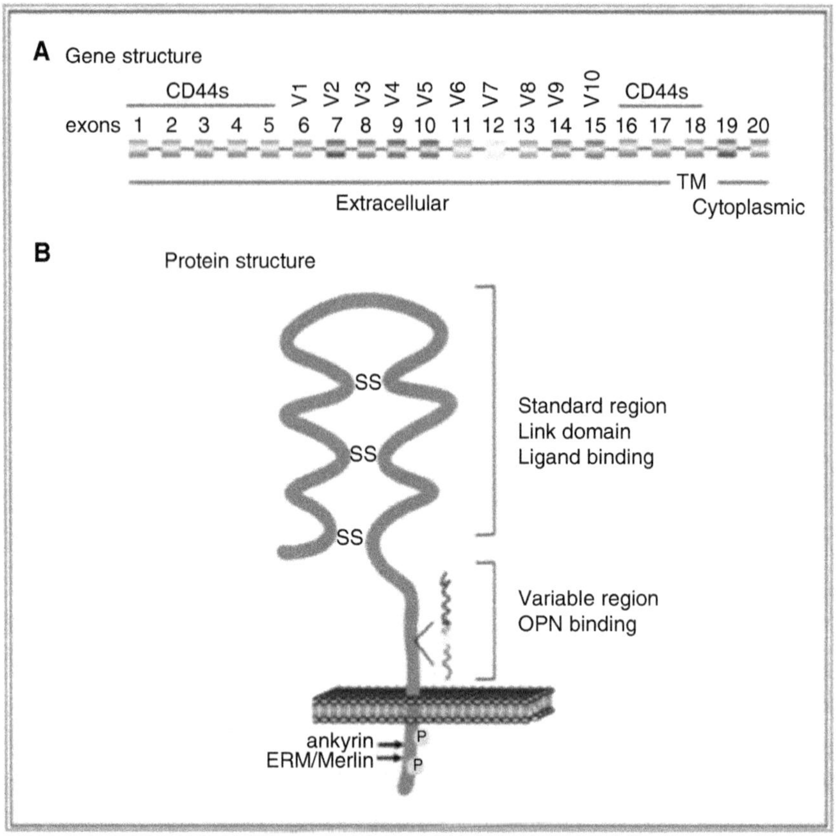

Fig. 2.3 Genetic encoding of CD44 (**a**) and its basic structure (**b**) [52]

as well as by nonhematopoietic cells such as neurons, epithelial cells, and epithelial stem cells [61–63].

The main role of CD24 in most cell types is unclear; however, certain immune regulatory functions of CD24 are known [64]. CD24 is broadly overexpressed by many types of tumor tissues, particularly those of the breast [65] and ovary [66]. For example, in breast cancer, cell surface and cytoplasmic expression of CD24 is associated with poor prognosis, histological grades, tumor size, and lymph node positivity [65, 67]. CD24 is expressed in epithelial ovarian cancer. Although most published work demonstrates an association of CD24 expression with advanced disease stage and poor prognosis, the association is controversial. $CD24^+$ cells exhibit increased tumor-forming and tumor-initiating capacities. Interestingly, $CD24^+$ or $CD24^-$ cells can initiate a tumor. This may be explained by in vitro and in vivo lineage tracking experiments showing the conversion of $CD24^-$ to $CD24^+$ cells. Therefore, $CD24^+$ may act as a transition phase between cancer stem cells and tumorigenesis [68].

2.5.3 CD133

CD133 is a surface marker that was identified in epithelial ovarian cancer [69], endometrial cancer, neuronal cancer, and colon cancer [70]. The role of CD133 in tumor progression is unclear. CD133 may serve as a prognostic marker of low-risk endometrial cancer [71]. The expression of CD133 may be associated with enhanced tumorigenesis in animal models of human melanoma and colon cancer [72, 73]. The role of CD133 in disease progression and resistance to chemotherapy is unclear.

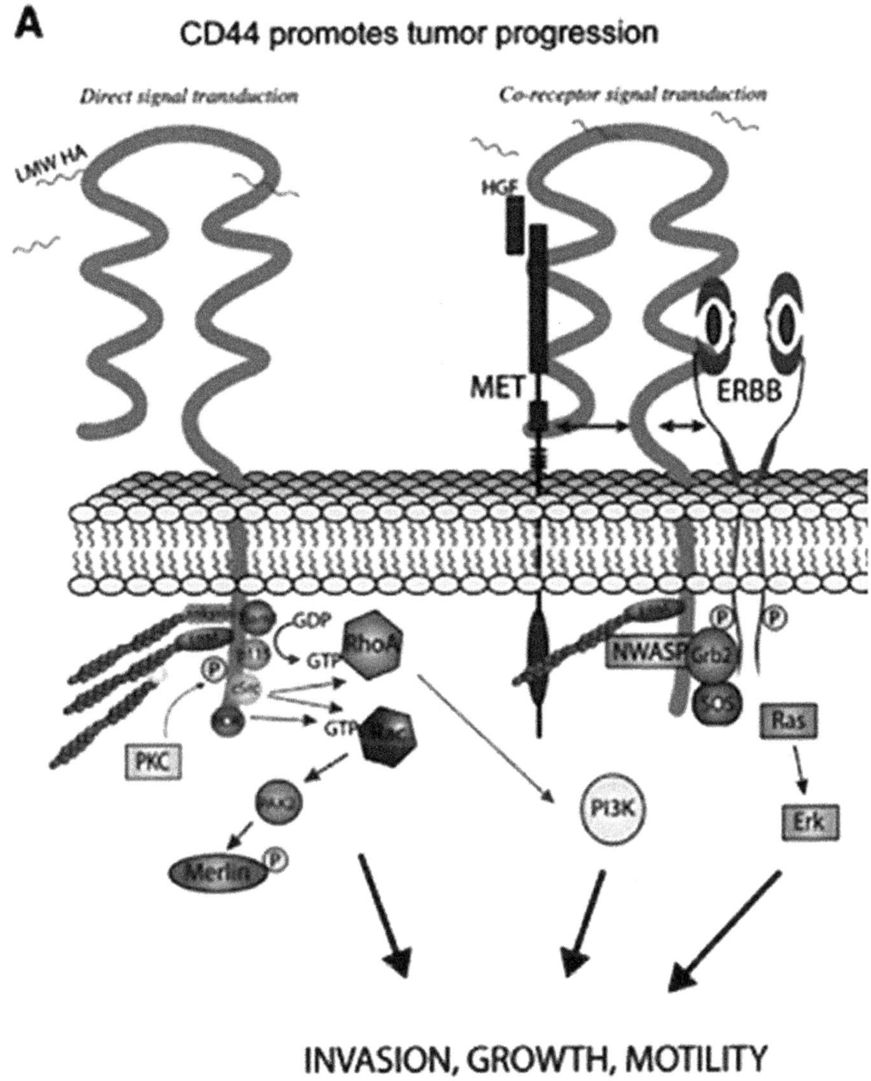

Fig. 2.4 Signal transduction of CD44 [58]

2.6 Resistance to Chemotherapy and CSCs

It is often suggested that CSCs are resistant to therapy in the same way that normal stem cells are protected against insult. These protections include the aforementioned quiescence as well as expression of drug pumps, high expression of antiapoptotic proteins, and resistance to DNA damage [74]. Some groups have started to determine if CSCs are more resistant to therapy than their progeny. For example, CD133-expressing glioma cells are more resistant to ionizing radiation compared with CD133^{negative} tumor cells [75]. CD44^{high}/CD24^{low} breast cancer CSCs appear intrinsically resistant to conventional chemotherapy and ionizing radiation [76], and chronic myeloid leukemia is sustained by leukemic stem cells that are relatively resistant to imatinib [77].

In EOC, platinum resistance is a very important issue because of the high recurrence rate of the disease. Several studies attempted to explain the development of platinum resistance in EOC, but there is no consensus regarding its

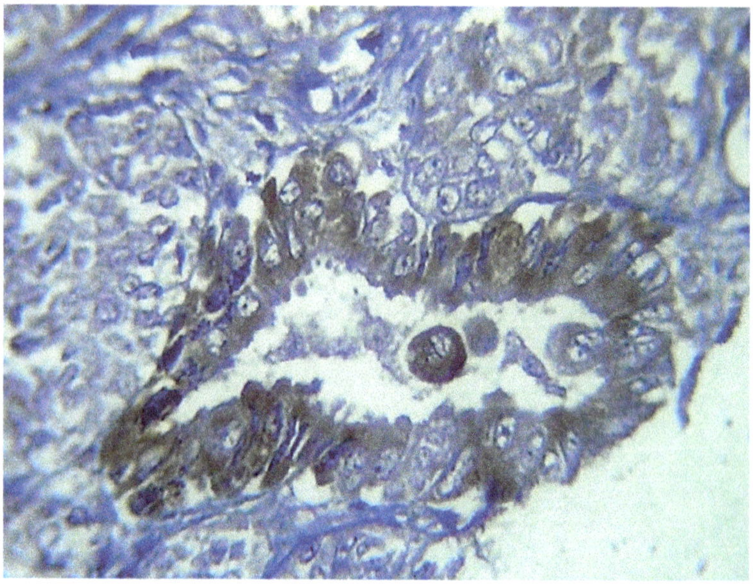

Fig. 2.5 Peculiar pattern of CD44 staining restricted to scattered tumor glands showing moderately intense staining. (Courtesy of Prof Dr Bassma El-Sabaa)

development. Patients with primary "platinum refractory" disease are intrinsically drug resistant and do not respond or progress very early following treatment. Primary platinum-refractory ovarian cancers are uncommon and usually occur with nonserous ovarian cancers such as clear cell carcinoma or mucinous carcinoma vs. the more common high-grade serous carcinoma. It is likely that the mechanisms of resistance among these various histotypes are very different.

Patients who experience an initial response to platinum chemotherapy may have tumors comprising populations of intrinsically platinum-resistant and platinum-sensitive cells. The sensitive cells undergo apoptosis following chemotherapy (tumor response), but the resistant subpopulation persists and expands, leading to early recurrence in platinum-resistant disease. Platinum-sensitive patients may respond repeatedly to platinum, because of the regrowth of the sensitive population. Ultimately however the sensitive cells may alter, rendering them resistant, or the resistant cell population will outgrow the sensitive population [30].

Another important characteristic of EOC is its heterogeneity. Heterogeneity exists spatially within the primary tumor and between the primary tumor and its metastases that are transient, as indicated by biopsies performed at different times [78, 79]. This heterogeneity significantly adds to the complexity of assessing or interpreting the response to treatment and patients' outcomes. This property is supported by anecdotal clinical observations of patients with differential responses to treatment, with progression at one site and responses at other sites. The mechanisms that explain how frequently this occurs are unknown, and there is no guidance or consensus on the appropriate management of these patients. In future studies, particularly of targeted therapies, repeat biopsies upon recurrence and after further treatment will be essential to gain a better understanding of the mechanisms of resistance. The CSC theory can explain this heterogeneity, where different subsets of CSCs proliferate in the same tumor and during the development of different metastasis, leading to different phenotypes of the same tumor.

Treatment of EOC recurrence is a dilemma, particularly for platinum-resistant patients. Chemotherapeutic agents such as doxorubicin, gemcitabine, topotecan, trabectedin, and paclitaxel as well as targeted therapies such as bevacizumab, olaparib, and niraparib were evaluated in clinical trials designed to develop a strategy to achieve an adequate response. Until recently, phase III trials did not reveal any significant improvement in the progression-free interval

(PFI) or OS. Two studies of chemotherapy plus an antiangiogenic agent achieved improved PFI, but not OS, in the platinum-resistant subset. The AURELIA study (involving chemotherapy combined with bevacizumab) achieved an approximate doubling of the PFI (3.4 vs. 6.7 months, HR 0.48, $p < 0.001$) vs. without bevacizumab, but no improvement in OS [80]. The TRINOVA 1 study (paclitaxel combined with the angiopoietin 1/2 inhibitor trebananib) achieved an improved PFI (5.4 vs. 7.2 months, $p < 0.001$) [81].

Poly ADP-ribose inhibitors achieved promising results in reducing the recurrence of EOC. Patients with platinum-sensitive recurrence with or without a BRCA mutation experience a slightly better response to olaparib and niraparib [82, 83]. An unanswered question is if the same effect will appear in the platinum-resistant subset of patients.

Recurrent EOC remains a significant treatment dilemma, mainly because of our limited understanding of the development of the resistance to chemotherapy. CSCs may contribute to recurrent EOC, and this will remain a hypothetical possibility until experimentally and clinically verified.

Acknowledgments We thank Prof Dr. Bassma El-Sabaa for her courtesy in providing material for Fig. 2.5.

We thank Edanz Group (www.edanzediting.com/ac) for editing a draft of this manuscript.

References

1. Torre, L. A., Bray, F., Siegel, R. L., Ferlay, J., Lortet-Tieulent, J., & Jemal, A. (2015). Global cancer statistics, 2012. *CA: A Cancer Journal for Clinicians, 65*, 87–108.
2. Sehouli, J., Senyuva, F., Fotopoulou, C., Neumann, U., Denkert, C., Werner, L., et al. (2009). Intra-abdominal tumor dissemination pattern and surgical outcome in 214 patients with primary ovarian cancer. *Journal of Surgical Oncology, 99*, 424–427.
3. Garcia, M., Jemal, A., Ward, E., Center, M., Hao, Y., Siegel, R., et al. (2007). *Global cancer facts & figures 2007*. American Cancer Society: Atlanta, GA.
4. National-Cancer-Comprehensive-Network. (2017). *NCCN clinical practice guidelines in oncology (NCCN Guidelines®) ovarian cancer*. NCCN.
5. AGO-Leitlinienkommission. (2013). S3-Leitlinie Diagnostik, Therapie und Nachsorge maligner Ovarialtumoren.
6. du Bois, A., Reuss, A., Pujade-Lauraine, E., Harter, P., Ray-Coquard, I., & Pfisterer, J. (2009). Role of surgical outcome as prognostic factor in advanced epithelial ovarian cancer: A combined exploratory analysis of 3 prospectively randomized phase 3 multicenter trials: By the Arbeitsgemeinschaft Gynaekologische Onkologie Studiengruppe Ovarialkarzinom (AGO-OVAR) and the Groupe d'Investigateurs Nationaux Pour les Etudes des Cancers de l'Ovaire (GINECO). *Cancer, 115*, 1234–1244.
7. Trimbos, J. B., Vergote, I., Bolis, G., Vermorken, J. B., Mangioni, C., Madronal, C., et al. (2003). Impact of adjuvant chemotherapy and surgical staging in early-stage ovarian carcinoma: European Organisation for Research and Treatment of Cancer-Adjuvant ChemoTherapy in Ovarian Neoplasm trial. *Journal of the National Cancer Institute, 95*, 113–125.
8. Hoskins, W. J., McGuire, W. P., Brady, M. F., Homesley, H. D., Creasman, W. T., Berman, M., et al. (1994). The effect of diameter of largest residual disease on survival after primary cytoreductive surgery in patients with suboptimal residual epithelial ovarian carcinoma. *American Journal of Obstetrics and Gynecology, 170*, 974–9; discussion 9–80.
9. Winter, W. E., III, Maxwell, G. L., Tian, C., Carlson, J. W., Ozols, R. F., Rose, P. G., et al. (2007). Prognostic factors for stage III epithelial ovarian cancer: A Gynecologic Oncology Group Study. *Journal of Clinical Oncology, 25*, 3621–3627.
10. Ataseven, B., Grimm, C., Harter, P., Heitz, F., Traut, A., Prader, S., et al. (2015). Prognostic impact of debulking surgery and residual tumor in patients with epithelial ovarian cancer FIGO stage IV. *Gynecologic Oncology, 140*(2), 215–220.
11. Piccart, M. J., Bertelsen, K., James, K., Cassidy, J., Mangioni, C., Simonsen, E., et al. (2000). Randomized intergroup trial of cisplatin-paclitaxel versus cisplatin-cyclophosphamide in women with advanced epithelial ovarian cancer: Three-year results. *Journal of the National Cancer Institute, 92*, 699–708.
12. McGuire, W. P., Hoskins, W. J., Brady, M. F., Kucera, P. R., Partridge, E. E., Look, K. Y., et al. (1996). Cyclophosphamide and cisplatin compared with paclitaxel and cisplatin in patients with stage III and stage IV ovarian cancer. *The New England Journal of Medicine, 334*, 1–6.
13. Omura, G., Blessing, J. A., Ehrlich, C. E., Miller, A., Yordan, E., Creasman, W. T., et al. (1986). A randomized trial of cyclophosphamide and doxorubicin with or without cisplatin in advanced ovarian carcinoma. A Gynecologic Oncology Group Study. *Cancer, 57*, 1725–1730.
14. du Bois, A., Luck, H. J., Meier, W., Adams, H. P., Mobus, V., Costa, S., et al. (2003). A randomized clinical trial of cisplatin/paclitaxel versus carboplatin/paclitaxel as first-line treatment of ovarian

cancer. *Journal of the National Cancer Institute, 95*, 1320–1329.

15. Ozols, R. F., Bundy, B. N., Greer, B. E., Fowler, J. M., Clarke-Pearson, D., Burger, R. A., et al. (2003). Phase III trial of carboplatin and paclitaxel compared with cisplatin and paclitaxel in patients with optimally resected stage III ovarian cancer: A Gynecologic Oncology Group study. *Journal of Clinical Oncology, 21*, 3194–3200.
16. Burger, R. A., Brady, M. F., Bookman, M. A., Fleming, G. F., Monk, B. J., Huang, H., et al. (2011). Incorporation of bevacizumab in the primary treatment of ovarian cancer. *The New England Journal of Medicine, 365*, 2473–2483.
17. Oza, A. M., Cook, A. D., Pfisterer, J., Embleton, A., Ledermann, J. A., Pujade-Lauraine, E., et al. (2015). Standard chemotherapy with or without bevacizumab for women with newly diagnosed ovarian cancer (ICON7): Overall survival results of a phase 3 randomised trial. *The Lancet Oncology, 16*, 928–936.
18. Kehoe, S., Hook, J., Nankivell, M., Jayson, G. C., Kitchener, H., Lopes, T., et al. (2015). Primary chemotherapy versus primary surgery for newly diagnosed advanced ovarian cancer (CHORUS): An open-label, randomised, controlled, non-inferiority trial. *The Lancet, 386*, 249–257.
19. Vergote, I., Amant, F., Kristensen, G., Ehlen, T., Reed, N. S., & Casado, A. (2011). Primary surgery or neoadjuvant chemotherapy followed by interval debulking surgery in advanced ovarian cancer. *European Journal of Cancer, 47*, S88–S92.
20. Fagotti, A., Ferrandina, G., Vizzielli, G., Fanfani, F., Gallotta, V., Chiantera, V., et al. (2016). Phase III randomised clinical trial comparing primary surgery versus neoadjuvant chemotherapy in advanced epithelial ovarian cancer with high tumour load (SCORPION trial): Final analysis of peri-operative outcome. *European Journal of Cancer, 59*, 22–33.
21. Wright, A. A., Bohlke, K., Armstrong, D. K., Bookman, M. A., Cliby, W. A., Coleman, R. L., et al. (2016). Neoadjuvant chemotherapy for newly diagnosed, advanced ovarian cancer: Society of Gynecologic Oncology and American Society of Clinical Oncology clinical practice guideline. *Journal of Clinical Oncology, 34*, 3460–3473.
22. Armstrong, D. K., Bundy, B., Wenzel, L., Huang, H. Q., Baergen, R., Lele, S., et al. (2006). Intraperitoneal cisplatin and paclitaxel in ovarian cancer. *The New England Journal of Medicine, 354*, 34–43.
23. Cannistra, S. A. (2004). Cancer of the ovary. *The New England Journal of Medicine, 351*, 2519–2529.
24. Markman, M., Rothman, R., Hakes, T., Reichman, B., Hoskins, W., Rubin, S., et al. (1991). Second-line platinum therapy in patients with ovarian cancer previously treated with cisplatin. *Journal of Clinical Oncology, 9*, 389–393.
25. Blackledge, G., Lawton, F., Redman, C., & Kelly, K. (1989). Response of patients in phase II studies of chemotherapy in ovarian cancer: Implications for patient treatment and the design of phase II trials. *British Journal of Cancer, 59*, 650–653.
26. Gore, M. E., Fryatt, I., Wiltshaw, E., & Dawson, T. (1990). Treatment of relapsed carcinoma of the ovary with cisplatin or carboplatin following initial treatment with these compounds. *Gynecologic Oncology, 36*, 207–211.
27. Balbi, G., Di Prisco, L., Musone, R., Menditto, A., Cassese, E., Balbi, C., et al. (2002). Second-line with paclitaxel and carboplatin for recurrent disease following first paclitaxel and platinum compounds in ovarian carcinoma. *European Journal of Gynaecological Oncology, 23*, 347–349.
28. Hoekstra, A. V., Hurteau, J. A., Kirschner, C. V., & Rodriguez, G. C. (2009). The combination of monthly carboplatin and weekly paclitaxel is highly active for the treatment of recurrent ovarian cancer. *Gynecologic Oncology, 115*, 377–381.
29. Rose, P. G., Fusco, N., Fluellen, L., & Rodriguez, M. (1998). Second-line therapy with paclitaxel and carboplatin for recurrent disease following first-line therapy with paclitaxel and platinum in ovarian or peritoneal carcinoma. *Journal of Clinical Oncology, 16*, 1494–1497.
30. Davis, A., Tinker, A. V., & Friedlander, M. (2014). "Platinum resistant" ovarian cancer: What is it, who to treat and how to measure benefit? *Gynecologic Oncology, 133*, 624–631.
31. Goodisman, J., Hagrman, D., Tacka, K. A., & Souid, A. K. (2006). Analysis of cytotoxicities of platinum compounds. *Cancer Chemotherapy and Pharmacology, 57*, 257–267.
32. Cohen, G. M. (1997). Caspases: The executioners of apoptosis. *The Biochemical Journal, 326*(Pt 1), 1–16.
33. Petit, P. X., Zamzami, N., Vayssiere, J. L., Mignotte, B., Kroemer, G., & Castedo, M. (1997). Implication of mitochondria in apoptosis. *Molecular and Cellular Biochemistry, 174*, 185–188.
34. Nagata, S. (2000). Apoptotic DNA fragmentation. *Experimental Cell Research, 256*, 12–18.
35. Tapia, G., & Diaz-Padilla, I. (2013). Molecular mechanisms of platinum resistance in ovarian cancer. In I. Diaz-Padilla (Ed.), *Ovarian cancer—A clinical and translational update*. Rijeka: InTech.
36. Siddik, Z. H. (2003). Cisplatin: Mode of cytotoxic action and molecular basis of resistance. *Oncogene, 22*, 7265–7279.
37. Dean, M., Fojo, T., & Bates, S. (2005). Tumour stem cells and drug resistance. *Nature Reviews. Cancer, 5*, 275–284.
38. Massard, C., Deutsch, E., & Soria, J. C. (2006). Tumour stem cell-targeted treatment: Elimination or differentiation. *Annals of Oncology, 17*, 1620–1624.
39. Mimeault, M., Hauke, R., Mehta, P. P., & Batra, S. K. (2007). Recent advances in cancer stem/progenitor cell research: Therapeutic implications for overcoming resistance to the most aggressive cancers. *Journal of Cellular and Molecular Medicine, 11*, 981–1011.
40. Helleman, J., Jansen, M. P., Burger, C., van der Burg, M. E., & Berns, E. M. (2010). Integrated genomics

of chemotherapy resistant ovarian cancer: A role for extracellular matrix, TGFbeta and regulating microRNAs. *The International Journal of Biochemistry & Cell Biology, 42*, 25–30.

41. Reya, T., Morrison, S. J., Clarke, M. F., & Weissman, I. L. (2001). Stem cells, cancer, and cancer stem cells. *Nature, 414*, 105–111.
42. Baum, C. M., Weissman, I. L., Tsukamoto, A. S., Buckle, A. M., & Peault, B. (1992). Isolation of a candidate human hematopoietic stem-cell population. *Proceedings of the National Academy of Sciences of the United States of America, 89*, 2804–2808.
43. Lapidot, T., Sirard, C., Vormoor, J., Murdoch, B., Hoang, T., Caceres-Cortes, J., et al. (1994). A cell initiating human acute myeloid leukaemia after transplantation into SCID mice. *Nature, 367*, 645–648.
44. Al-Hajj, M., Wicha, M. S., Benito-Hernandez, A., Morrison, S. J., & Clarke, M. F. (2003). Prospective identification of tumorigenic breast cancer cells. *Proceedings of the National Academy of Sciences of the United States of America, 100*, 3983–3988.
45. Ebben, J. D., Treisman, D. M., Zorniak, M., Kutty, R. G., Clark, P. A., & Kuo, J. S. (2010). The cancer stem cell paradigm: A new understanding of tumor development and treatment. *Expert Opinion on Therapeutic Targets, 14*, 621–632.
46. Sherman, L., Sleeman, J., Herrlich, P., & Ponta, H. (1994). Hyaluronate receptors: Key players in growth, differentiation, migration and tumor progression. *Current Opinion in Cell Biology, 6*, 726–733.
47. Slevin, M., Krupinski, J., Gaffney, J., Matou, S., West, D., Delisser, H., et al. (2007). Hyaluronan-mediated angiogenesis in vascular disease: Uncovering RHAMM and CD44 receptor signaling pathways. *Matrix Biology, 26*, 58–68.
48. Elzarkaa, A. A., Sabaa, B. E., Abdelkhalik, D., Mansour, H., Melis, M., Shaalan, W., et al. (2016). Clinical relevance of CD44 surface expression in advanced stage serous epithelial ovarian cancer: A prospective study. *Journal of Cancer Research and Clinical Oncology, 142*, 949–958.
49. Misra, S., Hascall, V. C., Berger, F. G., Markwald, R. R., & Ghatak, S. (2008). Hyaluronan, CD44, and cyclooxygenase-2 in colon cancer. *Connective Tissue Research, 49*, 219–224.
50. Louderbough, J. M., & Schroeder, J. A. (2011). Understanding the dual nature of CD44 in breast cancer progression. *Molecular Cancer Research, 9*, 1573–1586.
51. Screaton, G. R., Bell, M. V., Jackson, D. G., Cornelis, F. B., Gerth, U., & Bell, J. I. (1992). Genomic structure of DNA encoding the lymphocyte homing receptor CD44 reveals at least 12 alternatively spliced exons. *Proceedings of the National Academy of Sciences of the United States of America, 89*, 12160–12164.
52. Marhaba, R., & Zoller, M. (2004). CD44 in cancer progression: Adhesion, migration and growth regulation. *Journal of Molecular Histology, 35*, 211–231.
53. Tölg, C., Hofmann, M., Herrlich, P., & Ponta, H. (1993). Splicing choice from ten variant exons establishes CD44 variability. *Nucleic Acids Research, 21*, 1225–1229.
54. Haylock, D. N., & Nilsson, S. K. (2006). The role of hyaluronic acid in hemopoietic stem cell biology. *Regenerative Medicine, 1*, 437–445.
55. Aruffo, A., Stamenkovic, I., Melnick, M., Underhill, C. B., & Seed, B. (1990). CD44 is the principal cell surface receptor for hyaluronate. *Cell, 61*, 1303–1313.
56. Weber, G. F., Ashkar, S., Glimcher, M. J., & Cantor, H. (1996). Receptor-ligand interaction between CD44 and osteopontin (Eta-1). *Science, 271*, 509–512.
57. Stern, R., Asari, A. A., & Sugahara, K. N. (2006). Hyaluronan fragments: An information-rich system. *European Journal of Cell Biology, 85*, 699–715.
58. Springer, T., Galfre, G., Secher, D. S., & Milstein, C. (1978). Monoclonal xenogeneic antibodies to murine cell surface antigens: Identification of novel leukocyte differentiation antigens. *European Journal of Immunology, 8*, 539–551.
59. Kay, R., Takei, F., & Humphries, R. K. (1990). Expression cloning of a cDNA encoding M1/69-J11d heat-stable antigens. *Journal of Immunology, 145*, 1952–1959.
60. Hough, M. R., Rosten, P. M., Sexton, T. L., Kay, R., & Humphries, R. K. (1994). Mapping of CD24 and homologous sequences to multiple chromosomal loci. *Genomics, 22*, 154–161.
61. Rougon, G., Alterman, L. A., Dennis, K., Guo, X. J., & Kinnon, C. (1991). The murine heat-stable antigen: A differentiation antigen expressed in both the hematolymphoid and neural cell lineages. *European Journal of Immunology, 21*, 1397–1402.
62. Shackleton, M., Vaillant, F., Simpson, K. J., Stingl, J., Smyth, G. K., Asselin-Labat, M. L., et al. (2006). Generation of a functional mammary gland from a single stem cell. *Nature, 439*, 84–88.
63. Lawson, D. A., Xin, L., Lukacs, R. U., Cheng, D., & Witte, O. N. (2007). Isolation and functional characterization of murine prostate stem cells. *Proceedings of the National Academy of Sciences of the United States of America, 104*, 181–186.
64. Bai, X. F., Li, O., Zhou, Q., Zhang, H., Joshi, P. S., Zheng, X., et al. (2004). CD24 controls expansion and persistence of autoreactive T cells in the central nervous system during experimental autoimmune encephalomyelitis. *The Journal of Experimental Medicine, 200*, 447–458.
65. Kristiansen, G., Winzer, K. J., Mayordomo, E., Bellach, J., Schluns, K., Denkert, C., et al. (2003). CD24 expression is a new prognostic marker in breast cancer. *Clinical Cancer Research, 9*, 4906–4913.
66. Kristiansen, G., Denkert, C., Schluns, K., Dahl, E., Pilarsky, C., & Hauptmann, S. (2002). CD24 is expressed in ovarian cancer and is a new independent prognostic marker of patient survival. *The American Journal of Pathology, 161*, 1215–1221.
67. Athanassiadou, P., Grapsa, D., Gonidi, M., Athanassiadou, A. M., Tsipis, A., & Patsouris, E. (2009). CD24 expression has a prognostic impact in

breast carcinoma. *Pathology, Research and Practice, 205*, 524–533.
68. Burgos-Ojeda, D., Wu, R., McLean, K., Chen, Y. C., Talpaz, M., Yoon, E., et al. (2015). CD24+ ovarian cancer cells are enriched for cancer-initiating cells and dependent on JAK2 signaling for growth and metastasis. *Molecular Cancer Therapeutics, 14*, 1717–1727.
69. Szotek, P. P., Pieretti-Vanmarcke, R., Masiakos, P. T., Dinulescu, D. M., Connolly, D., Foster, R., et al. (2006). Ovarian cancer side population defines cells with stem cell-like characteristics and Mullerian inhibiting substance responsiveness. *Proceedings of the National Academy of Sciences of the United States of America, 103*, 11154–11159.
70. Ricci-Vitiani, L., Lombardi, D. G., Pilozzi, E., Biffoni, M., Todaro, M., Peschle, C., et al. (2007). Identification and expansion of human colon-cancer-initiating cells. *Nature, 445*, 111–115.
71. Mancebo, G., Sole-Sedeno, J. M., Pino, O., Miralpeix, E., Mojal, S., Garrigos, L., et al. (2017). Prognostic impact of CD133 expression in endometrial cancer patients. *Scientific Reports, 7*, 7687.
72. O'Brien, C. A., Pollett, A., Gallinger, S., & Dick, J. E. (2007). A human colon cancer cell capable of initiating tumour growth in immunodeficient mice. *Nature, 445*, 106–110.
73. Monzani, E., Facchetti, F., Galmozzi, E., Corsini, E., Benetti, A., Cavazzin, C., et al. (2007). Melanoma contains CD133 and ABCG2 positive cells with enhanced tumourigenic potential. *European Journal of Cancer, 43*, 935–946.
74. Zhou, B. B., Zhang, H., Damelin, M., Geles, K. G., Grindley, J. C., & Dirks, P. B. (2009). Tumour-initiating cells: Challenges and opportunities for anticancer drug discovery. *Nature Reviews. Drug Discovery, 8*, 806–823.
75. Bao, S., Wu, Q., McLendon, R. E., Hao, Y., Shi, Q., Hjelmeland, A. B., et al. (2006). Glioma stem cells promote radioresistance by preferential activation of the DNA damage response. *Nature, 444*, 756–760.
76. Li, X., Lewis, M. T., Huang, J., Gutierrez, C., Osborne, C. K., Wu, M. F., et al. (2008). Intrinsic resistance of tumorigenic breast cancer cells to chemotherapy. *Journal of the National Cancer Institute, 100*, 672–679.
77. O'Hare, T., Corbin, A. S., & Druker, B. J. (2006). Targeted CML therapy: Controlling drug resistance, seeking cure. *Current Opinion in Genetics & Development, 16*, 92–99.
78. Gerlinger, M., Rowan, A. J., Horswell, S., Math, M., Larkin, J., Endesfelder, D., et al. (2012). Intratumor heterogeneity and branched evolution revealed by multiregion sequencing. *The New England Journal of Medicine, 366*, 883–892.
79. Bashashati, A., Ha, G., Tone, A., Ding, J., Prentice, L. M., Roth, A., et al. (2013). Distinct evolutionary trajectories of primary high-grade serous ovarian cancers revealed through spatial mutational profiling. *The Journal of Pathology, 231*, 21–34.
80. Poveda, A. M., Selle, F., Hilpert, F., Reuss, A., Savarese, A., Vergote, I., et al. (2015). Bevacizumab combined with weekly paclitaxel, Pegylated liposomal doxorubicin, or topotecan in platinum-resistant recurrent ovarian cancer: Analysis by chemotherapy cohort of the randomized phase III AURELIA trial. *Journal of Clinical Oncology, 33*, 3836–3838.
81. Monk, B. J., Poveda, A., Vergote, I., Raspagliesi, F., Fujiwara, K., Bae, D.-S., et al. (2014). Anti-angiopoietin therapy with trebananib for recurrent ovarian cancer (TRINOVA-1): A randomised, multicentre, double-blind, placebo-controlled phase 3 trial. *The Lancet Oncology, 15*, 799–808.
82. Oza, A. M., Cibula, D., Benzaquen, A. O., Poole, C., Mathijssen, R. H. J., Sonke, G. S., et al. (2015). Olaparib combined with chemotherapy for recurrent platinum-sensitive ovarian cancer: A randomised phase 2 trial. *The Lancet Oncology, 16*, 87–97.
83. Mirza, M. R., Monk, B. J., Herrstedt, J., Oza, A. M., Mahner, S., Redondo, A., et al. (2016). Niraparib maintenance therapy in platinum-sensitive, recurrent ovarian cancer. *The New England Journal of Medicine, 375*, 2154–2164.

Ovarian Cancer: Therapeutic Strategies to Overcome Immune Suppression

3

Maureen L. Drakes and Patrick J. Stiff

3.1 Introduction

Approximately 85% of ovarian cancer cases are of ovarian and/or the fallopian tube epithelial origin, while others may arise from germ, epidermoid, stromal, and border cells [1]. Epithelial ovarian cancers consist of five different histological types: high-grade serous (HGSOC), low-grade serous, endometroid, clear cell, and mucinous cancer [2]. Features of HGSOC, the most frequently occurring group, include rapid progression, ascites formation in the peritoneal cavity, metastasis to distant organs, high recurrence rate, and resistance to conventional therapy. After diagnosis, cytoreductive surgery is usually performed to remove visible tumor masses. Tumors are staged (I–IV) by the International Federation of Gynecology and Obstetrics (FIGO) guidelines, where Stage IV tumors are those which have metastasized to distant sites. Low-grade tumors are slower growing and more genetically stable than high-grade tumors. In HGSOC the most common mutation is TP53 which is found in over 90% of patients [3]. Additionally, greater than 50% of HGSOC patients have DNA repair pathway deficiency in genes including those of BRCA1 and BRCA2 proteins [4]. This latter finding is harnessed in the development of novel therapy for the disease, as in the case of poly ADP-ribose polymerase (PARP) inhibitors [5, 6]. Other treatments which show promise and are currently in clinical trials include immune modulators such as immune checkpoint (IC) blockers, c-MET family inhibitors, dendritic cell vaccine therapy, and adoptive T cell-based therapies. Despite the fact that many preclinical studies are aimed at deciphering relevant immune and molecular biomarkers for therapy, there has only been minimal success with novel agents entering clinical practice to transform ovarian cancer outcome. Thus, statistics for this disease remain grim, and it is projected that in 2020 in the United States the numbers of ovarian cancer cases diagnosed will reach 21,750, and ovarian cancer deaths will be about 13,940 [7]. The text below will focus on immune resistance mechanisms which may favor ovarian cancer progression and impede the efficacy of novel therapies for the disease, and discuss the status of novel therapeutic strategies which have potential to improve survival in ovarian cancer.

M. L. Drakes (✉) · P. J. Stiff
Department of Medicine, Cardinal Bernardin Cancer Center, Loyola University Chicago, Maywood, IL, USA
e-mail: mdrakes@luc.edu

H. Schatten (ed.), *Ovarian Cancer: Molecular & Diagnostic Imaging and Treatment Strategies*, Advances in Experimental Medicine and Biology 1330, https://doi.org/10.1007/978-3-030-73359-9_3

3.2 Factors Regulating a Tumor-Promoting Equilibrium in Ovarian Cancer

The tumor microenvironment (TME) of HGSOC has revealed a network of cells and surrounding milieu which fosters tumorigenicity. These include T regulatory cells, exhausted T cells, immature dendritic cells (DC), plasmacytoid dendritic cells, tumor-associated macrophages, some natural killer (NK) cells, soluble molecules released by these cells, as well as immune checkpoint inhibitory molecules and other receptors on the surface of cells, which downregulate immunity. On the contrary, immunocompetent immune cells in the tumor such as subsets of CD8+ T cells and M1 macrophages correlate with survival in ovarian cancer. In this section we will discuss cell types and receptor-ligand interactions which confer inhibitory or suppressive properties to the TME, as well as those which augment antitumor immunity.

3.2.1 Immune Suppression by T Cells in Ovarian Cancer

Of the T cell subsets which are classified as immune down-regulatory, a well-studied cell subset is the forkhead box P3 (FoxP3) T regulatory cells (T regs). CD4+CD25 high FoxP3+ T regs can release interleukin-10 (IL-10) and transforming growth factor (TGF)-β, cytokines which are associated with poor outcome in ovarian cancer. These T regs are often considered to be detrimental to the prognosis of ovarian cancer, based on several published reports [8, 9]. It is also known that a range of other T cells contribute to the induction of pro-tumor immune responses in ovarian cancer. These include T cells expressing glucocorticoid-induced TNF receptor family-related protein (GITR), chemokine receptor CCR4, T cells low in CD28 expression, and exhausted T cells which express immune checkpoint inhibitory molecules including programmed death-1 (PD-1; CD279), cytotoxic T lymphocyte-associated protein-4 (CTLA-4), or lymphocyte activation gene-3 (LAG-3; CD223) [10–14].

3.2.2 Critical Role of Immune-Enhancing T Cells in Ovarian Cancer Outcome

The beneficial role of tumor-infiltrating lymphocytes (TILs) is substantiated in ovarian cancer. Immunocompetent TILs recognize cancer antigens or overexpressed self antigens which have been processed by antigen presenting cells (APC) and develop antitumor immune responses to these antigens. CD8+ and CD4+ TILs release interferon (IFN)-γ and interleukin-2 (IL-2), which are important mediators of antitumor immunity. CD4+ TILs also recruit DC which can stimulate T cells to secrete granzyme B or perforin, molecules which can kill tumor cells [15, 16].

A series of studies have shown that CD3+, CD4+, and CD8+ immunocompetent T cells infiltrating tumors (TILs) correlate with longer survival in ovarian cancer patients [17–19]. Furthermore, a large study of 3196 HGSOC patients which compared the density of CD8+ TILs in the tumor epithelium showed that medium survival rates were 2.8, 3.0, 3.8, and 5.1 years for patients with none, low, moderate, or high density of TILs, respectively. The presence of CD8+ TILs distinctly correlated with increased survival [20]. Other investigators reported that a high ratio of CD8 T cells/FoxP3 T cells was a relevant predictor of outcome in ovarian cancer [21].

CD103 is a marker for intraepithelial lymphocytes (IELs). TILs of the CD3+TCRαβ+CD8αβ+CD4− phenotype have been linked to prolonged survival in HGSOC [22, 23]. These beneficial TILs were primarily located in the tumor epithelium, whereas stromal TILs were not of benefit to survival [22]. Notably, epithelial CD8+ T cells and total CD103+ cells were of beneficial outcome only in patients who received primary surgery and adjuvant chemotherapy. These CD103+ TILs showed signs of recent activation and co-expressed PD-1 and

CD27. CD103+ TILs appeared to be induced as a result of an adaptive antitumor immune response against HGSOC [22]. The presence of PD-1 and CD27 on these CD103+ TILs may provide new avenues for immunotherapy of CD103+ high expressing patients, by blocking PD-1 and CD27 in combination.

Other subsets of T cells found in ovarian cancer which may be important in antitumor immunity are the Th17 CD4+ T cells. This cell type is reported to have different roles depending on the cancer type, but in ovarian cancer some reports indicate that these cells correlate inversely with the presence of T regs and directly with better survival [24, 25].

From the foregoing discussions, we see that TIL density at the time of diagnosis can predict disease outcome in ovarian cancer. It is also known that immunocompetent TILs in the TME also contribute to the optimum efficacy of novel immunotherapeutic strategies. In this respect, reports indicate that immunogenic or "hot" tumors consist of high numbers of TILs and "cold" tumors have a paucity of TILs. CD8+ T cells are critical antitumor effector cells, and "hot" tumors are generally more amenable to the effects of immunotherapy and yield better outcomes to immunotherapy than "cold" tumors [26, 27].

3.2.3 Immune Checkpoint Junctions and Ovarian Cancer

Immune checkpoint (IC) molecules can tip the balance between health and disease. Immune regulation at this level is determined by several co-inhibitory molecules. Generally, linkage of co-inhibitory receptor to ligand suppresses T cell receptor signaling and attenuates immune responses. Whereas, this dampening of immune function is crucial for the resolution of infections, and in the development of self-tolerance to limit autoimmune disease, high levels of IC molecules on T cells or on cancer cells are known to be a robust tumor-promoting mechanism [28]. The two most studied of these immune checkpoint molecules CTLA-4 and PD-1 are expressed on T cells and contribute significantly to immune resistance in cancer [29, 30].

Briefly, ligation of CD28 on T cells to B7-1/B7-2 (CD80/CD86) leads to heightened T cell antitumor responses. This is mediated by phosphoinositide-3-kinase (PI3K) and protein kinase B (AKT) pathways [31]. When T cells are activated, CTLA-4 is upregulated and this inhibits the CD28-B7 interaction, as CTLA-4 competes with CD28 to bind to B7 molecules. Binding of CTLA-4 to B7-1/B7-2 limits T cell function. Inhibition of CD28 mediated PI3K/AKT pathway is a critical mechanism of T cell suppression. The preferential binding of CTLA-4 to B7-1 or B7-2 on antigen presenting cells diminishes T cell proliferation, reduces tumor cell killing, and lessens Th1 cytokine secretion [32]. In an ovarian cancer cell culture system, antibody blocking of CTLA-4 during the initial phase of culture augmented the expansion of more potent CD8+ TILs [33].

Programmed death-1 (PD-1) is primarily expressed on T cells and is associated with T cell exhaustion [34]. It cross-links with PD-l ligand (PD-L1; CD274) which is usually displayed on tumor cells and on immunosuppressive macrophages. PD-1/PD-L1 interaction directly inhibits TCR signaling by dephosphorylation of downstream molecules [35]. PD-1/PD-L1 also increases T regs generation and function mediated by reduced signaling via the AKT/mTOR pathway. Interaction of PD-1 on CD4+ or CD8+ T cells with PD-L1 can result in diminished T cell proliferation and reduced T cell activation, with a skewing towards a T helper 2 (Th2) pro-tumor cytokine profile [36, 37]. In ovarian cancer, PD-L1 was expressed significantly more often in high-grade tumors (41.5%) than in low-grade tumors (7.7%) [38]. The development of antibodies to block IC in patients has opened a new and exciting avenue of immunotherapy.

Importantly, T cells in the ovarian TME also express other immune checkpoint inhibitory junctions including LAG-3, T cell immunoglobulin and mucin domain-containing protein 3 (TIM-3), B and T lymphocyte attenuator (BTLA), and V-domain Ig suppressor of T cell activation

(VISTA), which may all contribute to the immune suppressive networks in the ovarian cancer TME.

LAG-3 is closely related to CD4 but shares less than 20% homology at the amino acid level [28]. Like CD4 it also binds to major histocompatibility complex (MHC) class II molecules on APC, but with stronger affinity. Additionally, it binds to MHC-II on tumor cells and galectin-3 on stromal cells in the TME. The binding of LAG-3 to the MHC-II prevents the binding of MHC-II to the T cell receptor (TCR) and CD4 [28, 39]. LAG-3 is expressed on regulatory T cells, including activated natural T regs (nTregs) and induced CD4+ FoxP3+ T regs (iTregs). Blocking LAG-3 inhibits the suppressive function of T regs. In general, the expression of LAG-3 on CD8+, CD4+, and NK+ cells correlates with reduced immune function. Soluble LAG-3 impairs the differentiation of monocytes to DC and macrophages. Ligands for LAG-3 (such as MHC-II) are expressed on myeloid-derived suppressor cells (MDSC), APC and tumor cells, among others [39], and thus it is plausible that the inhibitory effects of this checkpoint molecule can have widespread effects on the TME. In an ovarian cancer model it is reported the LAG-3 and PD-1 act together to limit CD8+ T cell immunity [13].

Surface marker TIM-3 appears to have a negative role in ovarian cancer [40]. It binds to the C-type lectin galectin-9, ceacam-1, and high mobility group protein B1 (HMGB-1), among other ligands [28]. In ovarian cancer, in a sample size of 20 HGSOC patients, it was found that TILs co-expressing PD-1 and TIM-3 exhibited features of functional exhaustion and correlated with poor disease outcome [41]. Nonspecific stimulation with phorbol myristate acetate (PMA) and ionomycin induced IFN-γ production and CD107a (a marker of cytotoxic cell degranulation) in TIM-3− PD-1− cells, above that induced in TIM-3+ PD1+ cells. Furthermore, incubation with anti-TIM-3 and anti-PD-1 antibodies together (but not alone) increased the ability of tumor infiltrating CD8+ T cells to upregulate IFN-γ, and cytolytic molecules granzyme B and perforin, further indicating that these TIM-3+ PD-1+ cells are a subset of functionally impaired T cells [41]. In another study, blocking of TIM-3 was associated with improved antitumor responses in an ovarian cancer model [42].

BTLA (CD272) is an immune regulatory receptor which ligates with tumor necrosis factor (receptor) superfamily member 14 (TNFRSF14), also known as herpesvirus entry mediator (HVEM). BTLA is found on T cells, B cells, macrophages, DC, and NK cells. Complexes of BTLA-HVEM are associated with attenuated T cell immune responses and immune tolerance [28].

In ovarian cancer preclinical studies using the WF-3/Luc tumor model, the combination of chemotherapy and anti-BTLA antibody reduced peritoneal tumor volume and extended survival in tumor-bearing mice. This combination treatment resulted in a superior outcome to either treatment alone [43]. The investigators studied the expression of CD223, an activated T cell marker in these mice. The percentages of CD223+CD4+ and CD223+CD8+ cells in chemotherapy combined with anti-BTLA treatment were highest in the spleen cells of this group, in comparison with spleen cells of the monotherapy groups [43]. Studies in patients showed that of 254 diagnosed with epithelial ovarian cancer, 149 had detectable BTLA expression. Patients with BTLA positive tumors had significantly higher incidences of advanced disease, disease relapse, and disease-related deaths than patients not expressing BTLA [43].

VISTA, also called Differentiation of Embryonic Stem Cells 1 (Dies 1) and PD-1 homolog (PD-1H), among other names, belongs to the B7 family and shares sequence homology with PD-1 and PD-L1 [44]. It is highly expressed on CD11b high MDSC, and also on CD4+ and CD8+ regulatory T cells, and on TILs. VISTA suppresses T cell activation when expressed as a ligand on APC, or as a receptor on T cells [28]. The binding of VISTA to V-Set and Immunoglobulin domain containing-3 (VSIG-3) inhibits T cell function [45]. VISTA in tumor cells suppressed T cell proliferation and cytokine production in vitro and decreased the numbers of tumor-infiltrating CD8+ T cells in vivo. Anti-VISTA antibody treatment prolonged the survival of tumor-bearing mice [46]. This immune checkpoint molecule is highly expressed in human ovarian and endometrial cancers [47].

Future studies are needed to shed light on the array of immune suppressive mechanisms due to immune checkpoint synapses, and how we can more strategically target their emergence, to prevent acquired resistance to novel immunotherapy regimens.

3.2.4 Impact of Tumor-Associated Macrophages in the TME

Tumor-associated macrophages (TAMs) are derived from blood monocytes or from resident peritoneal macrophages [48–51]. Both of these subsets of TAMs express CD163 and CD206, and similar levels of genes which control phagocytosis and antigen presentation. Notably however TAMs in the TME have an upregulation of genes associated with extracellular matrix (ECM) remodeling, and these cells are known to alter the ECM [52]. In ovarian cancer, TAMs have received notoriety for their roles in proliferation of ovarian cancer cells, tumor cell invasion, angiogenesis, metastasis, and early relapse [53–56], but as we shall discuss, these myeloid progenitor cells are a dynamic entity which can change phenotype and/or function based on the cytokines, chemokines, and other soluble molecules present in the TME (reviewed in [57]).

Of the TAMs, M2 alternatively activated macrophages secrete IL-10 and TGF-β and contribute to tissue remodeling and tumor progression. High levels of IL-4, IL-13, IL-10, TGF-β, and colony-stimulating factor (CSF) foster this M2 macrophage lineage [58, 59]. On the contrary, CD4+ and CD8+ T cells may secrete IFN-γ in the TME, and this favors the generation of M1 polarized antitumor macrophages. M1 macrophages may also be stimulated by Toll-like receptor (TLR) ligands and Th1 cytokines such as IL-12, IL-23, and TNF-α. M1 macrophages are associated with survival in HGSOC [60].

A much studied group of TAMs in cancer are the myeloid-derived suppressor cells (MDSC). These cells enhance stemness and promote metastasis of ovarian cancer cells. MDSC in the TME are associated with heightened disease, increased tumor burden, and resistance to immunotherapy [61–65]. Based on phenotypic differences, MDSC in both mice and human are divided into two main subpopulations, the monocytic MDSC (m-MDSC) and the granulocytic MDSC (g-MDSC) [66–69]. The m-MDSC subpopulation is most often studied in the context of immune suppression in cancer.

MDSC limit the antitumor functions of T cells by several mechanisms. For example, activated MDSC produce high levels of nitric oxide (NO) and arginase-1 (ARG-1), which can contribute to cell cycle arrest in T cells by depletion of amino acid, L-arginine from the TME [70, 71]. They can inhibit the functioning of TCR and IL-2 signaling, reduce the numbers of T cells trafficking into LN and tumors, induce T cell apoptosis, and stimulate the generation of T regs [67, 72, 73]. MDSC also express high levels of PD-L1, the binding of which to PD-1 on T cells can lead to T cell exhaustion. They can also reduce the antitumor effect of T cells by secreting and releasing TGF-β and IL-10 [74], and impair NK cell function, a process mediated by NO [75]. MDSC stimulate tumor angiogenesis by secreting vascular endothelial growth factor (VEGF), and they also secrete metalloproteinases (MMP) such as MMP 9 which alters the ECM and basal membrane, and facilitates the entry of tumor cells to the bloodstream to enhance metastasis [76, 77].

In solid cancers, the presence of higher numbers of MDSC in peripheral blood has been found in patients with metastatic melanoma who did not respond well to immune checkpoint inhibitor anti-CTLA-4 antibody ipilimumab therapy. In poor responders to ipilimumab, these MDSC were more suppressive as compared with MDSC in ipilimumab responders. A similar observation was made in patients given anti-PD-1 antibody treatment nivolumab after ipilumimab progression, such that higher baseline MDSC number correlated with progression and poor overall survival [78–80].

Immune suppressive TAMs contribute significantly to the establishment of ovarian cancer. Blocking these cells or their functions with anti-CCL2 antibody, anti-CD52 antibody, anti-colony-stimulating factor 1 receptor (CSF-IR), and anti-CD11b antibody has been studied in preclinical mouse models of this disease, and in a limited number of clinical trials [81–85].

3.2.5 Opposing Roles of Dendritic Cells in Ovarian Cancer Immunity

Dendritic cells are professional antigen-presenting cells (APCs), which capture antigen, process and present antigen in the form of peptides to cells in the immune system [86]. Exogenous peptides are presented to CD4+ T cells via MHC class II, and endogenous peptides to CD8+ T cells via MHC class I. DC also present exogenously captured antigens as MHC class I associated peptides (cross presentation), thereby facilitating more efficient CD4+ and CD8+ T cell functions [87, 88].

Immature myeloid DC leave the bone marrow and enter the bloodstream, lymph nodes (LN), and tissues. These cells can only mount low immune responses. When these cells encounter antigen, DC migrate to lymph nodes from tissues and present the specific antigen to immune cells [89, 90]. In the presence of antigen, DC mature and upregulate co-stimulatory molecules (such as CD80, CD86, and CD40), and are now capable of efficiently activating CD8+ T cells, crosslinking with CD40 ligand on T cells, and secreting IL-12 [87, 89, 91, 92]. In the ovarian TME, myeloid DC are sparse and are generally immature. This is consistent with the abundance of pro-tumor soluble molecules such as TGF-β, IL-10, VEGF, ARG-1, indoleamine 2, 3-dioxygenase (IDO), as well as cellular components such as exhausted T cells and suppressive TAMs, which all favor tumor progression [93–95]. Reports indicate that depletion of DC in mice at later stages in ovarian cancer delayed tumor growth [96].

However, on a positive note regarding myeloid DC in cancer, the transcription factor basic leucine zipper transcription factor (Batf3) is essential for the development of mouse CD103+ DC [97, 98]. Batf3 lineage CD103+ DC correlated with increased levels of CXCL9, CXCL10, and CXCL11, chemokines which recruit effector T cells into tumors, and are associated with survival in cancer [99, 100]. These cells are major secretors of IL-12, a potent anticancer cytokine which drives Th1 antitumor immunity [101]. In preclinical studies, adoptive transfer of tumor-associated CD103+ DC resulted in tumor rejection when combined with PD-1 blockade to enhance CD8+ T cell activity. CD103+ DC are also identified in patient ascites [102], indicating the potential relevance of these DC in human ovarian cancer.

A frequently characterized DC in the ovarian TME is the plasmacytoid DC (pDC; CD4+CD123+BDCA2+), which has tolerogenic and pro-tumor properties. These cells secrete IDO, an enzyme which catalyzes tryptophan degradation, enhances tumor angiogenesis and metastasis and has an inhibitory effect on TIL function, and correlate with poor outcome in cancer [93, 103–105]. Antigen presentation by pDC is largely considered to induce T cell anergy and/or deletion, due to the ability of pDC to secrete IL-10, TGF-β, and IDO, or of these cells to express PD-L1, inducible T cell costimulator ligand (ICOS-L), or OX40-L [106, 107]. In ovarian cancer, pDC is also associated with tumor cell vascularization through the secretion of TNF-α and IL-8 [108].

3.2.6 Other Important Cellular Components in the Complex Ovarian Cancer Milieu

Natural killer cells (NK cells) mediate their lethal effects on tumor cells through the CD16 receptor, the NKG2D receptor, and the NKp30 receptor on their cell surface. Aberrant receptor/ligand expression, low numbers of NK cells, or inability of these cells to secrete cytotoxic molecules, can make these cells less effective at killing tumor cells [51]. For example, in ovarian cancer high expression of soluble B7-H6 (a ligand for the NKp30 receptor) was associated with reduced NKp30 levels on NK cells and diminished NK cell activity [109].

Efficient NK-DC interaction in tumors can upregulate CXCR3 and CCR5 on DC leading to the recruitment of CD8+ effector T cells into tumors [110], which can contribute to antitumor immune responses. Data from the Immunological Genome Project revealed that NK cells can secrete chemokines CCL5, CCL3, CXCL1, CCL4 and XCL1, and in tumors, NK cells potently induced DC chemo-attractants XCL1 and CCL5 [111, 112].

Polymorphonuclear neutrophils (PMNs, neutrophils) are a heterogeneous group of cells which can be classified as N1 (antitumor) and N2 (pro-tumor) [113, 114]. They migrate from blood into tissues guided primarily by chemokines CXCL1 and CXCL2. In a study of 213 HGSOC patients, some biopsies showed clustering of PMNs to ZEB1 (an epithelial-mesenchymal-transition (EMT) transcription factor) positive cells, primarily in areas of low E-cadherin [115]. The process of EMT identifies with the aggressive nature of cancers. Pro-tumor neutrophils contribute to angiogenesis, metastasis, and suppression of adaptive immune responses [116, 117]. On the contrary, antitumor N1 tumor-associated neutrophils may directly kill tumor cells, and promote CD8+ T cell recruitment and activation, by releasing chemokines CXCL9 and CXCL10 [118–120]. A recent meta analysis study in ovarian cancer revealed that a high neutrophil to lymphocyte ratio correlated with worse overall survival (O/S) in Asians but not in Caucasians [121].

Due to the immune suppressive nature of ovarian TME, several novel therapies are in clinical trials to target and diminish a variety of pro-tumor mechanisms in ovarian cancer.

3.3 Newer Therapies in Ovarian Cancer

HGSOC is marked by disease recurrence and metastasis. Conventional medical treatment using surgery and cytotoxic chemotherapy such as carboplatin and paclitaxel are not optimum to manage this disease. In this section, we will address the status of novel therapies which are in clinical trials and those which have been FDA approved for therapy, and will shed light on how we can better manage the clinical support of these patients.

With the administration of novel therapies such as most immunotherapy several considerations are important, since it has been found that the benefit of most of these recent approaches are only effective in a low percentage of ovarian cancer patients. A better selection of patients who can respond to immunotherapy needs to be identified so that patients are not given therapy that they do not have the potential to respond to, and in some cases are unnecessarily exposed to adverse effects of these treatments. Identifying biomarkers which may indicate that a patient is likely to respond to a particular therapy can be very useful in selecting patient cohorts. Some potential biomarkers are tumor PD-L1 expression, mismatch repair deficiency (dMMR) [122], immune cell infiltration, tumor mutational burden (TMB) [123], or neoantigen burden [124, 125]. The more information we can gather on patient immune status will aid in tailoring the best therapy for cancer patients.

Tumor mutational burden (TMB) is the total number of somatic (acquired) mutations in a tumor. This can be measured by whole exome sequencing (WES). TMB may vary 1000-fold between different cancer types [126]. Highly mutated tumors are more likely to have an abundance of tumor-specific mutant epitopes, which act as neoantigens and are recognized as nonself, leading to immune responses. In comparison with overall neoantigen load, TMB is easier and less expensive to measure and correlates with outcome to immunotherapy in some solid cancers.

Immune checkpoint inhibitors have demonstrated improved efficacy against hypermutated cancers such as melanomas and lung cancers. These cancers have more tumor-specific neoantigens that stimulate the recruitment of more immunocompetent TILs to potentiate antitumor immunity [123, 127]. Tumors with higher neoantigen load are associated with improved overall survival, and increased expression of TCR, IFN-γ, and TNF receptor pathway genes, parameters which are associated with tumor cell cytotoxicity [124]. Ovarian cancer is classified as a low TMB cancer, and this feature may be a contributing factor to the low response to most immunotherapy, as is currently observed.

In the preceding sections, we outlined several aspects of immune resistance in ovarian cancer (cells, soluble molecules, and IC junctions) which may negatively impact the course of disease, as well as limit the effectiveness of novel therapies. Table 3.1 summarizes several of these resistance mechanisms which may be reduced with conventional and newer therapies. Current thinking is that the way to improve ovarian cancer outcome is to target multiple facets of resistance mechanisms in combination therapy strategies. It will be evident in the future text that most clinical trials are currently investigating combination therapy regimens in patients.

3.3.1 Bevacizumab

Vascular endothelial growth factor (VEGF) is a key player in the progression of ovarian cancer, ascites formation, and the spread of tumor cells [128]. VEGF signaling reduces the trafficking of CTLs into the TME, promotes infiltration of T regs through a selective endothelium, as well as induces MDSC. VEGF can also reduce T cell proliferation and cytotoxic function and inhibit the maturation of DC [64, 129–131]. Bevacizumab (Avastin) is an anti-angiogenic agent which prevents VEGF from binding to its receptor, and significantly improves progression-free survival in some HGSOC patients. In 2014, the FDA approved bevacizumab in combination with paclitaxel, topotecan, or pegylated doxyrubicin (PLD) for the treatment of platinum-resistant epithelial ovarian cancer, fallopian tube, and primary peritoneal cancer. Bevacizumab and other anti-angiogenic agents are in clinical trials for ovarian and other cancers [132–134].

Table 3.1 **Regulation of immunity by cancer therapy.** A summary of mechanisms of action of anticancer treatments

Therapeutic agent	Biological actions
Chemotherapy: Carboplatin, paclitaxel, gemcitabine, cyclophosphamide, doxorubicin	Enhances antigen presentation and DC cross priming, reduces MDSC and T regs, increases CD8+ T cells, potentiates immunogenic cell death (ICD) leading to the release of tumor antigens and danger signals by dying cancer cells (damage-associated molecular patterns; DAMPS)
Radiation	Increases tumor antigen expression, allows the release of DAMPS, increases apoptosis and necrosis
Immune checkpoint (IC) inhibitors: Antibodies targeting CTLA-4, PD-1, LAG-3, TIM-3, BTLA, VISTA	Upregulates T cell activation, re-invigorates T cells, eliminates T regs, reduces alternate IC resistance
Inhibitors of:	
ARG-1	Decreases tumor metastasis and angiogenesis
CSF1-R	Reduces immune suppression by TAMS/MDSC
TGF-β	Upregulates immunocompetent TILs, improves T cell antitumor responses
IDO	Enhances T cell proliferation and migration to tumor
CXCR2	Decreases trafficking of pro-tumor TILs
CD73	Enhances T cell re-invigoration
Inhibitors of VEGF: bevacizumab	Diminishes angiogenesis and reduces T reg suppression
PARP inhibitors	Prevents cancer cells from repairing their damaged DNA leading to cancer cell death
Epigenetic regulators	Upregulates chemokines which recruit effector T cells to the TME. Stimulates the generation of cancer testis antigen (CTA)-specific antitumor T cells. Overcomes T cell exhaustion. Reduces MDSC function. Reactivates silenced immune genes

(continued)

Table 3.1 (continued)

Therapeutic agent	Biological actions
Blockers of oncologic signaling pathways	
c-MET	Decreases metastasis of tumor cells
MAPK	Upregulates IFN-γ signaling and MHC-1, and increases tumor cell lysis potential
PI3K-γ	Decreases MDSC
Vaccines	
DC vaccines	Increases antigen presentation and recognition, and upregulate T effector cell activity
Peptide vaccines	Elicits the generation of tumor antigen-specific CTLs
ACT therapy	
TCR	Targets immune-suppressed TME
TIL	Enhances tumor TIL infiltration and immune potency
CAR T cell	Uses engineered TCRs to attack cancer antigens
Probiotics	Augments the release of antitumor cytokines

3.3.2 Blocking the c-MET Axis

The c-mesenchymal-epithelial transition (c-MET, MET) axis has been proposed as a potential therapeutic target in ovarian cancer. MET is a proto-oncogene [135], which is associated with acquired resistance to some approved therapies such as anti-epithelial growth factor receptor (EGFR) therapies [136, 137]. The c-MET receptor is expressed by epithelial and endothelial cells [138, 139], and binds to its ligand hepatocyte growth factor (HGF), which is upregulated in ovarian cancer [140, 141]. A meta-analysis of 7 studies with about 568 patients showed that c-MET had no statistical correlation with FIGO stage or LN metastasis; however, high c-MET expression on patient tumor was significantly associated with poor prognosis in epithelial ovarian cancer [142]. A range of important agents targeting c-MET has been utilized in clinical trials for solid cancers, including Tivantinib, Crizotinib, Foretinib, and Capmatinib, as well as Rilotumumab which neutralizes HGF [143].

There are also several other critical oncogenic pathways which regulate the immune suppression of cancer. Indeed it is known that inhibitors of the mitogen activated protein kinase pathway (MAPK) enhance IFN-γ signaling and MHC class I expression, thereby promoting tumor cell cytotoxicity [144]. PI3k-γ inhibitors were shown to decrease MDSC in the TME and improve responses to immune checkpoint blockers in animal models [145].

3.3.3 PARP Inhibitors

The poly-ADP-ribose polymerase (PARP) proteins are a family of 17 enzymes which regulate varying cellular functions, of which PARP1 and PARP2 are engaged in DNA repair. DNA damage occurs constantly and the process requires a complex network of molecular repair pathways so as to maintain genomic integrity and prevent cell death [5, 6]. Homologous repair (HR) allows repair of double-stranded DNA breaks. It operates during the S and G2 phases of the cell cycle and relies on many proteins including BRCA1 and BRCA2. In cells with nonfunctioning HR, sometimes due to BRCA1 or BRCA2 deficiency, other repair pathways such as nonhomologous end-joining (NHEJ) which is more error prone occurs. This results in the accumulation of additional mutations and chromosomal instability, and a higher chance of a cell becoming malignant [6]. Germline mutations are inherited and BRCA1 and BRCA2 are the two most commonly known susceptibility genes. Somatic (acquired) mutations also occur in BRCA1/2.

PARP 1 and PARP 2 are excellent anticancer targets [146]. With germline testing on DNA extracted from healthy cells, patients may be considered for targeted therapy, such as PARP inhibitors for BRCA1/2 mutation positive patients. Patients with BRCA mutated tumors have an earlier age of diagnosis, improved survival, increased sensitivity to platinum-based chemotherapy, and sensitivity to PARP inhibitors. Olaparib (Lynparza) an inhibitor of PARP is effective in BRCA mutation positive ovarian cancer and is FDA approved for ovarian cancer with this mutation. Other FDA approved PARP inhibitors are Rucaparib (Rubraca) and Niraparib (Zejula). A detailed summary of the use of PARP inhibitors in ovarian and other cancers is given elsewhere [6, 30].

Studies suggest that HR-deficient BRCA1/2 mutated tumors may have higher neoantigen load and increased CD3+ and CD8+ TILs, as well as elevated PD-1 and PD-L1 expression, when compared with homologous recombination proficient tumors [147]. It has also been reported that with BRCA1 mutation in some patients, only the co-administration of anti-CTLA-4 antibody and PARP inhibitor showed significant treatment efficacy in HGSOC [148]. These findings suggest that combination therapy with PARP inhibitors and ICI may yield a good outcome in ovarian cancer.

3.3.4 Epigenetic Modulators in Ovarian Cancer

In addition to the immune resistance mechanisms in ovarian cancer (discussed in Sect. 3.3) which confer immune suppression, there are also changes in normal gene expression patterns, both of somatic and of epigenetic nature which can regulate immunity. Epigenetic changes are heritable changes in gene expression that do not involve changes to the underlying DNA sequence. Epigenetic changes can induce initiation and progression of cancers, and various epigenetic states can govern resistance to cancer therapies [149–151]. The benefits of epigenetic modifiers are currently being exploited in cancer since these agents function in dual roles, firstly in their traditionally ascribed role to impair tumor cell functions, and more recently reported, in a novel role to induce antitumor functions in immune cells [152–154].

Epigenetic modifying agents are FDA approved for the treatment of hematologic malignancies [155, 156]. Gene silencing caused by DNA hypermethylation can be reversed by DNA hypermethyltransferase (DNMT) inhibitors including azanucleosides such as 5-aza-2′-deoxycytidine (decitabine; DAC). DNMT inhibitors are nucleoside analogs that covalently bind and lead to degradation of DNMTs, resulting in reduced cellular methylation [149, 150]. Aberrant DNA methylation has been implicated in resistance to platinum-based therapy in ovarian cancer [151]. Another class of epigenetic modifiers is the histone deacetylase (HDAC) inhibitors. These are small molecules that enter to block active sites of HDAC to prevent removal of acetyl groups on histones and maintain a more open chromatic structure. There are about 18 HDAC enzymes, and these additionally act on over 1750 nonhistone proteins as well, regulating their biologic functions, with many nonspecific effects [153, 157].

Several preclinical studies and clinical trials have been conducted in ovarian cancer to test epigenetic modifiers in immune-oncology. It has been reported that decitabine (DAC) treatment can upregulate CCL2, CCL5, and CXCL10 by BR5FVB1-Akt cells in vitro, and increase the concentration of CXCL10 (which recruits CD8+ T cells) in peritoneal fluid of mice, concomitant with upregulated CD4 and CD8 T cells and NK cells in the TME [158].

Cancer testis antigens (CTAs) are well known for increasing tumor cell immunogenicity and potential recognition by T cells. CTAs are expressed during embryonic development but are silenced in adult somatic cells. DAC and azacitidine (AZA) can induce several CTAs on human ovarian cancer cell lines, a phenomenon which can be enhanced by addition of Trichostatin A (TSA), an HDAC inhibitor [159–162]. In vivo, this function of DAC and AZA can elicit the generation of CTA-specific antitumor T cells. DNMT inhibitors have also been shown to upregulate tumor cell expression of genes such as PSMB8, PSMB9, and transporter associated with antigen presentation 1 (TAP1) which are involved in antigen processing, as well as adhesion molecules such as intracellular adhesion molecule-1 (ICAM-1) which is involved in forming the T cell tumor cell synapse [163–165]. The endogenous antitumor immune responses induced by DNMT or HDAC inhibitors such as increased Th1 chemokine production, and upregulated IFN-γ signaling may prime the immune system for later benefit with other novel therapies such as immune checkpoint inhibitors in ovarian cancer (NCT02901899, NCT02915523, NCT02900560).

3.3.5 Immunotherapeutic Strategies in Ovarian Cancer

The concept of cancer immunotherapy has gained much attention over the past few decades. Briefly, this is treatment administered to enhance the patients' immune function to efficiently kill malignant cells. It may include several different categories of therapy such as cancer vaccines, adoptive immune cell therapies, chimeric antigen receptor (CAR) T cell therapy, and the most widely used of these treatments, immune checkpoint blockade therapy. Immunotherapy in ovarian cancer can potentially be a revolutionary treatment strategy, because tumor-specific T cells can target underlying immunosuppressive networks in the ovarian TME, giving improved outcome.

3.3.5.1 Immune Checkpoint Inhibitor Therapy

Immune checkpoints can confer a significant immune suppressive component in the ovarian cancer TME. Blocking of immune checkpoints for cancer therapy has had a surge in clinical trials over the past decade. Immune checkpoint inhibitors interrupt the negative receptor/ligand interactions in the TME and enable exhausted T cells to become functional. Some of these antibody blockers are FDA approved for melanoma, non-small cell lung cancer (NSCLC), bladder cancer, renal cell carcinoma, and Hodgkin's lymphoma among others. In many of these cancers, the outcome of this therapy have reached encouraging levels of success [166, 167], but this is not the case for ovarian cancer, in which the response rate is 10–15% for anti-PD-1 antibody therapy (such as Pembrolizumab and Nivolumab) [168–171], possibly due to failure of these treatments to overcome the intricate levels of immune suppression in the ovarian TME.

Ipilimumab, targeting CTLA-4 was the first immune checkpoint inhibitor (ICI) to be approved for cancer therapy. Today there are hundreds of ongoing clinical trials to evaluate Ipilimumab as monotherapy or in combination with chemotherapy or other agents. Most cancer clinical trials with ICI investigated anti-PD-1 (Pembrolizimab, Nivolumab) or anti-PD-L1 (Avelumab, Atezolizumab, Durvalumab) antibodies, and fewer using ipilimumab. The interested reader can refer to a summary [172] for a detailed account of immune checkpoint inhibitors in phase III and IV clinical trials. There are also several clinical trials with ICI therapy in ovarian cancer with some reported results [173].

Some examples for ovarian cancer ICI combination clinical trials are as follows. There is an ongoing Phase I/II clinical trial with PARP inhibition (Olaparib) and CTLA-4 blockade (Tremelimumab) in BRACA-deficient ovarian cancer patients (NCT02571725). In another trial, a combination of Nivolumab and Ipilimumab resulted in better response rates in patients with recurrent or persistent ovarian cancer than with Nivolumab alone (NCT02498600). In a phase II clinical trial with 19 patients using combination treatment of Pembrolizumab plus chemotherapy in recurrent platinum-resistant ovarian cancer, with a combination of cisplatin and gemcitabine with Pembrolizumab, the overall response rate was 57%, with 7% showing a complete response and 50% of patients achieving a partial response (NCT02608684). Four patients (29%) achieved stable disease as the best response. Other ongoing studies of ICI in combination trials are summarized elsewhere [173]. Some clinical trials blocking LAG-3 are also in progress for multiple solid cancers (NCT03459222, NCT01968109, NCT02966548, and NCT03335540).

An illustrative summary of therapeutic agents used in clinical trials for HGSOC patients and other cancers is shown in Fig. 3.1, with the understanding that in translational medicine, combinations can be made across many of these treatment groups.

3.3.5.2 Dendritic Cell Vaccine Therapy

DC were discovered in 1973 by Zanvil Cohn and Ralph Steinman [174]. These cells are attractive candidates for immunotherapy because they are superb antigen-presenting cells which regulate

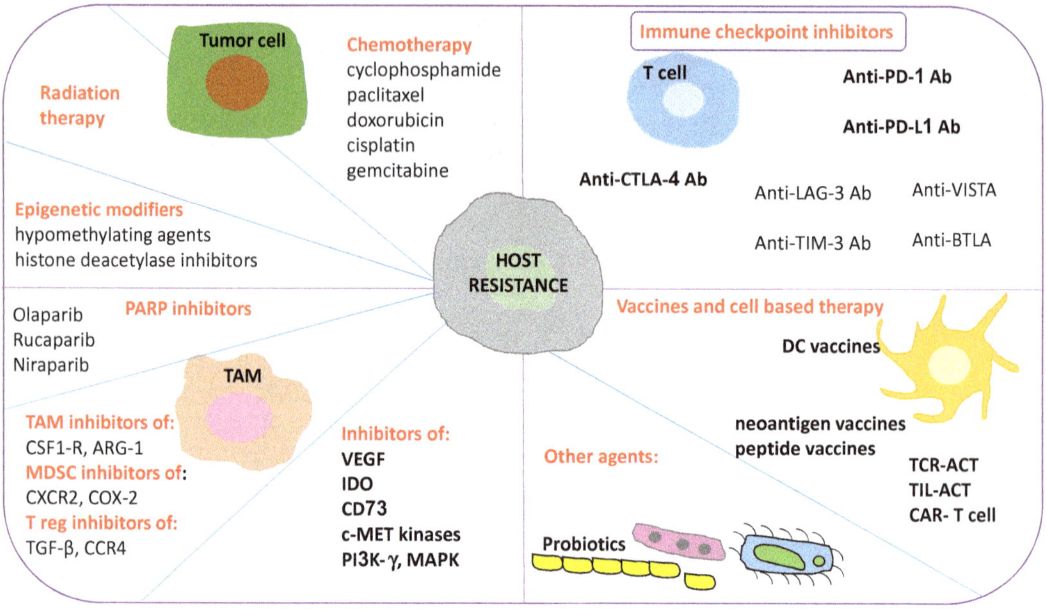

Fig. 3.1 **Overview of classes of agents used in combination therapy in ovarian cancer**. A multiplicity of clinical trials are ongoing in solid cancers targeting immune suppression using combination therapies from groups represented in figure panels. Strategies to reduce tumor burden such as chemotherapy or radiation can be combined with immune checkpoint inhibitors to potentiate antitumor T cell responses. Agents which target TAMS or MDSC have the ability to improve T cell antitumor responses. Combination therapy with MDSC inhibitors and ICI can further lead to re-invigoration of T cells and improved outcome in ovarian cancer. Dual treatments with ICI such as anti-PD-1 and anti-LAG-3 can potentially reduce emerging resistance due to upregulation of IC with anti-PD-1 monotherapy. DC vaccines increase antigen presentation and potentiate T effector cell activity. Combining DC vaccines with epigenetic modifiers which stimulate the generation of TAAs can elicit potent antitumor T cell immunity and disease improvement. VEGF inhibitors improve the maturation of DC. Combining this treatment with DC vaccines and/or with ICI can boost antitumor T cell immunity. The possibilities for combination therapy with various classes of agents are evident by the plethora of ongoing clinical trials

innate and adaptive immunity, and are stimulators of antitumor responses. Due to their efficacy at cross presentation, DC have been used in vaccines to induce antitumor cytotoxic T lymphocyte (CTL) immune responses [86, 88, 175].

DC prepared for vaccine therapy is often generated from CD14+ monocytes. Autologous blood mononuclear cells are collected in the clinic by leukapheresis and DC propagated under sterile conditions in a current Good Manufacturing Practices (cGMP) facility. Mature cells such as the alpha-DC-1 (α-DC-1) [176–178] are often used for DC vaccine therapy. In the DC generation process, antigens are usually added to make the final product more immunogenic. Undefined antigens may include tumor lysate, total genomic DNA, total RNA from tumor cells, or apoptotic tumor cells. Alternatively, DC made be loaded with peptides, proteins, cDNA, and RNA, which allow specific immune responses to be measured after vaccination. Other immunogenic antigens include p53, or NY-ESO-1, melanoma-associated antigen (MAGE), Wilms' tumor-1 (WT-1), and mucin 1 (MUC-1), which are human leukocyte antigen (HLA-A2) restricted [176, 179].

An important landmark in DC vaccine therapy was reached in 2010, with the FDA approval of Sipuleucil-T as a DC vaccine for prostate cancer. This monotherapy resulted in 4.1 months in improvement in overall survival in subjects [180]. Many other DC vaccines have been in clinical trials but the efficacy of this treatment has not reached the level as was anticipated, as far as inducing significant objective clinical responses in cancers. This may be attributed to the fact that patients in clinical trials had advanced metastatic disease, had a high level of immune suppression, the DC were not at the peak of maturity, or the best immunogenic antigen for the patient was not known for use in the preparation of the DC vaccines. There may also be a reduction of TAAs by cancer cells, a mechanism of immune evasion. Thus, like most other novel therapies we are now in the phase of exploiting suitable therapy combinations which may be synergistic with DC vaccine regimens [181, 182].

DC vaccines are generally used in combination with other therapies such as cyclophosphamide (NCT00703105), ICI therapy, or agents such as bevacizumab in an effort to attenuate several immune suppressive parameters in the TME simultaneously and boost antitumor immunity. In a recent Phase I clinical trial, ovarian cancer patients who had recurrent, measurable disease following surgery and chemotherapy was administered an autologous DC vaccine pulsed with oxidized tumor lysate. The cohort of patients who received a combination of DC vaccine, bevacizumab, and cyclophosphamide had significantly higher overall survival rates than those who received DC vaccine and bevacizumab (with no cyclophosphamide prior to each vaccine) [183].

An exhaustive list of dendritic cell vaccine clinical trials in combination treatment strategies in immuno-oncology is reported in Trial Watch: dendritic cell vaccination for cancer immunotherapy, 2019 [184].

3.3.5.3 Adoptive Immunotherapy

Adoptive cell transfer (ACT) immunotherapy is based on the intravenous infusion of autologous immune cells after modification and expansion in culture to improve antitumor responses when injected into patients. In 1965, it was demonstrated that adoptive immunotherapy had an improvement on acute leukemia in preclinical studies and in a clinical trial [185, 186]. Adoptive immunotherapy includes the transfusion of lymphokine activator killer cells (LAK), natural killer cells (NK), cytokine-induced killer cells (CK), hematopoietic stem cell transplants, tumor-infiltrating lymphocyte therapy (TIL), chimeric antigen receptor therapy (CAR) T cell and T cell receptor (TCR) T cell therapy. We will focus our discussion on those cellular therapies which may be of direct interest in ovarian cancer treatment in the near future.

Due to complex immune suppression mechanisms in ovarian cancer, like most therapies, the success of ACT may depend on the use of additional treatment agents used in combination. Clinical trials with TIL therapy in metastatic ovarian cancer in combination with chemotherapy or with immune checkpoint inhibitors include NCT02482090 and NCT03287674.

Chimeric antigen receptor therapy (CAR) T cell and T cell receptor (TCR) T cell therapy is becoming increasingly popular for the treatment of cancer. TCRs on T cells recognize specific antigens. TCR T cells express a genetically engineered antigen TCR alpha and beta chain pair that can recognize tumor-specific antigens. While this TCR T cell treatment has reached some level of success in some colorectal cancer, metastatic melanoma, and multiple myeloma patients, there has not yet been much success in ovarian cancer patients with this treatment [187].

So far the success of CD19 CAR T cells is primarily attributed to their performance in hematologic cancers [188]. In ovarian cancer, the main targets for CAR T cell therapy include MUC 16 [189], mesothelin [190], and folate receptor-alpha (NCT03585764). MUC-16 plays an important role in the progression and metastasis of ovarian cancer [189]. Like TCR T cell therapy however there may be serious side effects with CAR T cell therapy, since the CAR T cell antigen is expressed in some normal tissue resulting in immune-mediated rejection, known as an "on-target, off tumor response," which can cause damage to vital organs such as the liver and lungs [191]. This problem may be overcome with novel CAR T cell designs where T cells are transduced with a CAR offering suboptimal activation upon binding of one antigen, and a chimeric co-stimulatory receptor (CCR) that recognizes a second antigen. Antigens such as FR-α and mesothelin are highly expressed in ovarian cancer tissue compared with normal tissue, and trans-signaling CAR can identify tumor cells based on this parameter and diminish damage to normal tissues. Yet other problems may include cytokine release syndrome [192, 193] or the general immune suppression in the ovarian TME, which may prevent T cells from infiltrating into tumors. Clinical trials with ACT in ovarian cancer are summarized by Yang and coworkers, 2019 [187].

3.4 Looking Forward

There is an urgent need for effective therapy options which will impact the care and survival of HGSOC patients. As we have covered in the foregoing text, owing to the diverse networks of immune suppression in the ovarian cancer TME, it is evident that monotherapy is met with several existing and emerging resistance mechanisms. Improved outcomes may also be diminished by a lack of understanding of the best dosages, optimum timing of administration of therapy agents, duration of each therapy for maximal effectiveness, and which patients will respond best to each therapy.

There is an agreement in the scientific community that the idea of combination therapy to alleviate the multiplicity of immune suppressive networks at the same time is both attractive and promising, yet additional questions arise such as: how will the best combination agents be chosen for additive or synergistic benefits while avoiding dual toxicity (adverse effects) or emerging resistance? How will the greatest anticancer effects be achieved with minimal treatment?

A useful model was proposed by Blank and colleagues [194], the immunogram model, describing interactions between solid cancers and the immune system, and factors which should be taken into account. These authors considered that in some patients overcoming T cell inhibition may be the only factor to be addressed for disease improvement, yet in other patients, there may be a multiplicity of factors. These include general immune status, immune cell infiltration, levels of checkpoint molecules, the presence of soluble inhibitors such as IDO, inhibitory tumor mechanisms, and tumor sensitivity to immune effectors (tumor evasion mechanisms). A recent study also proposed that mathematical models will be helpful in choosing feasible and effective combinations for ovarian cancer therapy [195].

The molecular composition of the tumor at the time of surgery may also be important to determining disease course and response to novel therapy. This can be studied by using genomics to understand gene expression signatures in the tumors of HGSOC patients, or by combinations of proteomics and other genomic data studies [196, 197]. This should enable a better selection of patients with the potential to respond to a particular therapy.

In this chapter, we outlined the mechanisms of resistance which need to be overcome for better outcome in HGSOC. Even so, the dynamics of combination treatment and response remains complicated and challenging. We need to continue to delineate the nature of the patient tumor, underlying mutations, and the relevance of these to the disease, develop additional useful biomarkers, understand immune cell and tumor cell resistance in ovarian cancer, and investigate the efficacy of novel agents in preclinical animal models, and in clinical trials. We anticipate that success with these measures will signal a new era where there are effective treatment options for HGSOC, resulting in improved quality of life and a significant improvement in survival of patients.

Acknowledgments This work was supported by the Cronk Foundation funds to M.L.D. and the Perritt Charitable Foundation funds to P.J.S.

Funding: This research received no external funding.

Conflicts of InterestThe authors declare no conflicts of interest.

References

1. Perets, R., Wyant, G. A., Muto, K. W., Bijron, J. G., Poole, B. B., Chin, K. T., et al. (2013). Transformation of the fallopian tube secretory epithelium leads to high-grade serous ovarian cancer in Brca; Tp53; Pten models. *Cancer Cell, 24*(6), 751–765.
2. Prat, J. (2012). New insights into ovarian cancer pathology. *Annals of Oncology, 23*(Suppl 10), 111.
3. Cole, A. J., Dwight, T., Gill, A. J., Dickson, K. A., Zhu, Y., Clarkson, A., et al. (2016). Assessing mutant

p53 in primary high-grade serous ovarian cancer using immunohistochemistry and massively parallel sequencing. *Scientific Reports, 6*, 26191.

4. Cancer Genome Atlas Research Network. (2011). Integrated genomic analyses of ovarian carcinoma. *Nature, 474*(7353), 609–615.
5. Lord, C. J., Tutt, A. N., & Ashworth, A. (2015). Synthetic lethality and cancer therapy: Lessons learned from the development of PARP inhibitors. *Annual Review of Medicine, 66*, 455–470.
6. Yi, M., Dong, B., Qin, S., Chu, Q., Wu, K., & Luo, S. (2019). Advances and perspectives of PARP inhibitors. *Experimental Hematology & Oncology, 8*, 29. eCollection 2019.
7. Siegel, R. L., Miller, K. D., & Jemal, A. (2020). Cancer statistics, 2020. *CA: A Cancer Journal for Clinicians, 70*(1), 7–30.
8. Curiel, T. J., Coukos, G., Zou, L., Alvarez, X., Cheng, P., Mottram, P., et al. (2004). Specific recruitment of regulatory T cells in ovarian carcinoma fosters immune privilege and predicts reduced survival. *Nature Medicine, 10*(9), 942–949.
9. Wolf, D., Wolf, A. M., Rumpold, H., Fiegl, H., Zeimet, A. G., Muller-Holzner, E., et al. (2005). The expression of the regulatory T cell-specific forkhead box transcription factor FoxP3 is associated with poor prognosis in ovarian cancer. *Clinical Cancer Research, 11*(23), 8326–8331.
10. Filaci, G., Fenoglio, D., Fravega, M., Ansaldo, G., Borgonovo, G., Traverso, P., et al. (2007). CD8+ CD28− T regulatory lymphocytes inhibiting T cell proliferative and cytotoxic functions infiltrate human cancers. *Journal of Immunology, 179*(7), 4323–4334.
11. Tanaka, A., & Sakaguchi, S. (2017). Regulatory T cells in cancer immunotherapy. *Cell Research, 27*(1), 109–118.
12. Wang, R. F. (2008). CD8+ regulatory T cells, their suppressive mechanisms, and regulation in cancer. *Human Immunology, 69*(11), 811–814.
13. Huang, R. Y., Eppolito, C., Lele, S., Shrikant, P., Matsuzaki, J., & Odunsi, K. (2015). LAG3 and PD1 co-inhibitory molecules collaborate to limit CD8+ T cell signaling and dampen antitumor immunity in a murine ovarian cancer model. *Oncotarget, 6*(29), 27359–27377.
14. Rodriguez, G. M., Galpin, K. J. C., McCloskey, C. W., & Vanderhyden, B. C. (2018). The tumor microenvironment of epithelial ovarian cancer and its influence on response to immunotherapy. *Cancers (Basel), 10*(8). https://doi.org/10.3390/cancers10080242.
15. Hamai, A., Benlalam, H., Meslin, F., Hasmim, M., Carre, T., Akalay, I., et al. (2010). Immune surveillance of human cancer: If the cytotoxic T-lymphocytes play the music, does the tumoral system call the tune? *Tissue Antigens, 75*(1), 1–8.
16. Kaech, S. M., & Cui, W. (2012). Transcriptional control of effector and memory CD8+ T cell differentiation. *Nature Reviews. Immunology, 12*(11), 749–761.
17. Santoiemma, P. P., & Powell, D. J. (2015). Tumor infiltrating lymphocytes in ovarian cancer. *Cancer Biology & Therapy, 16*(6), 807–820.
18. Fridman, W. H., Pagès, F., Saut's-Fridman, C., & Galon, J. (2012). The immune contexture in human tumours: Impact on clinical outcome. *Nature Reviews. Cancer, 12*(4), 298–306.
19. Zhang, L., Conejo-Garcia, J., Katsaros, D., Gimotty, P. A., Massobrio, M., Regnani, G., et al. (2003). Intratumoral T cells, recurrence, and survival in epithelial ovarian cancer. *The New England Journal of Medicine, 348*(3), 203–213.
20. Ovarian Tumor Tissue Analysis (OTTA) Consortium, Goode, E. L., Block, M. S., Kalli, K. R., Vierkant, R. A., Chen, W., et al. (2017). Dose-response association of CD8+ tumor-infiltrating lymphocytes and survival time in high-grade serous ovarian cancer. *JAMA Oncology, 3*(12), e173290.
21. Sato, E., Olson, S. H., Ahn, J., Bundy, B., Nishikawa, H., Qian, F., et al. (2005). Intraepithelial CD8+ tumor-infiltrating lymphocytes and a high CD8+/regulatory T cell ratio are associated with favorable prognosis in ovarian cancer. *Proceedings of the National Academy of Sciences of the United States of America, 102*(51), 18538–18543.
22. Komdeur, F. L., Wouters, M. C. A., Workel, H. H., Tijans, A. M., Terwindt, A. L. J., Brunekreeft, K. L., et al. (2016). CD103+ intraepithelial T cells in high-grade serous ovarian cancer are phenotypically diverse TCRaβ+ CD8aβ+ T cells that can be targeted for cancer immunotherapy. *Oncotarget, 7*(46), 75130–75144.
23. Webb, J. R., Milne, K., Watson, P., Deleeuw, R. J., & Nelson, B. H. (2014). Tumor-infiltrating lymphocytes expressing the tissue resident memory marker CD103 are associated with increased survival in high grade serous ovarian cancer. *Clinical Cancer Research, 20*(2), 434–444.
24. Zeng, Y., Zhang, Q., Wang, H., Lu, M., Kong, H., Zhang, Y., et al. (2015). Prognostic significance of interleukin-17 in solid tumors: A meta-analysis. *International Journal of Clinical and Experimental Medicine, 8*(7), 10515–10536.
25. Fialova, A., Partlova, S., Sojka, L., Hromadkova, H., Brtnicky, T., Fucikova, J., et al. (2013). Dynamics of T-cell infiltration during the course of ovarian cancer: The gradual shift from a Th17 effector cell response to a predominant infiltration by regulatory T-cells. *International Journal of Cancer, 132*(5), 1070–1079.
26. Haanen, J. B. A. G. (2017). Converting cold into hot tumors by combining immunotherapies. *Cell, 170*(6), 1055–1056.
27. Ros, X. R., & Vermeulen, L. (2018). Turning cold tumors hot by blocking TGF-beta. *Trends Cancer, 4*(5), 335–337.
28. Qin, S., Xu, L., Yi, M., Yu, S., Wu, K., & Luo, S. (2019). Novel immune checkpoint targets: Moving beyond PD-1 and CTLA-4. *Molecular Cancer, 18*(1), 155–152.

29. Alsaab, H. O., Sau, S., Alzhrani, R., Tatiparti, K., Bhise, K., Kashaw, S. K., et al. (2017). PD-1 and PD-L1 checkpoint signaling inhibition for cancer immunotherapy: Mechanism, combinations, and clinical outcome. *Frontiers in Pharmacology, 8*, 561.
30. Zhang, G., Liu, C., Bai, H., Cao, G., Cui, R., & Zhang, Z. (2019). Combinatorial therapy of immune checkpoint and cancer pathways provides a novel perspective on ovarian cancer treatment. *Oncology Letters, 17*(3), 2583–2591.
31. Miller, J., Baker, C., Cook, K., Graf, B., Sanchez-Lockhart, M., Sharp, K., et al. (2009). Two pathways of costimulation through CD28. *Immunologic Research, 45*(2–3), 159–172.
32. Rowshanravan, B., Halliday, N., & Sansom, D. M. (2018). CTLA-4: A moving target in immunotherapy. *Blood, 131*(1), 58–67.
33. Friese, C., Harbst, K., Borch, T. H., Westergaard, M. C. W., Pedersen, M., Kverneland, A., et al. (2020). CTLA-4 blockade boosts the expansion of tumor-reactive CD8(+) tumor-infiltrating lymphocytes in ovarian cancer. *Scientific Reports, 10*(1), 3914–3914.
34. Ahmadzadeh, M., Johnson, L. A., Heemskerk, B., Wunderlich, J. R., Dudley, M. E., White, D. E., et al. (2009). Tumor antigen-specific CD8 T cells infiltrating the tumor express high levels of PD-1 and are functionally impaired. *Blood, 114*(8), 1537–1544.
35. Sun, C., Mezzadra, R., & Schumacher, T. N. (2018). Regulation and function of the PD-L1 checkpoint. *Immunity, 48*(3), 434–452.
36. Qin, W., Hu, L., Zhang, X., Jiang, S., Li, J., Zhang, Z., et al. (2019). The diverse function of PD-1/PD-L pathway beyond cancer. *Frontiers in Immunology, 10*, 2298.
37. Iwai, Y., Hamanishi, J., Chamoto, K., & Honjo, T. (2017). PMC5381059; Cancer immunotherapies targeting the PD-1 signaling pathway. *Journal of Biomedical Science, 24*(1), 1–11.
38. Drakes, M. L., Mehrotra, S., Aldulescu, M., Potkul, R. K., Liu, Y., Grisoli, A., et al. (2018). Stratification of ovarian tumor pathology by expression of programmed cell death-1 (PD-1) and PD-ligand-1 (PD-L1) in ovarian cancer. *Journal of Ovarian Research, 11*(1), 43.
39. Solinas, C., Migliori, E., De Silva, P., & Willard-Gallo, K. (2019). LAG3: The biological processes that motivate targeting this immune checkpoint molecule in human cancer. *Cancers (Basel), 11*(8), 1213. https://doi.org/10.3390/cancers11081213.
40. Xu, Y., Zhang, H., Huang, Y., Rui, X., & Zheng, F. (2017). Role of TIM-3 in ovarian cancer. *Clinical & Translational Oncology, 19*(9), 1079–1083.
41. Fucikova, J., Rakova, J., Hensler, M., Kasikova, L., Belicova, L., Hladikova, K., et al. (2019). TIM-3 dictates functional orientation of the immune infiltrate in ovarian cancer. *Clinical Cancer Research, 25*(15), 4820–4831.
42. Guo, Z., Cheng, D., Xia, Z., Luan, M., Wu, L., Wang, G., et al. (2013). Combined TIM-3 blockade and CD137 activation affords the long-term protection in a murine model of ovarian cancer. *Journal of Translational Medicine, 11*, 215.
43. Chen, Y. L., Lin, H. W., Chien, C. L., Lai, Y. L., Sun, W. Z., Chen, C. A., et al. (2019). BTLA blockade enhances Cancer therapy by inhibiting IL-6/IL-10-induced CD19(high) B lymphocytes. *Journal for Immunotherapy of Cancer, 7*(1), 313–314.
44. Ceeraz, S., Nowak, E. C., & Noelle, R. J. (2013). B7 family checkpoint regulators in immune regulation and disease. *Trends in Immunology, 34*(11), 556–563.
45. Wang, J., Wu, G., Manick, B., Hernandez, V., Renelt, M., Erickson, C., et al. (2019). VSIG-3 as a ligand of VISTA inhibits human T-cell function. *Immunology, 156*(1), 74–85.
46. Le Mercier, I., Chen, W., Lines, J. L., Day, M., Li, J., Sergent, P., et al. (2014). VISTA regulates the development of protective antitumor immunity. *Cancer Research, 74*(7), 1933–1944.
47. Mulati, K., Hamanishi, J., Matsumura, N., Chamoto, K., Mise, N., Abiko, K., et al. (2019). VISTA expressed in tumour cells regulates T cell function. *British Journal of Cancer, 120*(1), 115–127.
48. Franklin, R. A., Liao, W., Sarkar, A., Kim, M. V., Bivona, M. R., Liu, K., et al. (2014). The cellular and molecular origin of tumor-associated macrophages. *Science, 344*(6186), 921–925.
49. Liu, Y., & Cao, X. (2015). The origin and function of tumor-associated macrophages. *Cellular & Molecular Immunology, 12*(1), 1–4.
50. Elliott, L. A., Doherty, G. A., Sheahan, K., & Ryan, E. J. (2017). Human tumor-infiltrating myeloid cells: Phenotypic and functional diversity. *Frontiers in Immunology, 8*, 86.
51. Worzfeld, T., von Strandmann, E. P., Huber, M., Adhikary, T., Wagner, U., Reinartz, S., et al. (2017). The unique molecular and cellular microenvironment of Ovarian cancer. *Frontiers in Oncology, 7*, 24.
52. Finkernagel, F., Reinartz, S., Lieber, S., Adhikary, T., Wortmann, A., Hoffmann, N., et al. (2016). The transcriptional signature of human ovarian carcinoma macrophages is associated with extracellular matrix reorganization. *Oncotarget, 7*(46), 75339–75352.
53. Reinartz, S., Schumann, T., Finkernagel, F., Wortmann, A., Jansen, J. M., Meissner, W., et al. (2014). Mixed-polarization phenotype of ascites-associated macrophages in human ovarian carcinoma: Correlation of CD163 expression, cytokine levels and early relapse. *International Journal of Cancer, 134*(1), 32–42.
54. Yin, M., Li, X., Tan, S., Zhou, H. J., Ji, W., Bellone, S., et al. (2016). Tumor-associated macrophages drive spheroid formation during early transcoelomic metastasis of ovarian cancer. *The Journal of Clinical Investigation, 126*(11), 4157–4173.
55. Pollard, J. W. (2004). Tumour-educated macrophages promote tumour progression and metastasis. *Nature Reviews. Cancer, 4*(1), 71–78.

56. Aras, S., & Zaidi, M. R. (2017). TAMeless traitors: Macrophages in cancer progression and metastasis. *British Journal of Cancer, 117*(11), 1583–1591.
57. Drakes, M. L., & Stiff, P. J. (2018). Regulation of ovarian cancer prognosis by immune cells in the tumor microenvironment. *Cancers (Basel), 10*(9). https://doi.org/10.3390/cancers10090302.
58. Lan, C., Huang, X., Lin, S., Huang, H., Cai, Q., Wan, T., et al. (2013). Expression of M2-polarized macrophages is associated with poor prognosis for advanced epithelial ovarian cancer. *Technology in Cancer Research & Treatment, 12*(3), 259–267.
59. Carroll, M. J., Kapur, A., Felder, M., Patankar, M. S., & Kreeger, P. K. (2016). M2 macrophages induce ovarian cancer cell proliferation via a heparin binding epidermal growth factor/matrix metalloproteinase 9 intercellular feedback loop. *Oncotarget, 7*(52), 86608–86620.
60. Zhang, M., He, Y., Sun, X., Li, Q., Wang, W., Zhao, A., et al. (2014). A high M1/M2 ratio of tumor-associated macrophages is associated with extended survival in ovarian cancer patients. *Journal of Ovarian Research, 7*, 19.
61. Baert, T., Vankerckhoven, A., Riva, M., Van Hoylandt, A., Thirion, G., Holger, G., et al. (2019). Myeloid derived suppressor cells: Key drivers of immunosuppression in ovarian cancer. *Frontiers in Immunology, 10*, 1273.
62. Okla, K., Wertel, I., Polak, G., Surowka, J., Wawruszak, A., & Kotarski, J. (2016). Tumor-associated macrophages and myeloid-derived suppressor cells as immunosuppressive mechanism in ovarian cancer patients: Progress and challenges. *International Reviews of Immunology, 35*(5), 372–385.
63. Wu, L., Deng, Z., Peng, Y., Han, L., Liu, J., Wang, L., et al. (2017). Ascites-derived IL-6 and IL-10 synergistically expand CD14(+)HLA-DR(−/low) myeloid-derived suppressor cells in ovarian cancer patients. *Oncotarget, 8*(44), 76843–76856.
64. Horikawa, N., Abiko, K., Matsumura, N., Hamanishi, J., Baba, T., Yamaguchi, K., et al. (2017). Expression of vascular endothelial growth factor in ovarian cancer inhibits tumor immunity through the accumulation of myeloid-derived suppressor cells. *Clinical Cancer Research, 23*(2), 587–599.
65. Cui, T. X., Kryczek, I., Zhao, L., Zhao, E., Kuick, R., Roh, M. H., et al. (2013). Myeloid-derived suppressor cells enhance stemness of cancer cells by inducing microRNA101 and suppressing the corepressor CTBP2. *Immunity, 39*(3), 611–621.
66. Gabrilovich, D. I. (2017). Myeloid-derived suppressor cells. *Cancer Immunology Research, 5*(1), 3–8.
67. Weber, R., Fleming, V., Hu, X., Nagibin, V., Groth, C., Altevogt, P., et al. (2018). Myeloid-derived suppressor cells hinder the anti-cancer activity of immune checkpoint inhibitors. *Frontiers in Immunology, 9*, 1310.
68. Solito, S., Marigo, I., Pinton, L., Damuzzo, V., Mandruzzato, S., & Bronte, V. (2014). Myeloid-derived suppressor cell heterogeneity in human cancers. *Annals of the New York Academy of Sciences, 1319*, 47–65.
69. Bronte, V., Brandau, S., Chen, S. H., Colombo, M. P., Frey, A. B., Greten, T. F., et al. (2016). Recommendations for myeloid-derived suppressor cell nomenclature and characterization standards. *Nature Communications, 7*, 12150.
70. Rodriguez, P. C., Quiceno, D. G., Zabaleta, J., Ortiz, B., Zea, A. H., Piazuelo, M. B., et al. (2004). Arginase I production in the tumor microenvironment by mature myeloid cells inhibits T-cell receptor expression and antigen-specific T-cell responses. *Cancer Research, 64*(16), 5839–5849.
71. Rodriguez, P. C., Quiceno, D. G., & Ochoa, A. C. (2007). L-Arginine availability regulates T-lymphocyte cell-cycle progression. *Blood, 109*(4), 1568–1573.
72. Nagaraj, S., Gupta, K., Pisarev, V., Kinarsky, L., Sherman, S., Kang, L., et al. (2007). Altered recognition of antigen is a mechanism of CD8+ T cell tolerance in cancer. *Nature Medicine, 13*(7), 828–835.
73. Pan, P. Y., Ma, G., Weber, K. J., Ozao-Choy, J., Wang, G., Yin, B., et al. (2010). Immune stimulatory receptor CD40 is required for T-cell suppression and T regulatory cell activation mediated by myeloid-derived suppressor cells in cancer. *Cancer Research, 70*(1), 99–108.
74. Sinha, P., Clements, V. K., Bunt, S. K., Albelda, S. M., & Ostrand-Rosenberg, S. (2007). Cross-talk between myeloid-derived suppressor cells and macrophages subverts tumor immunity toward a type 2 response. *Journal of Immunology, 179*(2), 977–983.
75. Stiff, A., Trikha, P., Mundy-Bosse, B., McMichael, E., Mace, T. A., Benner, B., et al. (2018). Nitric oxide production by myeloid-derived suppressor cells plays a role in impairing Fc receptor-mediated natural killer cell function. *Clinical Cancer Research, 24*(8), 1891–1904.
76. Baniyash, M. (2016). Myeloid-derived suppressor cells as intruders and targets: Clinical implications in cancer therapy. *Cancer Immunology, Immunotherapy, 65*(7), 857–867.
77. Jacob, A., & Prekeris, R. (2015). The regulation of MMP targeting to invadopodia during cancer metastasis. *Frontiers in Cell and Development Biology, 3*, 4.
78. Meyer, C., Cagnon, L., Costa-Nunes, C., Baumgaertner, P., Montandon, N., Leyvraz, L., et al. (2014). Frequencies of circulating MDSC correlate with clinical outcome of melanoma patients treated with ipilimumab. *Cancer Immunology, Immunotherapy, 63*(3), 247–257.
79. Sade-Feldman, M., Kanterman, J., Klieger, Y., Ish-Shalom, E., Olga, M., Saragovi, A., et al. (2016). Clinical significance of circulating CD33+CD11b+HLA−DR− myeloid cells in patients with stage IV melanoma treated with ipilimumab. *Clinical Cancer Research, 22*(23), 5661–5672.

80. Gebhardt, C., Sevko, A., Jiang, H., Lichtenberger, R., Reith, M., Tarnanidis, K., et al. (2015). Myeloid cells and related chronic inflammatory factors as novel predictive markers in melanoma treatment with ipilimumab. *Clinical Cancer Research, 21*(24), 5453–5459.
81. Khan, A. N. H., Kolomeyevskaya, N., Singel, K. L., Grimm, M. J., Moysich, K. B., Daudi, S., et al. (2015). Targeting myeloid cells in the tumor microenvironment enhances vaccine efficacy in murine epithelial ovarian cancer. *Oncotarget, 6*(13), 11310–11326.
82. Pulaski, H. L., Spahlinger, G., Silva, I. A., McLean, K., Kueck, A. S., Reynolds, R. K., et al. (2009). Identifying alemtuzumab as an anti-myeloid cell antiangiogenic therapy for the treatment of ovarian cancer. *Journal of Translational Medicine, 7*, 49.
83. Moughon, D. L., He, H., Schokrpur, S., Jiang, Z. K., Yaqoob, M., David, J., et al. (2015). Macrophage blockade using CSF1R inhibitors reverses the vascular leakage underlying malignant ascites in late-stage epithelial ovarian cancer. *Cancer Research, 75*(22), 4742–4752.
84. Brana, I., Calles, A., LoRusso, P. M., Yee, L. K., Puchalski, T. A., Seetharam, S., et al. (2015). Carlumab, an anti-C-C chemokine ligand 2 monoclonal antibody, in combination with four chemotherapy regimens for the treatment of patients with solid tumors: An open-label, multicenter phase 1b study. *Targeted Oncology, 10*(1), 111–123.
85. Sawa-Wejksza, K., & Kandefer-Szerszen, M. (2018). Tumor-associated macrophages as target for antitumor therapy. *Archivum Immunologiae et Therapiae Experimentalis, 66*(2), 97–111.
86. Banchereau, J., Briere, F., Caux, C., Davoust, J., Lebecque, S., Liu, Y. J., et al. (2000). Immunobiology of dendritic cells. *Annual Review of Immunology, 18*, 767–811.
87. Dudek, A. M., Martin, S., Garg, A. D., & Agostinis, P. (2013). Immature, semi-mature, and fully mature dendritic cells: Toward a DC-cancer cells interface that augments anticancer immunity. *Frontiers in Immunology, 4*, 438.
88. Sabado, R. L., Balan, S., & Bhardwaj, N. (2017). Dendritic cell-based immunotherapy. *Cell Research, 27*(1), 74–95.
89. Volovitz, I., Melzer, S., Amar, S., Bocsi, J., Bloch, M., Efroni, S., et al. (2016). Dendritic cells in the context of human tumors: Biology and experimental tools. *International Reviews of Immunology, 35*(2), 116–135.
90. Strioga, M., Schijns, V., Powell, D. J., Pasukoniene, V., Dobrovolskiene, N., & Michalek, J. (2013). Dendritic cells and their role in tumor immunosurveillance. *Innate Immunity, 19*(1), 98–111.
91. Braun, A., Worbs, T., Moschovakis, G. L., Halle, S., Hoffmann, K., Bolter, J., et al. (2011). Afferent lymph-derived T cells and DCs use different chemokine receptor CCR7-dependent routes for entry into the lymph node and intranodal migration. *Nature Immunology, 12*(9), 879–887.
92. Comerford, I., Harata-Lee, Y., Bunting, M. D., Gregor, C., Kara, E. E., & McColl, S. R. (2013). A myriad of functions and complex regulation of the CCR7/CCL19/CCL21 chemokine axis in the adaptive immune system. *Cytokine & Growth Factor Reviews, 24*(3), 269–283.
93. Ino, K. (2011). Indoleamine 2,3-dioxygenase and immune tolerance in ovarian cancer. *Current Opinion in Obstetrics & Gynecology, 23*(1), 13–18.
94. Cai, D. L., & Jin, L.-P. (2017). Immune cell population in ovarian tumor microenvironment. *Journal of Cancer, 8*(15), 2915–2923.
95. Chen, F., Hou, M., Ye, F., Lv, W., & Xie, X. (2009). Ovarian cancer cells induce peripheral mature dendritic cells to differentiate into macrophage like cells in vitro. *International Journal of Gynecological Cancer, 19*(9), 1487–1493.
96. Scarlett, U. K., Rutkowski, M. R., Rauwerdink, A. M., Fields, J., Escovar-Fadul, X., Baird, J., et al. (2012). Ovarian cancer progression is controlled by phenotypic changes in dendritic cells. *The Journal of Experimental Medicine, 209*(3), 495–506.
97. Hildner, K., Edelson, B. T., Purtha, W. E., Diamond, M., Matsushita, H., Kohyama, M., et al. (2008). Batf3 deficiency reveals a critical role for CD8a+ dendritic cells in cytotoxic T cell immunity. *Science, 322*(5904), 1097–1100.
98. Spranger, S., Dai, D., Horton, B., & Gajewski, T. F. (2017). Tumor-residing Batf3 dendritic cells are required for effector T cell trafficking and adoptive T cell therapy. *Cancer Cell, 31*(5), 711–723.e4.
99. Bronger, H., Singer, J., Windmuller, C., Reuning, U., Zech, D., Delbridge, C., et al. (2016). CXCL9 and CXCL10 predict survival and are regulated by cyclooxygenase inhibition in advanced serous ovarian cancer. *British Journal of Cancer, 115*(5), 553–563.
100. Lieber, S., Reinartz, S., Raifer, H., Finkernagel, F., Dreyer, T., Bronger, H., et al. (2018). Prognosis of ovarian cancer is associated with effector memory CD8(+) T cell accumulation in ascites, CXCL9 levels and activation-triggered signal transduction in T cells. *Oncoimmunology, 7*(5), e1424672.
101. Mittal, D., Vijayan, D., Putz, E. M., Aguilera, A. R., Markey, K. A., Straube, J., et al. (2017). Interleukin-12 from CD103(+) Batf3-dependent dendritic cells required for NK-cell suppression of metastasis. *Cancer Immunology Research, 5*(12), 1098–1108.
102. Flies, D. B., Higuchi, T., Harris, J. C., Jha, V., Gimotty, P. A., & Adams, S. F. (2016). Immune checkpoint blockade reveals the stimulatory capacity of tumor-associated CD103(+) dendritic cells in late-stage ovarian cancer. *Oncoimmunology, 5*(8), e1185583.
103. Inaba, T., Ino, K., Kajiyama, H., Yamamoto, E., Shibata, K., Nawa, A., et al. (2009). Role of the immunosuppressive enzyme indoleamine

2,3-dioxygenase in the progression of ovarian carcinoma. *Gynecologic Oncology, 115*(2), 185–192.
104. Tanizaki, Y., Kobayashi, A., Toujima, S., Shiro, M., Mizoguchi, M., Mabuchi, Y., et al. (2014). Indoleamine 2,3-dioxygenase promotes peritoneal metastasis of ovarian cancer by inducing an immunosuppressive environment. *Cancer Science, 105*(8), 966–973.
105. Labidi-Galy, S., Treilleux, I., Goddard-Leon, S., Combes, J. D., Blay, J. Y., Ray-Coquard, I., et al. (2012). Plasmacytoid dendritic cells infiltrating ovarian cancer are associated with poor prognosis. *Oncoimmunology, 1*(3), 380–382.
106. Wylie, B., Macri, C., Mintern, J. D., & Waithman, J. (2019). Dendritic cells and cancer: From biology to therapeutic intervention. *Cancers (Basel), 11*(4), 521. https://doi.org/10.3390/cancers11040521.
107. Matta, B. M., Castellaneta, A., & Thomson, A. W. (2010). Tolerogenic plasmacytoid DC. *European Journal of Immunology, 40*(10), 2667–2676.
108. Curiel, T. J., Cheng, P., Mottram, P., Alvarez, X., Moons, L., Evdemon-Hogan, M., et al. (2004). Dendritic cell subsets differentially regulate angiogenesis in human ovarian cancer. *Cancer Research, 64*(16), 5535–5538.
109. Pesce, S., Tabellini, G., Cantoni, C., Patrizi, O., Coltrini, D., Rampinelli, F., et al. (2015). B7-H6-mediated downregulation of NKp30 in NK cells contributes to ovarian carcinoma immune escape. *OncoImmunology, 4*(4), e1001224.
110. Wong, J. L., Berk, E., Edwards, R. P., & Kalinski, P. (2013). IL-18-primed helper NK cells collaborate with dendritic cells to promote recruitment of effector CD8+ T cells to the tumor microenvironment. *Cancer Research, 73*(15), 4653–4662.
111. Heng, T. S., Painter, M. W., & Immunological Genome Project Consortium. (2008). The Immunological Genome Project: Networks of gene expression in immune cells. *Nature Immunology, 9*(10), 1091–1094.
112. Böttcher, J. P., Bonavita, E., Chakravarty, P., Blees, H., Cabeza-Cabrerizo, M., Sammicheli, S., et al. (2018). NK cells stimulate recruitment of cDC1 into the tumor microenvironment promoting cancer immune control. *Cell, 172*(5), 1022, 1028.e14.
113. Fridlender, Z. G., Sun, J., Mishalian, I., Singhal, S., Cheng, G., Kapoor, V., et al. (2012). Transcriptomic analysis comparing tumor-associated neutrophils with granulocytic myeloid-derived suppressor cells and normal neutrophils. *PLoS One, 7*(2), e31524.
114. Fridlender, Z. G., Sun, J., Kim, S., Kapoor, V., Cheng, G., Ling, L., et al. (2009). Polarization of tumor-associated neutrophil phenotype by TGF-beta: "N1" versus "N2" TAN. *Cancer Cell, 16*(3), 183–194.
115. Mayer, C., Darb-Esfahani, S., Meyer, A. S., Hubner, K., Rom, J., Sohn, C., et al. (2016). Neutrophil granulocytes in ovarian cancer—Induction of epithelial-to-mesenchymal-transition and tumor cell migration. *Journal of Cancer, 7*(5), 546–554.
116. Loukinova, E., Dong, G., Enamorado-Ayalya, I., Thomas, G. R., Chen, Z., Schreiber, H., et al. (2000). Growth regulated oncogene-alpha expression by murine squamous cell carcinoma promotes tumor growth, metastasis, leukocyte infiltration and angiogenesis by a host CXC receptor-2 dependent mechanism. *Oncogene, 19*(31), 3477–3486.
117. Fridlender, Z. G., & Albelda, S. M. (2012). Tumor-associated neutrophils: Friend or foe? *Carcinogenesis, 33*(5), 949–955.
118. Lichtenstein, A., Seelig, M., Berek, J., & Zighelboim, J. (1989). Human neutrophil-mediated lysis of ovarian cancer cells. *Blood, 74*(2), 805–809.
119. Scapini, P., Lapinet-Vera, J. A., Gasperini, S., Calzetti, F., Bazzoni, F., & Cassatella, M. A. (2000). The neutrophil as a cellular source of chemokines. *Immunological Reviews, 177*, 195–203.
120. Beauvillain, C., Delneste, Y., Scotet, M., Peres, A., Gascan, H., Guermonprez, P., et al. (2007). Neutrophils efficiently cross-prime naive T cells in vivo. *Blood, 110*(8), 2965–2973.
121. Chen, S., Zhang, L., Yan, G., Cheng, S., Fathy, A. H., Yan, N., et al. (2017). Neutrophil-to-lymphocyte ratio is a potential prognostic biomarker in patients with ovarian cancer: A meta-analysis. *BioMed Research International, 2017*, 7943467.
122. Zhao, P., Li, L., Jiang, X., & Li, Q. (2019). Mismatch repair deficiency/microsatellite instability-high as a predictor for anti-PD-1/PD-L1 immunotherapy efficacy. *Journal of Hematology & Oncology, 12*(1), 54–51.
123. Samstein, R. M., Lee, C. H., Shoushtari, A. N., Hellmann, M. D., Shen, R., Janjigian, Y. Y., et al. (2019). Tumor mutational load predicts survival after immunotherapy across multiple cancer types. *Nature Genetics, 51*(2), 202–206.
124. Strickland, K. C., Howitt, B. E., Shukla, S. A., Rodig, S., Ritterhouse, L. L., Liu, J. F., et al. (2016). Association and prognostic significance of BRCA1/2-mutation status with neoantigen load, number of tumor-infiltrating lymphocytes and expression of PD-1/PD-L1 in high grade serous ovarian cancer. *Oncotarget, 7*(12), 13587–13598.
125. Jiang, T., Shi, T., Zhang, H., Hu, J., Song, Y., Wei, J., et al. (2019). Tumor neoantigens: From basic research to clinical applications. *Journal of Hematology & Oncology, 12*(1), 93–95.
126. Chalmers, Z. R., Connelly, C. F., Fabrizio, D., Gay, L., Ali, S. M., Ennis, R., et al. (2017). Analysis of 100,000 human cancer genomes reveals the landscape of tumor mutational burden. *Genome Medicine, 9*(1), 34–32.
127. Hellmann, M. D., Ciuleanu, T. E., Pluzanski, A., Lee, J. S., Otterson, G. A., Audigier-Valette, C., et al. (2018). Nivolumab plus ipilimumab in lung cancer with a high tumor mutational burden. *The New England Journal of Medicine, 378*(22), 2093–2104.
128. Chen, Y., Zhang, L., Liu, W. X., & Wang, K. (2018). VEGF and SEMA4D have synergistic effects on the promotion of angiogenesis in epithelial ovarian

cancer. *Cellular & Molecular Biology Letters, 23*, 2. eCollection 2018.
129. Mulligan, J. K., Day, T. A., Gillespie, M. B., Rosenzweig, S. A., & Young, M. R. (2009). Secretion of vascular endothelial growth factor by oral squamous cell carcinoma cells skews endothelial cells to suppress T-cell functions. *Human Immunology, 70*(6), 375–382.
130. Oyama, T., Ran, S., Ishida, T., Nadaf, S., Kerr, L., Carbone, D. P., et al. (1998). Vascular endothelial growth factor affects dendritic cell maturation through the inhibition of nuclear factor-kappa B activation in hemopoietic progenitor cells. *Journal of Immunology, 160*(3), 1224–1232.
131. Yang, J., Yan, J., & Liu, B. (2018). Targeting VEGF/VEGFR to modulate antitumor immunity. *Frontiers in Immunology, 9*, 978.
132. Rossi, L., Verrico, M., Zaccarelli, E., Papa, A., Colonna, M., Strudel, M., et al. (2017). Bevacizumab in ovarian cancer: A critical review of phase III studies. *Oncotarget, 8*(7), 12389–12405.
133. Aravantinos, G., & Pectasides, D. (2014). Bevacizumab in combination with chemotherapy for the treatment of advanced ovarian cancer: A systematic review. *Journal of Ovarian Research, 7*, 57. eCollection 2014.
134. Cortez, A. J., Tudrej, P., Kujawa, K. A., & Lisowska, K. M. (2018). Advances in ovarian cancer therapy. *Cancer Chemotherapy and Pharmacology, 81*(1), 17–38.
135. Chang, K., Karnad, A., Zhao, S., & Freeman, J. W. (2015). Roles of c-Met and RON kinases in tumor progression and their potential as therapeutic targets. *Oncotarget, 6*(6), 3507–3518.
136. Bardelli, A., Corso, S., Bertotti, A., Hobor, S., Valtorta, E., Siravegna, G., et al. (2013). Amplification of the MET receptor drives resistance to anti-EGFR therapies in colorectal cancer. *Cancer Discovery, 3*(6), 658–673.
137. Wilson, T. R., Fridlyand, J., Yan, Y., Penuel, E., Burton, L., Chan, E., et al. (2012). Widespread potential for growth-factor-driven resistance to anticancer kinase inhibitors. *Nature, 487*(7408), 505–509.
138. Hass, R., Jennek, S., Yang, Y., & Friedrich, K. (2017). c-Met expression and activity in urogenital cancers—Novel aspects of signal transduction and medical implications. *Cell Communication and Signaling: CCS, 15*(1), 10–12.
139. Sierra, J. R., & Tsao, M. S. (2011). c-MET as a potential therapeutic target and biomarker in cancer. *Therapeutic Advances in Medical Oncology, 3*(1 Suppl), 21.
140. Viticchie, G., & Muller, P. A. J. (2015). c-Met and other cell surface molecules: Interaction, activation and functional consequences. *Biomedicines, 3*(1), 46–70.
141. Parrott, J. A., & Skinner, M. K. (2000). Expression and action of hepatocyte growth factor in human and bovine normal ovarian surface epithelium and ovarian cancer. *Biology of Reproduction, 62*(3), 491–500.
142. Kim, J. H., Jang, H. J., Kim, H. S., Kim, B. J., & Park, S. H. (2018). Prognostic impact of high c-Met expression in ovarian cancer: A meta-analysis. *Journal of Cancer, 9*(19), 3427–3434.
143. Hughes, V. S., & Siemann, D. W. (2018). Have clinical trials properly assessed c-Met inhibitors? *Trends in Cancer, 4*(2), 94–97.
144. Loi, S., Dushyanthen, S., Beavis, P. A., Salgado, R., Denkert, C., Savas, P., et al. (2016). RAS/MAPK activation is associated with reduced tumor-infiltrating lymphocytes in triple-negative breast cancer: Therapeutic cooperation between MEK and PD-1/PD-L1 immune checkpoint inhibitors. *Clinical Cancer Research, 22*(6), 1499–1509.
145. De Henau, O., Rausch, M., Winkler, D., Campesato, L. F., Liu, C., Cymerman, D. H., et al. (2016). Overcoming resistance to checkpoint blockade therapy by targeting PI3Kgamma in myeloid cells. *Nature, 539*(7629), 443–447.
146. Murai, J., Huang, S. Y., Das, B. B., Renaud, A., Zhang, Y., Doroshow, J. H., et al. (2012). Trapping of PARP1 and PARP2 by clinical PARP inhibitors. *Cancer Research, 72*(21), 5588–5599.
147. Nolan, E., Savas, P., Policheni, A. N., Darcy, P. K., Vaillant, F., Mintoff, C. P., et al. (2017). Combined immune checkpoint blockade as a therapeutic strategy for BRCA1-mutated breast cancer. *Science Translational Medicine, 9*(393), eaal4922. https://doi.org/10.1126/scitranslmed.aal4922.
148. Higuchi, T., Flies, D. B., Marjon, N. A., Mantia-Smaldone, G., Ronner, L., Gimotty, P. A., et al. (2015). CTLA-4 blockade synergizes therapeutically with PARP inhibition in BRCA1-deficient ovarian cancer. *Cancer Immunology Research, 3*(11), 1257–1268.
149. Gnyszka, A., Jastrzebski, Z., & Flis, S. (2013). DNA methyltransferase inhibitors and their emerging role in epigenetic therapy of cancer. *Anticancer Research, 33*(8), 2989–2996.
150. Stresemann, C., & Lyko, F. (2008). Modes of action of the DNA methyltransferase inhibitors azacytidine and decitabine. *International Journal of Cancer, 123*(1), 8–13.
151. Fang, F., Zuo, Q., Pilrose, J., Wang, Y., Shen, C., Li, M., et al. (2014). Decitabine reactivated pathways in platinum resistant ovarian cancer. *Oncotarget, 5*(11), 3579–3589.
152. Topper, M. J., Vaz, M., Marrone, K. A., Brahmer, J. R., & Baylin, S. B. (2020). The emerging role of epigenetic therapeutics in immuno-oncology. *Nature Reviews. Clinical Oncology, 17*(2), 75–90.
153. McCaw, T. R., Randall, T. D., & Arend, R. C. (2019). Overcoming immune suppression with epigenetic modification in ovarian cancer. *Translational Research, 204*, 31–38.
154. Giannakakis, A., Karapetsas, A., Dangaj, D., Lanitis, E., Tanyi, J., Coukos, G., et al. (2014). Overexpression of SMARCE1 is associated with CD8+ T-cell infiltration in early stage ovarian can-

cer. *The International Journal of Biochemistry & Cell Biology, 53*, 389–398.

155. Shu, H. E., & Fang, T. (2020). Therapeutic efficacy of decitabine on low and moderate-risk MDS patients and the prognostic factors. *Zhongguo Shi Yan Xue Ye Xue Za Zhi, 28*(1), 214–217.
156. Hong, M., Zhu, H., Sun, Q., Zhu, Y., Miao, Y., Yang, H., et al. (2020). Decitabine in combination with low-dose cytarabine, aclarubicin and G-CSF tends to improve prognosis in elderly patients with high-risk AML. *Aging (Albany NY), 12*(7), 5792–5811.
157. Choudhary, C., Kumar, C., Gnad, F., Nielsen, M. L., Rehman, M., Walther, T. C., et al. (2009). Lysine acetylation targets protein complexes and co-regulates major cellular functions. *Science, 325*(5942), 834–840.
158. Wang, L., Amoozgar, Z., Huang, J., Saleh, M. H., Xing, D., Orsulic, S., et al. (2015). Decitabine enhances lymphocyte migration and function and synergizes with CTLA-4 blockade in a murine ovarian cancer model. *Cancer Immunology Research, 3*(9), 1030–1041.
159. Adair, S. J., & Hogan, K. T. (2009). Treatment of ovarian cancer cell lines with 5-aza-2′-deoxycytidine upregulates the expression of cancer-testis antigens and class I major histocompatibility complex-encoded molecules. *Cancer Immunology, Immunotherapy, 58*(4), 589–601.
160. Woloszynska-Read, A., Mhawech-Fauceglia, P., Yu, J., Odunsi, K., & Karpf, A. R. (2008). Intertumor and intratumor NY-ESO-1 expression heterogeneity is associated with promoter-specific and global DNA methylation status in ovarian cancer. *Clinical Cancer Research, 14*(11), 3283–3290.
161. Link, P. A., Zhang, W., Odunsi, K., & Karpf, A. R. (2013). BORIS/CTCFL mRNA isoform expression and epigenetic regulation in epithelial ovarian cancer. *Cancer Immunity, 13*, 6.
162. Zhang, W., Barger, C. J., Link, P. A., Mhawech-Fauceglia, P., Miller, A., Akers, S. N., et al. (2015). DNA hypomethylation-mediated activation of Cancer/Testis Antigen 45 (CT45) genes is associated with disease progression and reduced survival in epithelial ovarian cancer. *Epigenetics, 10*(8), 736–748.
163. Li, H., Chiappinelli, K. B., Guzzetta, A. A., Easwaran, H., Yen, R. W., Vatapalli, R., et al. (2014). Immune regulation by low doses of the DNA methyltransferase inhibitor 5-azacitidine in common human epithelial cancers. *Oncotarget, 5*(3), 587–598.
164. Siebenkäs, C., Chiappinelli, K. B., Guzzetta, A. A., Sharma, A., Jeschke, J., Vatapalli, R., et al. (2017). Inhibiting DNA methylation activates cancer testis antigens and expression of the antigen processing and presentation machinery in colon and ovarian cancer cells. *PLoS One, 12*(6), e0179501.
165. Srivastava, P., Paluch, B. E., Matsuzaki, J., James, S. R., Collamat-Lai, G., Taverna, P., et al. (2015). Immunomodulatory action of the DNA methyltransferase inhibitor SGI-110 in epithelial ovarian cancer cells and xenografts. *Epigenetics, 10*(3), 237–246.
166. Topalian, S. L., Hodi, F. S., Brahmer, J. R., Gettinger, S. N., Smith, D. C., McDermott, D. F., et al. (2012). Safety, activity, and immune correlates of anti-PD-1 antibody in cancer. *The New England Journal of Medicine, 366*(26), 2443–2454.
167. Rothlin, C. V., & Ghosh, S. (2020). Lifting the innate immune barriers to antitumor immunity. *Journal for Immunotherapy of Cancer, 8*(1), e000695. https://doi.org/10.1136/jitc-000695.
168. Chester, C., Dorigo, O., Berek, J. S., & Kohrt, H. (2015). Immunotherapeutic approaches to ovarian cancer treatment. *Journal for Immunotherapy of Cancer, 3*, 1–10.
169. Hamanishi, J., Mandai, M., & Konishi, I. (2016). Immune checkpoint inhibition in ovarian cancer. *International Immunology, 28*(7), 339–348.
170. Disis, M. L., Taylor, M. H., Kelly, K., Beck, J. T., Gordon, M., Moore, K. M., et al. (2019). Efficacy and safety of avelumab for patients with recurrent or refractory ovarian cancer: Phase 1b results from the JAVELIN solid tumor trial. *JAMA Oncology, 5*(3), 393–401.
171. Homicsko, K., & Coukos, G. (2015). Targeting programmed cell death 1 in ovarian cancer. *Journal of Clinical Oncology, 33*(34), 3987–3989.
172. Darvin, P., Toor, S. M., Sasidharan Nair, V., & Elkord, E. (2018). Immune checkpoint inhibitors: Recent progress and potential biomarkers. *Experimental & Molecular Medicine, 50*(12), 1–11.
173. Gaillard, S. L., Secord, A. A., & Monk, B. (2016). The role of immune checkpoint inhibition in the treatment of ovarian cancer. *Gynecologic Oncology Research and Practice, 3*, 1–14.
174. Steinman, R. M., & Cohn, Z. A. (1973). Identification of a novel cell type in peripheral lymphoid organs of mice. I. Morphology, quantitation, tissue distribution. *The Journal of Experimental Medicine, 137*(5), 1142–1162.
175. Palucka, K., & Banchereau, J. (2012). Cancer immunotherapy via dendritic cells. *Nature Reviews. Cancer, 12*(4), 265–277.
176. Stiff, P. J., Czerlanis, C., & Drakes, M. L. (2013). Dendritic cell immunotherapy in ovarian cancer. *Expert Review of Anticancer Therapy, 13*(1), 43–53.
177. Lee, J. J., Foon, K. A., Mailliard, R. B., Muthuswamy, R., & Kalinski, P. (2008). Type 1-polarized dendritic cells loaded with autologous tumor are a potent immunogen against chronic lymphocytic leukemia. *Journal of Leukocyte Biology, 84*(1), 319–325.
178. Mailliard, R. B., Wankowicz-Kalińska, A., Cai, Q., Wesa, A., Hilkens, C. M., Kapsenberg, M. L., et al. (2004). alpha-type-1 polarized dendritic cells: A novel immunization tool with optimized CTL-inducing activity. *Cancer Research, 64*(17), 5934–5937.
179. Whiteside, T. L. (2008). Evaluation of dendritic cell products generated for human therapy and post-treatment immune monitoring. *BioPharm International, 21*(3).

180. Kantoff, P. W., Higano, C. S., Shore, N. D., Berger, E. R., Small, E. J., Penson, D. F., et al. (2010). Sipuleucel-T immunotherapy for castration-resistant prostate cancer. *The New England Journal of Medicine, 363*(5), 411–422.
181. Belderbos, R. A., Aerts, J. G. J. V., & Vroman, H. (2019). Enhancing dendritic cell therapy in solid tumors with immunomodulating conventional treatment. *Molecular Therapy Oncolytics, 13*, 67–81.
182. van Gulijk, M., Dammeijer, F., Aerts, J. G. J. V., & Vroman, H. (2018). Combination strategies to optimize efficacy of dendritic cell-based immunotherapy. *Front in Immunology, 9*, 2759.
183. Tanyi, J. L., Bobisse, S., Ophir, E., Tuyaerts, S., Roberti, A., Genolet, R., et al. (2018). Personalized cancer vaccine effectively mobilizes antitumor T cell immunity in ovarian cancer. *Science Translational Medicine, 10*(436). https://doi.org/10.1126/scitranslmed.aao5931.
184. Sprooten, J., Ceusters, J., Coosemans, A., Agostinis, P., De Vleeschouwer, S., Zitvogel, L., et al. (2019). Trial watch: Dendritic cell vaccination for cancer immunotherapy. *Oncoimmunology, 8*(11), e1638212.
185. Mathé, G., Amiel, J. L., Schwarzenberg, L., Cattan, A., & Schneider, M. (1965). Adoptive immunotherapy of acute leukemia: Experimental and clinical results. *Cancer Research, 25*(9), 1525–1531.
186. Mathé, G., Amiel, J. L., Schwarzenberg, L., et al. (2016). Successful allogenic bone marrow transplantation in man: Chimerism, induced specific tolerance, and possible anti-leukemic effects. Blood. 1965;25(2):179–196. *Blood, 128*(2), 147–167.
187. Yang, S., Yin, X., Yue, Y., & Wang, S. (2019). Application of adoptive immunotherapy in ovarian cancer. *Oncotargets and Therapy, 12*, 7975–7991.
188. Park, J. H., Rivière, I., Gonen, M., Wang, X., Sénéchal, B., Curran, K. J., et al. (2018). Long-term follow-up of CD19 CAR therapy in acute lymphoblastic leukemia. *The New England Journal of Medicine, 378*(5), 449–459.
189. Felder, M., Kapur, A., Gonzalez-Bosquet, J., Horibata, S., Heintz, J., Albrecht, R., et al. (2014). MUC16 (CA125): Tumor biomarker to cancer therapy, a work in progress. *Molecular Cancer, 13*, 129.
190. Hassan, R., Kreitman, R. J., Pastan, I., & Willingham, M. C. (2005). Localization of mesothelin in epithelial ovarian cancer. *Applied Immunohistochemistry & Molecular Morphology, 13*(3), 243–247.
191. Morgan, R. A., Yang, J. C., Kitano, M., Dudley, M. E., Laurencot, C. M., & Rosenberg, S. A. (2010). Case report of a serious adverse event following the administration of T cells transduced with a chimeric antigen receptor recognizing ERBB2. *Molecular Therapy, 18*(4), 843–851.
192. Tanyi, J. L., Stashwick, C., Plesa, G., Morgan, M. A., Porter, D., Maus, M. V., et al. (2017). Possible compartmental cytokine release syndrome in a patient with recurrent ovarian cancer after treatment with mesothelin-targeted CAR-T cells. *Journal of Immunotherapy, 40*(3), 104–107.
193. Gödel, P., Shimabukuro-Vornhagen, A., & von Bergwelt-Baildon, M. (2018). Understanding cytokine release syndrome. *Intensive Care Medicine, 44*(3), 371–373.
194. Blank, C. U., Haanen, J. B., Ribas, A., & Schumacher, T. N. (2016). The "cancer immunogram". *Science, 352*(6286), 658–660.
195. Kozłowska, E., Vallius, T., Hynninen, J., Hietanen, S., Färkkilä, A., & Hautaniemi, S. (2019). Virtual clinical trials identify effective combination therapies in ovarian cancer. *Scientific Reports, 9*(1), 18678.
196. Reinartz, S., Finkernagel, F., Adhikary, T., Rohnalter, V., Schumann, T., Schober, Y., et al. (2016). A transcriptome-based global map of signaling pathways in the ovarian cancer microenvironment associated with clinical outcome. *Genome Biology, 17*(1), 108.
197. Zhang, H., Liu, T., Zhang, Z., Payne, S. H., Zhang, B., McDermott, J. E., et al. (2016). Integrated proteogenomic characterization of human high-grade serous ovarian cancer. *Cell, 166*(3), 755–765.

4 Pharmacological Effects of Natural Components Against Ovarian Cancer and Mechanisms

Huidi Liu and Shu-Lin Liu

4.1 Introduction

Ovarian cancer is among the three leading causes of female cancers but ranks first in mortality worldwide [1]. In the United States, 22,820 new ovarian cancer cases and 14,240 deaths were reported in 2016 [2]. The five most common ovarian cancer types include high-grade and low-grade serous, endometrioid, mucinous, and clear cell carcinoma. All ovarian cancers have an insidious onset with hardly noticeable progress until advanced stages (Stages III and IV). As a result, at the time when the disease is finally diagnosed, peritoneal metastasis often has already occurred. The standard treatment comprises of surgery to remove all macroscopic tumors and systemic chemotherapy to clear or suppress remaining cancer cells [3]. Although ovarian cancers are generally sensitive to platinum agents, and so taxane/platinum combined regimens are often used as first-line chemotherapy, resistance to platinum reagents is common at advanced stages. Conventional chemotherapy is usually cytotoxic with a myriad of side effects. Therefore, more effective and less cytotoxic therapies to treat ovarian cancers are urgently required [4].

Many natural compounds provide health and anticancer benefits. Paclitaxel, a natural antitumor agent, is in the standard front-line treatment and has significant effects on advanced malignancies including ovarian cancer [12]. Substances like paclitaxel are good examples of how natural compounds may be used to treat cancers, inspiring the discovery of safe and effective approaches in ovarian cancer prevention and therapy.

In this chapter, we summarize the available evidence about the effects of plant components from selected fruits, vegetables, and herbs their potential applications as alternative therapeutics against ovarian cancer, with a focus on our recent work in this field.

H. Liu
Systemomics Center, College of Pharmacy, and Genomics Research Center (State-Province Key Laboratories of Biomedicine-Pharmaceutics of China), Harbin Medical University, Harbin, China

HMU-UCCSM Centre for Infection and Genomics, Harbin, China

Department of Biochemistry and Molecular Biology, University of Calgary, Calgary, Canada

S.-L. Liu (✉)
Systemomics Center, College of Pharmacy, and Genomics Research Center (State-Province Key Laboratories of Biomedicine-Pharmaceutics of China), Harbin Medical University, Harbin, China

HMU-UCCSM Centre for Infection and Genomics, Harbin, China

Department of Microbiology, Immunology and Infectious Diseases, University of Calgary, Calgary, Canada
e-mail: slliu@hrbmu.edu.cn

H. Schatten (ed.), *Ovarian Cancer: Molecular & Diagnostic Imaging and Treatment Strategies*, Advances in Experimental Medicine and Biology 1330, https://doi.org/10.1007/978-3-030-73359-9_4

4.1.1 Lignans

Phytoestrogens, especially lignans, are abundant in food materials and are considered to have preventive and therapeutic effects against various cancers [5, 6]. Enterodiol (END) and enterolactone (ENL) are extensively investigated lignans for their potential medical uses [7, 8]. We and other authors have reported the production of END and ENL by human intestinal microbiota through biotransformation from flaxseeds (seeds of *Linum usitatissimum* L.) [9–12]. END and ENL both can reduce the risk of hormone-dependent cancers in the breast [9, 13], uterus [14], and prostate [15]. The anticancer activities of flaxseed lignans have been attributed to two mechanisms, i.e., antioxidant and hormone receptor modulating effects [16, 17]. END and ENL act as antioxidants against DNA damage and lipid peroxidation in cancer and probably also contribute to the reduction of hypercholesterolemia, hyperglycemia, and atherosclerosis [18]. Of specific significance, END and ENL can mimic the structure of human estrogens to upregulate or downregulate the functions of estrogen receptors (ERs) [19]. At relatively low doses, END and ENL exhibit the estrogenic activity, while at higher doses they appear to be anti-estrogenic. The "biphasic effects" might be caused by protein kinase inhibitors at low doses and the topoisomerase activity at higher doses, respectively [7, 20].

There is a considerable body of evidence from epidemiological studies correlating high concentrations of lignans in body fluids with a low incidence of hormone-dependent tumors, in particular breast cancer [21, 22]. For example, a follow-up study showed that postmenopausal breast cancer patients having high enterolignan levels may have a better survival [23]. In another study on serum concentrations in correlation with dietary intake of flaxseed, postmenopausal women consuming flaxseeds had decreased serum 17β-estradiol and estrone sulfate concentrations and lowered breast cancer risks [24]. Additionally, numerous in vitro studies and in vivo animal experiments have demonstrated potent anticancer effects of END and ENL, such as work on breast cancer cell lines MCF-7 and MDA MB 231, which demonstrated the anti-metastatic activity of ENL, probably by inhibiting cell adhesion, cell invasion, and cell motility through downregulating MMP2, MMP9, and MMP14 gene expression [25]. Researchers measured the urinary ENL level in postmenopausal women as well as in breast cancer patients, who were treated with breast cancer removal surgery, and found that breast cancer patients had significantly lower ENL levels compared to the control group, suggesting that ENL might be involved in reducing the risk of breast cancer [26]. In another study, flaxseed, which is a rich source of END and ENL, administered in a basal high-fat diet reduced the nuclear aberration and epithelial proliferation in female rat mammary gland, suggesting a protective effect of flaxseed against breast cancer [27]. Similar results have been found in colon cancer, in which lignans inhibited cell proliferation and induced apoptosis [28].

Nude mouse models have been used to evaluate the therapeutic effects of END, ENL, and other phytoestrogens. A study based on a model of human breast cancers in nude mice showed that cancer animals treated with tamoxifen and fed with flaxseeds or ENL exhibited decreased IL-1β levels compared to controls, which would suppress tumor angiogenesis and reduce microvessel density in vivo [29]. Another breast cancer mouse model with MCF-7 cells showed that ENL had potent effects against breast cancer growth, whereas GEN (Genistein) as the control did not [30]. Additionally, compared to genistein, END and ENL are more suitable for prolonged treatment [9]. The effects of ENL on colon cancer growth and the involved mechanisms of action have been investigated by detecting apoptosis- and proliferation-related proteins and establishing colon cancer mouse models [31]. ENL at a dose of 10 mg/kg could suppress human colon cancer cell growth both in vitro and in vivo [31].

Numerous findings have been reported on END and ENL with different types of tissues or cancers, such as those of lignans on hen ovaries [32, 33], but work about the effects of END and ENL on human ovarian cancers is lacking. We found that both ENL and END performed excellent anticancer effects, although ENL exhibited higher efficacy and less side effects than END in ovarian cancer treatments [34].

4.1.2 Main Components of Pomegranate, Ellagic Acid, and Luteolin

Pomegranate has been used as medicine in many cultures throughout history but is usually consumed as fresh fruit or commercial fruit juice. It possesses many pharmacological effects, including anti-inflammatory, antioxidant, antibacterial, and estrogenic activities [35, 36]. All biological activities are generally attributed to the high phenol, flavonoid, anthocyanin, and tannin contents of the juice, seed, and peel [37]. Recent studies have demonstrated that pomegranate is a potent anti-carcinogenic agent that inhibits multiple signaling pathways, inducing apoptosis and cell cycle arrest [38–41]. Additionally, pomegranate can significantly inhibit angiogenesis and metastasis in cancer development progress [42].

Luteolin (2-(3,4-dihydroxyphenyl) chromenylium-5,7-diol, L) is a nontoxic flavonoid compound that has been used in Chinese traditional medicine to treat various pathologies [43]. In different cancers, luteolin can act as an MMPs inhibitor, attenuating MMPs expression by suppressing the ERK/NF-κB pathway or directly inhibiting its activity [44, 45]. Ellagitannins, one subclass of hydrolyzable tannins, are broken down into free ellagic acid (2,3,7,8-Tetrahydroxy-chromeno[5,4,3-cde] chromene-5,10-dione, EA), which can be absorbed by stomach [46]. When the pomegranate juice is processed, each fruit produces a minimum of 2 g/L of ellagitannins [47]. EA has shown anti-proliferation activity in breast cancer and antioxidant activity through inhibiting inflammatory factors such as TNF-α [48]. However, controversial results have been reported on liver cancer, in which EA was demonstrated to promote hepatocarcinogenesis or perform no effect on hepatocarcinoma [49, 50].

While it is confirmed that pomegranate has significant effects on breast, prostate, and colon cancers [51–53], there are few detailed reports on ovarian cancer. The pharmacological effects and anticancer mechanisms of pomegranate fruit juice (PFJ), along with two of its main components, EA and L, on ovarian cancer provide theoretical basis for new anticancer drug development. We recently compared the efficacy of EA and L and found that EA performed stronger effects than L on ovarian cancer [54].

4.1.3 Mangiferin

Mangiferin (1,3,6,7-tetrohydroxyxanthone-c2-β-D-glucoside) is a kind of polyphenol extracted from the *Anacardiaceae* and *Gentianaceae* species [55], abundant in the leaves, bark of *Mangiferin indica* L. [56], and the roots of *Salacia chinensis* [57], and is commercially utilized in food and natural pharmaceutical industries [58]. Mangiferin has been shown to have promising chemotherapeutic and chemo-preventative potentials, such as antioxidant, anti-inflammatory, immunomodulatory, and anti-virial effects [59]. It also could mitigate the malignant progress of various cancers by suppressing proliferation, migration, and invasion. Furthermore, mangiferin could reverse epithelial-mesenchymal transition to exert anticancer activity in MCF7 breast cancer cell line by inhibiting Wnt/β-catenin pathway [60]. Mangiferin also suppressed expression levels of lung cancer associated enzymes (AHH, γ-GT, and 5′ND) in animal models [61]. Moreover, in some leukemia cases, mangiferin could suppress cyclin B1 and Akt phosphorylation levels, leading to cell arrest in G2/M phase, and activate Nrf2-reduced ROS reaction at a higher concentration [62, 63]. For other cancers, mangiferin could downregulate the Bcl-2/Bax ratio, which is involved in promoting cell apoptosis in nasopharyngeal carcinoma cells [64]. Additionally, it could also block methylmercury-induced DNA damage and oxidative stress in human neuroblastoma IMR-32 cells [65].

4.1.4 *Acanthopanax senticosus*

Acanthopanax senticosus is a small woody shrub that belongs to the Araliaceae family, distributed mainly over China, Korea, Japan, and Russia [66]. *A. senticosus* has a multitude of other names such as Siberian Ginseng, *Eleutherococcus senticosus,* and Ciwujia in China [66–69]. *A. senticosus* has been used in eastern Asia for over 2000 years [70], playing a vital role in traditional Chinese medicine. It is popular with its remarkable performance in the treatment of human cardiovascular diseases, diabetes, and neurasthenia [66, 67]. Recently, the possible anticancer activities of *A. senticosus* have attracted much interest in research, along with some other specific pharmacological effects such as immunostimulatory, immunomodulatory, radiation protection, and antioxidant functions [66]. The most active constituents of this plant are believed to come from the roots and stem, and the bioactivity of *A. senticosus* is attributed to the secondary compounds it synthesizes, such as lignans, saponins, coumarins, minerals, triterpenoid saponins, and various sugars [66]. This plant is processed into Herbal Tea and capsules and then dissolved in hot water. These compounds are suspected to interact with cancerous cells, the immune system, and protective enzymes to provoke anticancer and protective effects, which have been continually supported by experimental studies [71–73].

4.1.5 MMPs

Matrix metalloproteinases (MMPs) are a family of Ca^{2+}-dependent Zn^{2+}-containing endopeptidases, which are capable of degrading extracellular matrix proteins to promote cancer cell migration, invasion, and metastasis [74]. Among more than 20 members of MMPs, MMP2 and MMP9 are correlated with the aggressiveness of cancer [75]. MMPs are regulated by hormones, growth factors, and cytokines, which are all involved in ovarian cancer [76–78]. Thus, MMPs have been considered as significant targets for ovarian cancer therapy.

4.1.6 Natural Components in Ovarian Cancer Treatments

Ovarian cancer remains an overwhelming threat to the health and lives of women due to its high morbidity and mortality. Basic and clinical researchers are currently seeking effective antineoplastic agents without side effects for more accurate and efficient treatment of ovarian cancer and natural products from plants and other organisms provide hope. In this study, we demonstrated the activities of Pomegranate fruit juice (PFJ) and two of its main components, Ellagic acid (EA) and Luteolin (L), to suppress the migration and progression of ovarian cancer through downregulating the expression of MMP-2 and MMP-9.

4.2 Materials and Methods

4.2.1 Reagents

DMEM and McCoy's 5A media were purchased from GE Healthcare Life Sciences, HyClone Laboratories. Histostain-Plus Kits (SP-9001, SP-9002) and Mouse Anti-β actin mAb (TA-09) were purchased from ZSGB-Bio, Beijing, China. Luteolin (≥98%, L9283) and ellagic acid (≥95%, E2250) were purchased from Sigma-Aldrich (USA). DAB Horseradish Peroxidase Color Development Kit (P0203), BeyoECL Plus (P0018), and Hematoxylin Staining Solution (C0107) were purchased from Beyotime Institute of Biotechnology. MMP2 (BMO569) and MMP9 (PB0709) were both purchased as primary antibodies from Boster Biological Technology Co., LTD [79]. HRP-linked rabbit- and mouse-anti IgG (7074s, 7076s) were chosen from CST (USA). Mouse MMPs ELISA Kit (DM-X6142, DM-X6008) was purchased from Baomanbio, Shanghai, China.

4.2.2 Cell Culture

A2780 and ES-2, two human epithelial ovarian cancer cell lines, were purchased from Procell, Wuhan, China. A2780 cells were cultured in

DMEM supplemented with 10% FBS at 37 °C and 5% CO_2 in a humidified incubator. Cells were passaged twice weekly using 0.05% trypsin. Similarly, ES-2 cells were cultured in McCoy's 5A supplemented with 10% FBS at 37 °C and 5% CO_2 in a humidified incubator.

4.2.3 MTT Assay

3-(4,5-Dimethylthiazol-2-yl)-2,5-diphenyltetrazolium bromide (MTT), used to estimate the cytotoxicity of drugs, is a standard colorimetric assay for measuring cellular proliferation. For the cytotoxicity assay, cells were passaged into 96-well plates at 5000 (A2780) cells per well and grown to >80% confluence, before being treated with EA, L (5 μg/mL, 10 μg/mL, 15 μg/mL), or PFJ (5%, 10%). The viable cells were determined 12, 24, or 48 h later by the MTT assay. A total of 20 μL of MTT was added to each well at the indicated time points and 150 μL of DMSO was added to dissolve the formed formazan crystals. MTT has been validated to be an accurate measure of the viable cell population. DMSO at the concentrations used had no effect on cell viability.

4.2.4 Crystal Violet Assay

Crystal violet staining is a colorimetric indirect method to detect maintained adherence of cells. A2780 cells at 45,000 cells per 500 μL were seeded in 24-well plates and treated with different concentrations of PFJ (5%, 10%), EA (5 μg/mL, 10 μg/mL, 15 μg/mL), or L (5 μg/mL, 10 μg/mL, 15 μg/mL) overnight. After incubation for 48 h, the medium was aspirated from the wells and 300 μL 4% PFA was added per well to fix the cells. To remove the remaining liquid, invert the plate on filter paper, then dye cells with 1% crystal violet 300 μL/well for 5 min, and wash with a gentle stream of tap water. After added 1% SDS 300 μL per well, the plate was incubated at room temperature for 1–3 h on a bench rocker with a frequency of 20 oscillations/min, then measured the optical density of each well at 570 nm a plate reader.

4.2.5 Wound Healing Assay

In order to evaluate the migration ability, cells were passaged into 24-well plates at 300,000 (A2780) cells per well and grown to >80% confluence. Twenty microliter pipette tips were used to make a straight, 1-mm-wide scratch and the scattered cells were washed away by PBS twice. Next, cells were treated with serum-free medium (control), EA (5 μg/mL, 10 μg/mL, 15 μg/mL), L (5 μg/mL, 10 μg/mL, 15 μg/mL), or PFJ (5%, 10%) and cultured for another 24 h. The scratch gaps were photographed at time points of 0, 24 h under a light microscope and analyzed using the Digimizer Version 4.6.1 software.

4.2.6 Western Blot Analysis

The expression level of MMP2 and MMP9 in ovarian cancer cells were determined by western blot analysis, proteins extracted from A2780 cells were heat-inactivated and transferred to the PVDF membrane by electrophoresis (150 mA for 80 and 70 min for MMP2 and MMP9, respectively). 5% skim milk was used to block the other interfering proteins. The PVDF membrane was then washed with 1×TBST three times, dyed in a darkroom, and imaged. A 1:400 dilution of primary antibody in 5% skim milk and a 1:500 dilution of secondary antibody in 1×TBST were used. Finally, proteins grey-level was measured by Quantity One (Bio-Rad Quantity One version 4.6.2).

4.2.7 Nude Mice In Vivo Experiments

With the animal experiment approved by the Institutional Ethics Committee of Harbin Medical University, 24 female nude mice, weighing 16 ± 18 g and 4–8 weeks old, were purchased from VRL, Beijing, China. All mice were raised on purified, laminar air flow shelves in the sterile laboratory, at a constant temperature of 25 ± 2 °C. Humidity was maintained between 45% and 50%. We injected human ovarian cancer ES-2 cells (80 μL, 4.09 × 10^6/μL) into the right hind leg and monitored body weight every alternate day. All tumor mass could be touched 15 days after inoculation. The mice were randomly divided into four groups, to be treated with 50 mg/kg EA ($n = 7$), 50 mg/kg L ($n = 7$), 20 mL/kg PFJ ($n = 5$), or 1×PBS ($n = 5$). All mice were executed to examine tissue invasion around the tumor cells and the metastasis of superficial lymph node and viscera. The tumor tissues were used for histological examinations by HE and IHC staining. We also cut mice tails to draw blood for ELISA to detect the concentration of MMP2 and MMP9.

4.2.8 ELISA Analysis

Seven different concentrations of standard MMP2 and MMP9 were pipetted into pre-coated plates. Supernatant from A2780 cell culture and serum from nude mice were used as antigens. Next, biotin-labeled anti-human MMP2 and MMP9 were added to the plates and the reaction was allowed to continue for 60 min at 37 °C. After washing three times, Avidin-Biotin-enzyme Complex (ABC) was added to the stain. After a 15-min incubation at room temperature, a stop buffer was added and the optical density was measured at 450 nm.

4.2.9 HE Staining

Paraffin sections were dipped in xylene, followed by submersion in 100% and then in 80% ethanol solutions. Slides were washed with distilled water for 5 min, before being dyed with hematoxylin and eosin. The slides then underwent ethanol dehydration, being submerged in 85% and 90% ethanol, and then by carbol xylol. Finally, the slides were mounted using neutral balsam.

4.2.10 Immunohistochemical Staining

Previously prepared paraffin-fixed nude mice tissue sections (3 μm) (normal and tumor) were processed for peroxidase (DAB) immunohistochemistry. After deparaffinization and rehydration using xylene and a series of weakening concentrations of ethanol (95%, 80%, 70%), 50 μL of 1:200 dilution MMP2 and MMP9 primary antibody was added to each sample. The samples were stored overnight at 4 °C. After being washed with water for 5 min, addition of peroxidase-labeled polymer and substrate allowed the brown staining of the target proteins to be observed. The samples were counterstained by hematoxylin for 30 s.

4.2.11 Statistical Analysis

Data are presented as the mean of triplicate or quadruplicate determinants with standard error (s.e.). Assays were repeated at least three times. Statistical analysis was performed to assess the difference between the means of the untreated and treated samples using Student's *t*-test, Chi-square test, and Spearman's Rank correlation analysis with SPSS statistical software version 17.0 and GraphPad Prism software. *P*-value <0.05 was considered statistically significant.

4.3 Results

4.3.1 PFJ, EA, and L Could Inhibit the Proliferation of Human Ovarian Carcinoma Cell Line A2780 Cells

We examined different concentrations of PFJ, EA, and L to establish whether they might have the ability to inhibit the proliferation of cancer cells. As shown in Fig. 4.1, they (control, 5 μM/mL, 10 μM/mL, 15 μM/mL) all significantly suppressed the growth of A2780 cells in 12, 24, and 48 h, compared to the control group. Among the treatments ($n = 3$), dose- and time-dependent

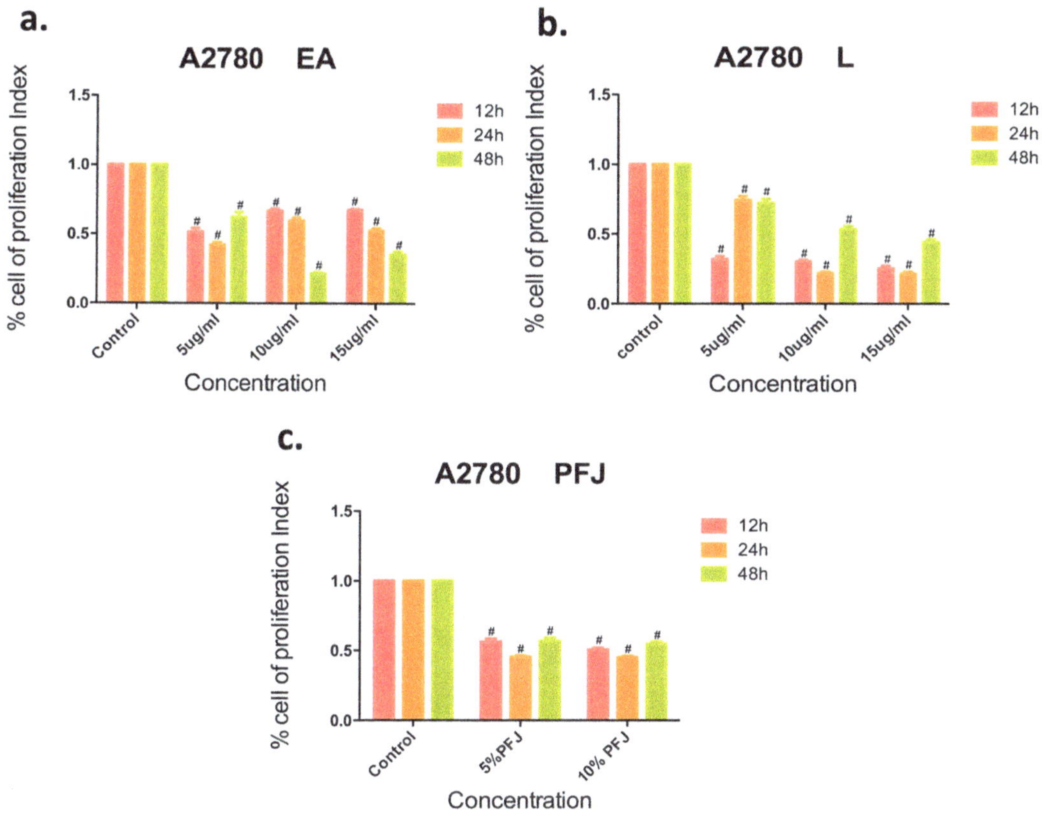

Fig. 4.1 Different concentrations of PFJ, EA, and L showed an inhibition effect on cell proliferation according to the MTT values. The inhibiting cancer cell proliferation activity of different concentration of EA (**a**), L (**b**), and PFJ (**c**). Treated after 12, 24, and 48 h, all of the three show obvious suppression features, both EA and L presented a desired dose- and time-dependent manner at 48 h. Results were obtained from three separate experiments. Student's *t*-test was used for statistical tests, # represents $P < 0.01$ and * represents $P < 0.05$ when compared with ctrl

responses were observed in L (Fig. 4.1b), while EA showed a dose-dependent response only at 48 h (Fig. 4.1a). In order to confirm the results of MTT assays, we performed crystal violet assays to verify the proliferation inhibitive effects of PFJ, EA, and L. Cells were stained by crystal violet after 48 h treatments with PFJ, EA, or L at different concentrations and OD values were compared. We found that PFJ, EA, and L all could decrease the cell number of A2780 in a dose-dependent manner (Fig. 4.2). Furthermore, by triple repeats of each treatment, we found that the inhibitive effect of EA was more obvious than L after 48 h treatment (Fig. 4.2a, b), which was consistent and more favorable with the results from MTT assays at the 48 h time point.

4.3.2 PFJ, EA, and L Could Inhibit the Migration of Human Ovarian Carcinoma Cell Line A2780 Cells

Migration is an initial step for a malignant tumor to make the disease rapidly deteriorating. As shown in Fig. 4.3, EA (Fig. 4.3a), L (Fig. 4.3c), and PFJ (Fig. 4.3e) significantly inhibited tumor migration in a dose-dependent manner. Consistent

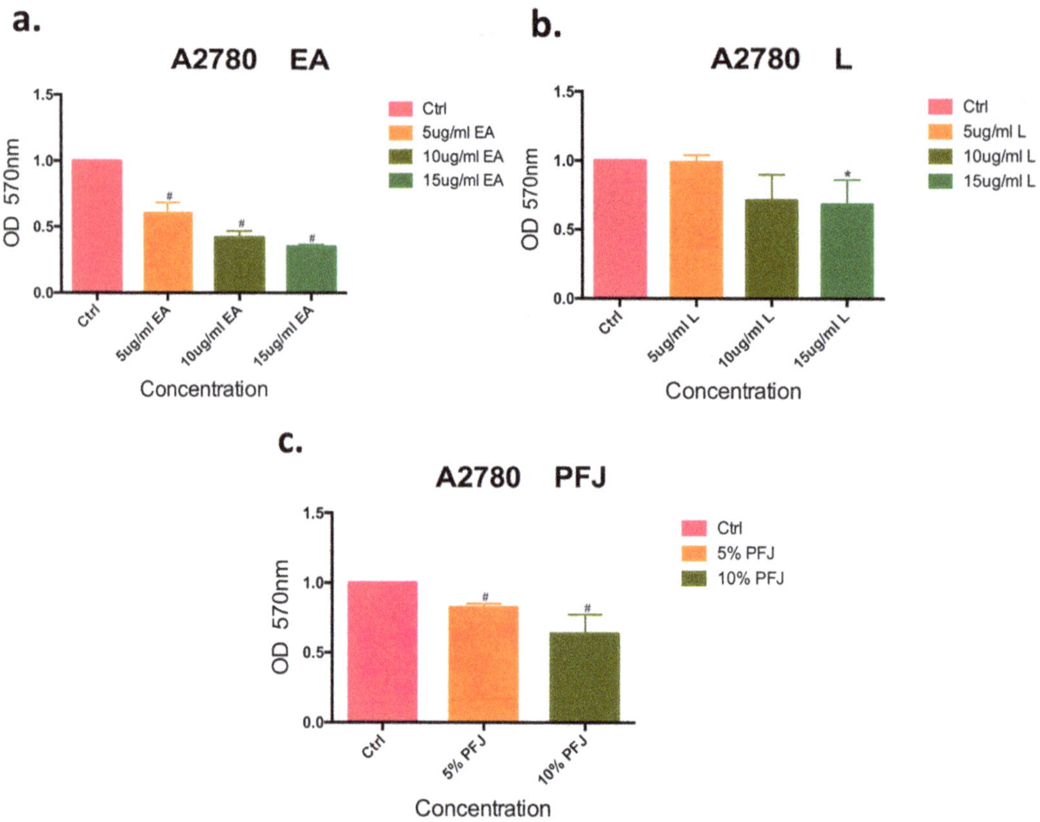

Fig. 4.2 Cell numbers inhibition by different concentrations of PFJ, EA, and Lin A2780 cell line by Crystal Violet assay. After treating with different concentrations of EA, L, and PFJ, A2780 cells were stained with crystal violet (**a**–**c**). Cell number was determined by OD570 value after treating with different concentration of EA, L, and PFJ. After data analysis, each of the three compounds could reduce cell number remarkably; moreover, EA performed a most effective and does-dependent inhibit function compared to L. PFJ could also reduce cell numbers as the concentrations increased. Results were obtained from three independent experiments, # represents $P < 0.01$ and * represents $P < 0.05$ when compared with control

with MTT and crystal violet assay results, treatments with PFJ, EA, and L all inhibited cell motility into a wounded area of confluent cultures in a dose-dependent manner (Fig. 4.3b–f).

4.3.3 The Expression Levels of MMP2 and MMP9 Were Markedly Downregulated by PFJ, EA, and L

To elucidate the mechanisms of the three compounds to inhibit cancer cell migration, we used MMP2 and MMP9 as markers of cancer metastasis. MMP2 (72 kDa type IV collagenase) is intimately linked with the invasion and metastasis of ovarian cancer, while MMP9 (68 kDa type IV collagenase) is a useful serum marker of ovarian cancer [80]. To determine whether EA or L might regulate the expression of MMPs, we conducted western blot analysis and evaluated the expression intensity of MMPs in A2780 cells after treatment with the products for 24 h. Compared with the control (Fig. 4.4a–c), EA at concentrations of 10–15 μg/mL markedly downregulated MMP2 and MMP9 expression in a dose-dependent manner. L could slightly reduce MMP2 and MMP9 expression at the concentration of 5 μg/mL, but at increased concentrations (10–15 μg/mL) the inhibitory effects became much higher. Decreased

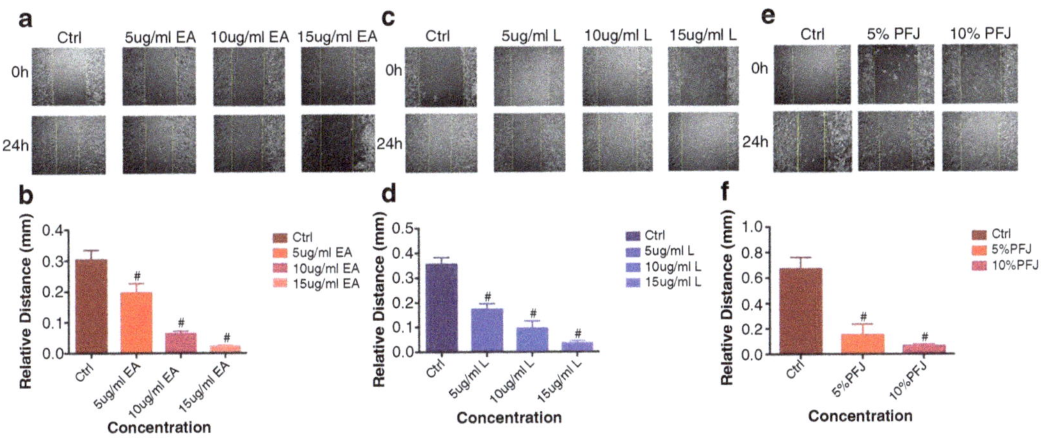

Fig. 4.3 The migration of ovarian cancer cells, quantified by Wound Healing, was significantly suppressed by EA and L. Wound healing in the cell vitro experiments is typically characterized by the remaining distance of the scar after treated with three compounds in 24 h cells comparing to 0 h. EA (**a**, **b**), L (**c**, **d**), and PFJ (**e**, **f**) show dose- and time-dependent and preliminary demonstrate our hypothesis. Results were obtained from three separate experiments. Student's *t*-test was used for statistical tests, # represents $P < 0.01$ and * represents $P < 0.05$ when compared with Ctrl

expression levels of MMP2 (Fig. 4.4b) and MMP9 (Fig. 4.4c) were observed when the cells were treated with 5% and 10% PFJ. Decreased MMP2 and MMP9 expression was also observed as PFJ at different concentrations in western blot analysis (Fig. 4.5a–c).

The contents of MMP2 and MMP9 in the cell supernatant decreased in a dose-dependent manner upon treatment with EA, L, and PFJ as examined using ELISA assays (Fig. 4.6a).

4.3.4 PFJ, EA, and L Inhibited Tumor Growth In Vivo

To better understand the impacts of EA and L on ovarian cancer, we injected ES-2 cells into the right hind leg of female nude mice. Two weeks later, all mice could be found bearing a tumor and were randomly assigned into four experimental groups (EA, L, PFJ, and PBS). As the body weight curve shows, all animals gained body weight gradually (Fig. 4.7a). Interestingly, the body weight of PBS group suddenly dropped during 15–20 days and recovered as we improved the living environment. Compared with that in the PBS group, the tumor volumes increased more slowly in the other three groups (Fig. 4.7b). At the end of the experiment, all mice were sacrificed for histological examinations of the tumor tissues. We found that all three treatments reduced both tumor weight and volume (Fig. 4.8a, b, f, g) with no effects on body weight (Fig. 4.8d) or spleen weight (Fig. 4.8c, e). EA showed a greater reduction of tumor weight and volume than L, suggesting that it could be a better candidate for a future anticancer drug.

4.3.5 PFJ, EA, and L Inhibited MMP2 and MMP9 Expression: Histological and Biochemical Evidence

Hematoxylin-Eosin staining showed dark and basophilic materials in the cytoplasm of most tumor cells compared to cells of control tissues (Fig. 4.9a). The HE staining results indicate that the three products all had anticancer effects through transforming cell structures. We then quantified the expression of MMPs in solid tumor paraffin sections by immunohistochemis-

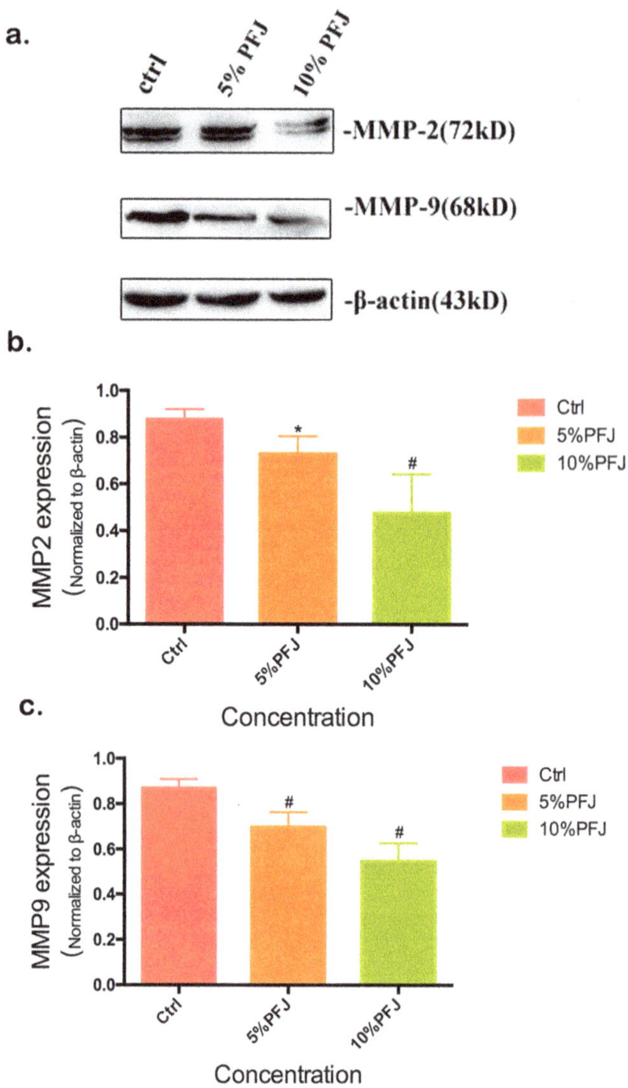

Fig. 4.4 The amount of MMP2 and MMP9 was significantly reduced by EA, L. To illustrate the metastasis inhibition mechanism of EA (**a**, **b**) and L (**a**, **c**), western blot show the downregulation of MMP2 and MMP9 in a dose-dependent tendency. But at the lower concentration (5 μg/mL) of EA (**b**), the inhibiting activity was not markedly compared with control group. Results were obtained from three separate experiments. Student's *t*-test was used for statistical tests, # represents $P < 0.01$ and * represents $P < 0.05$ when compared with Ctrl

try and found that the expression levels of MMP2 and MMP9 were significantly reduced compared to the PBS group (Fig. 4.9b, c). These results indicate that EA, L, and PFJ all had anticancer activities through downregulating MMPs expression. Staining of MMPs was strongest in the PBS groups but either "weak" or "moderate" after treatment, suggesting that all three treatments had therapeutic effects (Fig. 4.9c, d). Moreover, serum ELISA analysis demonstrated that EA, L, and PFJ had suppressive activities on MMP9 and MMP2 (Fig. 4.6b), further confirming the antitumor characteristics of the three products.

4.4 Discussion and Conclusion

Ovarian cancer remains a serious threat to the health and lives of women due to its high mortality. Basic and clinical researchers are currently seeking effective antineoplastic agents without side effects for more accurate and efficient use in diagnosis, treatment, or prognosis of ovarian cancer. We demonstrated the ability of pomegranate fruit juice (PFJ) and two of its main components, ellagic acid (EA) and luteolin (L), to suppress the proliferation, migration, and progression of ovarian cancer through downregulating the expression of MMP2 and MMP9.

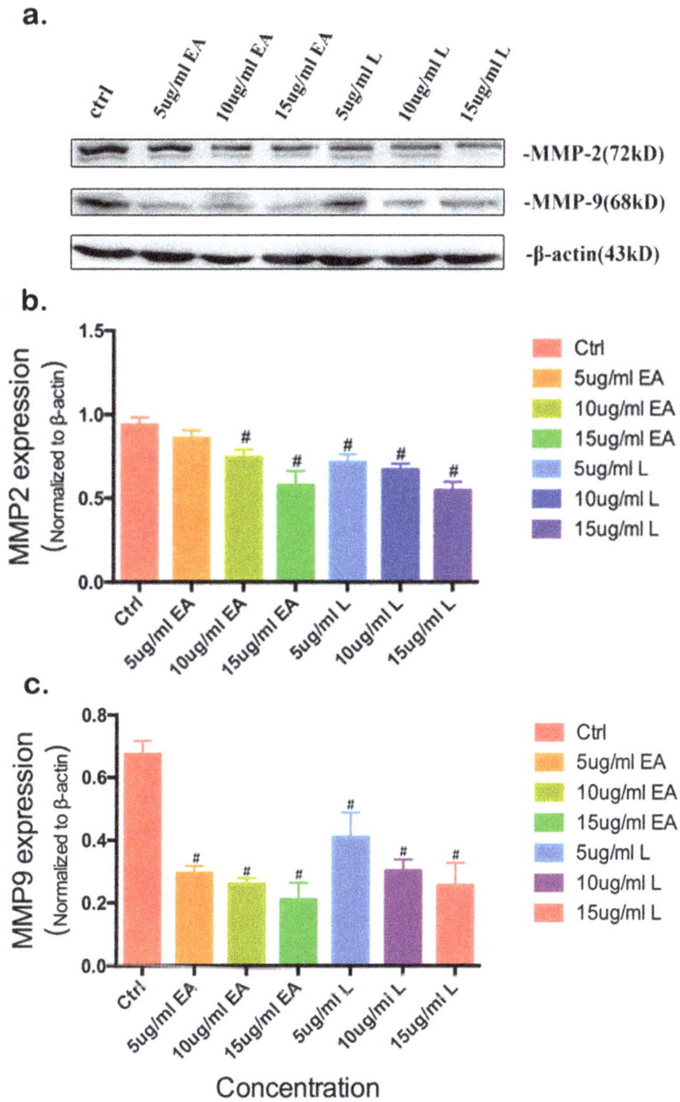

Fig. 4.5 The expression level of MMP2 and MMP9 was inhibited by PFJ. At the same time, different concentrations of PFJ also exert an influence on inhibiting the expression of MMPs. Compared with corresponding ctrl group, both MMP2 and MMP9 expression levels were restrained by 5% and 10% PFJ at does-dependent manners. Results were obtained from three separate experiments. Student's t-test was used for statistical tests, # represents $P < 0.01$ and * represents $P < 0.05$ when compared with Ctrl

A previous study in prostate cancer showed that pomegranate could exert anticancer activity, which was attributed to its high content of polyphenols [81]. Another research group confirmed that EA, L, and ursolic acid extracted from pomegranate caused a concentration-dependent decrease in PANC-1 cell proliferation [82]. Consistently, a study in prostate cancer indicated that PFJ components EA, L, and punicic acid together inhibited the growth of both hormone-dependent and -independent prostate cancer cells and inhibited their migration, progression, and metastasis. Similarly, EA has also been demonstrated to exert in vivo anti-angiogenic effect and inhibit MMP2 activity, both obviously contributing to antitumor activities [83]. L acts as an anti-metastatic agent by suppressing MMP2 and MMP9 production and downregulating expression in azoxymethane-induced colorectal cancer [45]. Yuan-Chiang and colleagues first investigated the effects of EA on ovarian cancer and pointed out that EA may be a potential novel chemoprevention and treatment assistant agent for human ovarian carcinoma [84]. We sought to clarify the antitumor mechanism of EA, L, and PFJ in ovarian cancer; more-

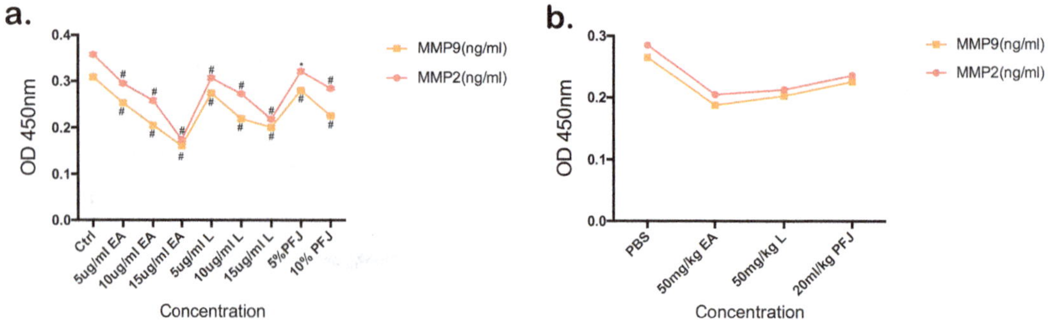

Fig. 4.6 The contents of MMP2 and MMP9 in the supernatant fluid of cultured cancer cells and mice serum were detected by ELISA. In cell supernatant ELISA assay (**a**), we measured the OD value at 450 nm, and it shows the same trend with western blot. In nude mice serum ELISA assay (**b**), each sample contains a corresponding number of mice group. And it also presents a trend that EA, L, and PFJ can decrease the expression level of MMP2 and MMP9. Cell supernatant ELISA assays were performed by three separate experiments. Student's *t*-test was used for statistical tests, # represents $P < 0.01$ and * represents $P < 0.05$ when compared with Ctrl

over, the efficacy of each treatment was compared as well.

We found EA and L to significantly reduce the proliferation, migration, and invasion of ovarian cancer both in vivo and in vitro. The growth of tumor cells was suppressed by PFJ, EA, and L and the inhibitory effect became even stronger with increasing concentrations of the fruit products. Both EA and L showed a time- and dose-dependent manner, EA performed a more obviously cell inhibitive effect comparing to L. Additionally, Wound Healing assays showed PFJ, EA, and L to have a dose-dependent inhibition of cell migration. MMP2 and MMP9, both important markers in tumor migration and invasion, showed the effects of treatments on protein levels. Intensity of MMP2 and MMP9 expression decreased with increasing concentrations of the compounds tested.

During our in vitro experiments, we noticed that EA seemed to have superior anticancer effects over L. To date, there is no publication comparing the antitumor activities of EA and L. Our publication may guide further study of the role played by EA in resisting ovarian cancer. In our experiment, PFJ was squeezed directly from fresh fruit, presumably containing anticancer substances such as anthocyanins. Therefore, the anticancer effect of PFJ should not be simply attributed to the effects of EA and L together. Because neither splenomegaly nor intense changes in body weight were observed, EA, L, and PFJ did not induce severe side effects in nude mice.

In addition to pomegranate, EA and L can be extracted from many other plants, including various berries, pineapple, broccoli, bird chili, and onion leaves [85, 86]. For further usages of these two compounds, our work encourages their dietary and medicinal applications.

Finally, the research demonstrated that EA, L, and PFJ suppressed the proliferation and migration of ovarian cancer through downregulating the expression of MMP2 and MMP9, both in vivo and in vitro. We reported for the first time that EA had greater effects than L, suggesting that EA may be a promising candidate for further preclinical testing for the treatment of human ovarian cancer.

4.5 Further Direction of Natural Compounds in Ovarian Cancer

Natural plants or fruit-derived metabolites are of great resources for adjunct therapies to complement conventional treatment. Natural products markedly inhibited the metastasis of ovarian cancer cells by downregulating the expression of MMPs and slowed down the growth of solid tumors in our in vivo experiments. Our results

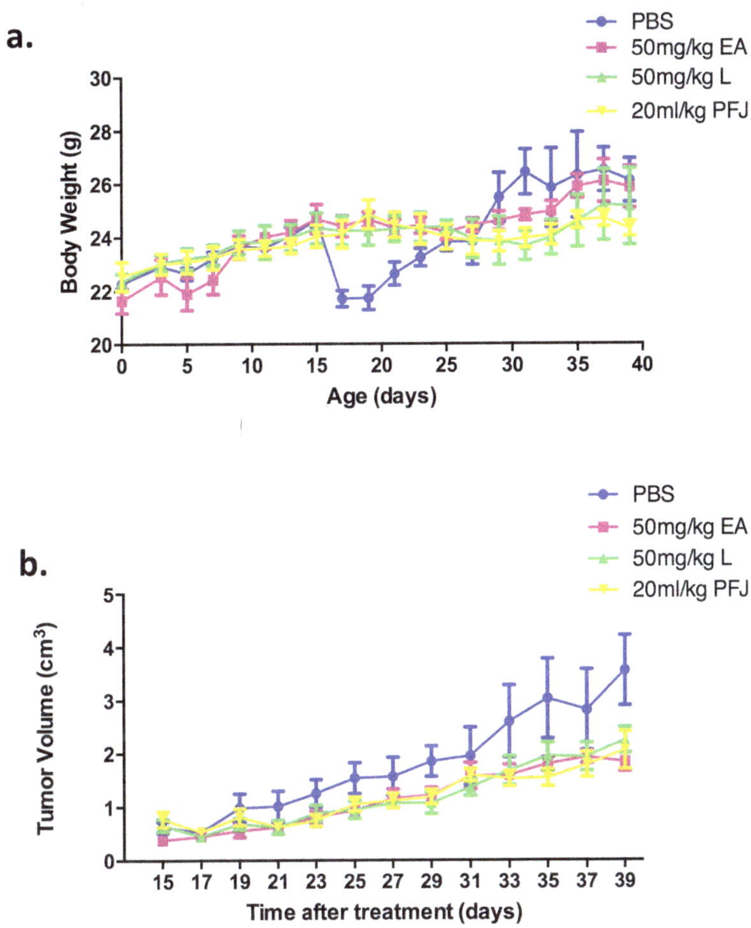

Fig. 4.7 Tumor growth and body weight changes as time during EA, L, and PFJ treatments after the tumor mass could be touched. Treated three compounds bearing ovarian cancer nude mice to clarify the in vivo effect of them. In the everlasting 40 days, nude mice were treated with PBS, 50 mg/kg EA, 50 mg/kg L, and 20 mL/kg PFJ as experimental design. Body weight and tumor volume were started to measure after the tumor mass could be touched. At 15th–19th (**a**), due to the surrounding was worse, the body weight of PBS group dropped drastically before the living environment was improved, subsequently, the body weight of PBS group was recovered and performed as our result shown. All groups were stably increased during the 40 days. Tumor volume were measured once every 2 days and the tumor sizes were calculated according to the formula $V = 0.5\ ab^2$. As shown by the tumor volume curve (**b**), the tumor volumes were increasing slowly in these three compound groups compared with PBS

indicate great potentials of using a broad variety of natural products to improve the prognosis of ovarian cancer. In conclusion, natural products are well on its way to improve the prognosis of ovarian cancer and have great potentials for further application as effective pharmacological treatments for ovarian cancer.

Acknowledgments This work was supported by grants of the National Natural Science Foundation of China (NSFC30970119, 81030029, 81271786, NSFC-NIH 81161120416, 81671980, 81871623) and College students' Innovation & Entrepreneurship project in Heilongjiang Province (201410226047 J.J.K, D.Y.; 201510226020 D.S.L, L.Y., T.L., L.Q., L.L.G.; 201610226095 H.Y.W., Z.H.S. T.T.G., S.J.H., S.G.; 201610226094 Y.Y.Q., M.Y.; 201710226073 S.J.H., S.G.). H.D.L. is supported by a scholarship from China Scholarship Council, CSC No. 201508230143, for an academic visit to the University of Calgary (Univ. of Calgary ID number: 30016355). We also thank the Health and Family Planning Commission of Heilongjiang Province

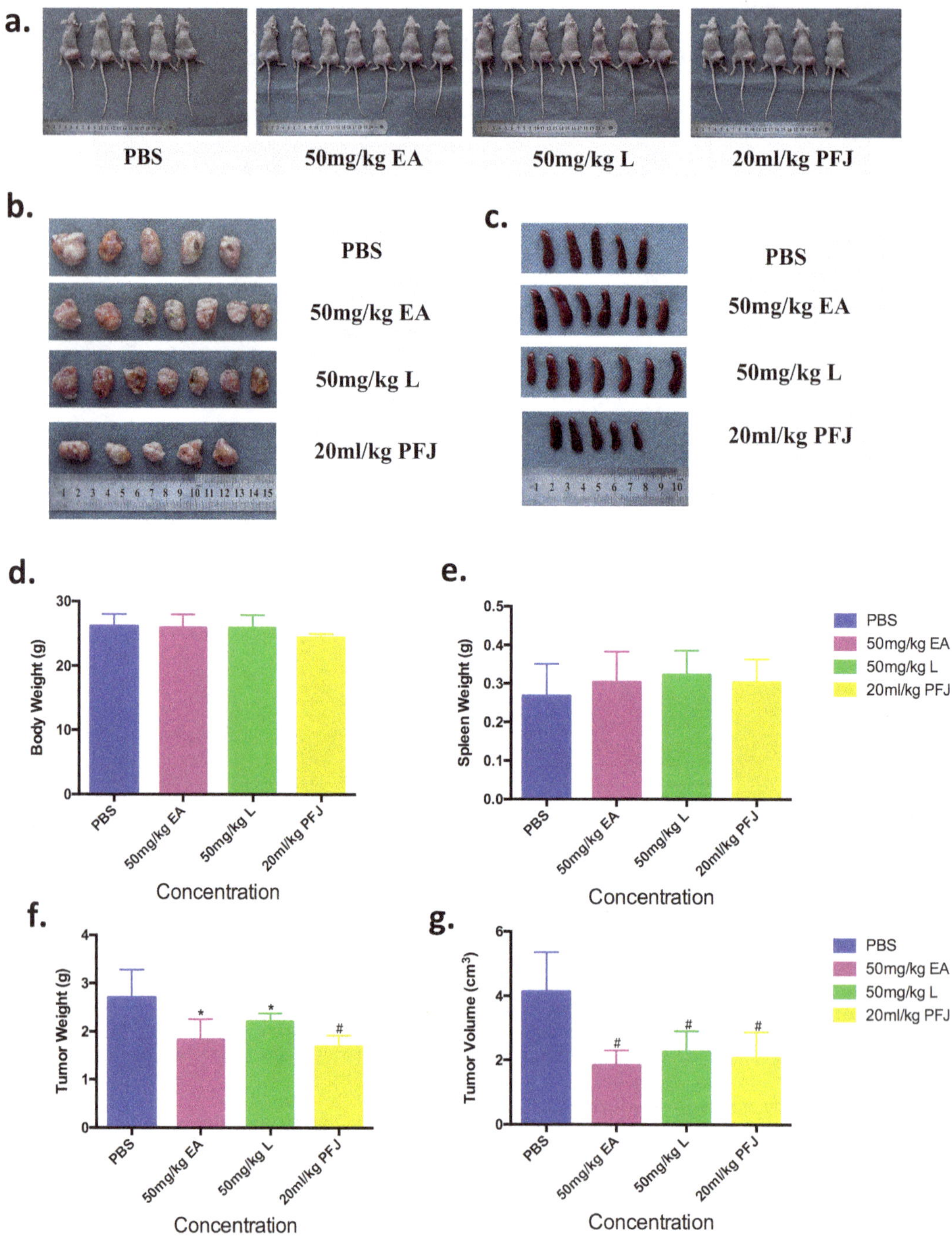

Fig. 4.8 The final body weight, spleen weight, tumor weight, and tumor volume were measured and recorded at necroscopy (the 40th day). Nude mice were sacrificed then gathered the solid tumor and spleen at the end of the experiment (**a**–**c**). And contrast to the value of body weight, tumor weight, and tumor volume in different groups (**d**–**g**), EA shows more advantages and less side effect in vivo. It seems that EA could be used as a medicine in future treatment

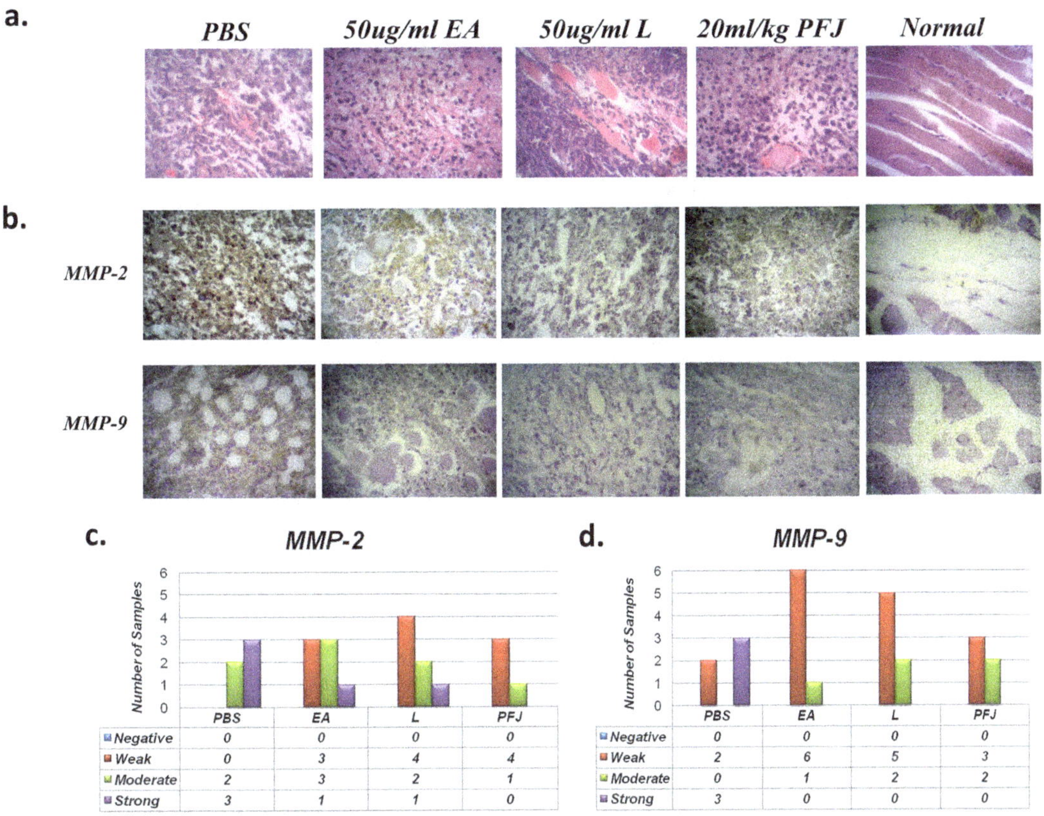

Fig. 4.9 Immunohistochemistry staining for MMP2 and MMP9 and Hematoxylin-Eosin staining in different nude mice ovarian carcinoma tissues. The HE and immunohistochemistry staining in solid tumor paraffin sections. The result of MMP2: PBS ($n = 5$; moderate: 2, strong: 3), EA ($n = 7$; weak: 3, moderate: 3, strong: 1), L ($n = 7$; weak: 4, moderate: 2, strong: 1), and PFJ ($n = 5$; weak: 4, moderate: 1). And MMP9: PBS ($n = 5$; weak: 2, strong: 3), EA ($n = 7$; weak: 6, moderate: 1), L ($n = 7$; weak: 5, moderate: 2), and PFJ ($n = 5$; weak: 3, moderate: 2). It seems that all three compounds may inhibit MMP9 expression strongly than MMP2

(2016-188), the Fundamental Research Funds for the Provincial Universities (2017JCZX57), University Nursing Program for Young Scholars with Creative Talents in Heilongjiang Province (UNPYSCT-2018064), the China Postdoctoral Science Foundation (2018M630380), and the Heilongjiang Postdoctoral Financial Assistance (LBH-Z18198) program for their support.

References

1. Collins, Y., Holcomb, K., Chapman-Davis, E., Khabele, D., & Farley, J. H. (2014). Gynecologic cancer disparities: A report from the Health Disparities Taskforce of the Society of Gynecologic Oncology. *Gynecologic Oncology, 133*(2), 353–361.
2. Siegel, R. L., Miller, K. D., & Jemal, A. (2016). Cancer statistics, 2016. *CA: A Cancer Journal for Clinicians, 66*(1), 7–30.
3. Bhatt, A., & Glehen, O. (2016). The role of cytoreductive surgery and hyperthermic intraperitoneal chemotherapy (HIPEC) in ovarian cancer: A review. *Indian Journal of Surgical Oncology, 7*(2), 188–197.
4. Saucier, J. M., Yu, J., Gaikwad, A., Coleman, R. L., Wolf, J. K., & Smith, J. A. (2007). Determination of the optimal combination chemotherapy regimen for treatment of platinum-resistant ovarian cancer in nude mouse model. *Journal of Oncology Pharmacy Practice, 13*(1), 39–45.
5. Al-Anazi, A. F., Qureshi, V. F., Javaid, K., & Qureshi, S. (2011). Preventive effects of phytoestrogens against postmenopausal osteoporosis as compared to the available therapeutic choices: An overview. *Journal of Natural Science Biology and Medicine, 2*(2), 154–163.

6. Wietrzyk, J., Grynkiewicz, G., & Opolski, A. (2005). Phytoestrogens in cancer prevention and therapy—Mechanisms of their biological activity. *Anticancer Research, 25*(3c), 2357–2366.
7. Wang, L. Q. (2002). Mammalian phytoestrogens: Enterodiol and enterolactone. *Journal of Chromatography. B, Analytical Technologies in the Biomedical and Life Sciences, 777*(1–2), 289–309.
8. Martinchik, A. N., & Zubtsov, V. V. (2012). [Phytoestrogenis properties of flaxseed lignans]. *Voprosy Pitaniia, 81*(6):61–66.
9. Power, K. A., Saarinen, N. M., Chen, J. M., & Thompson, L. U. (2006). Mammalian lignans enterolactone and enterodiol, alone and in combination with the isoflavone genistein, do not promote the growth of MCF-7 xenografts in ovariectomized athymic nude mice. *International Journal of Cancer, 118*(5), 1316–1320.
10. Tao, Y. L., Yang, D. H., Zhang, Y. T., Zhang, Y., Wang, Z. Q., Wang, Y. S., Cai, S. Q., & Liu, S. L. (2014). Cloning, expression, and characterization of the beta-glucosidase hydrolyzing secoisolariciresinol diglucoside to secoisolariciresinol from Bacteroides uniformis ZL1. *Applied Microbiology and Biotechnology, 98*(6), 2519–2531.
11. Zhu, H.-Y., Li, M.-X., Yang, D.-H., Tao, Y.-L., Zhang, Y., & Liu, S.-L. (2014). Biotransformation of the SDG in defatted flaxseed into END co-cultured by three single bacterial colonies. *Process Biochemistry, 49*(1), 19–24.
12. Wang, C. Z., Ma, X. Q., Yang, D. H., Guo, Z. R., Liu, G. R., Zhao, G. X., Tang, J., Zhang, Y. N., Ma, M., Cai, S. Q., et al. (2010). Production of enterodiol from defatted flaxseeds through biotransformation by human intestinal bacteria. *BMC Microbiology, 10*, 115.
13. Mousavi, Y., & Adlercreutz, H. (1992). Enterolactone and estradiol inhibit each other's proliferative effect on MCF-7 breast cancer cells in culture. *The Journal of Steroid Biochemistry and Molecular Biology, 41*(3–8), 615–619.
14. Power, K. A., Ward, W. E., Chen, J. M., Saarinen, N. M., & Thompson, L. U. (2006). Genistein alone and in combination with the mammalian lignans enterolactone and enterodiol induce estrogenic effects on bone and uterus in a postmenopausal breast cancer mouse model. *Bone, 39*(1), 117–124.
15. Denis, L., Morton, M. S., & Griffiths, K. (1999). Diet and its preventive role in prostatic disease. *European Urology, 35*(5–6), 377–387.
16. Hallund, J., Ravn-Haren, G., Bugel, S., Tholstrup, T., & Tetens, I. (2006). A lignan complex isolated from flaxseed does not affect plasma lipid concentrations or antioxidant capacity in healthy postmenopausal women. *The Journal of Nutrition, 136*(1), 112–116.
17. Zhou, Y., Liu, Y. E., Cao, J., Zeng, G., Shen, C., Li, Y., Zhou, M., Chen, Y., Pu, W., Potters, L., et al. (2009). Vitexins, nature-derived lignan compounds, induce apoptosis and suppress tumor growth. *Clinical Cancer Research, 15*(16), 5161–5169.
18. Prasad, K. (2000). Antioxidant activity of secoisolariciresinol diglucoside-derived metabolites, secoisolariciresinol, enterodiol, and enterolactone. *International Journal of Angiology, 9*(4), 220–225.
19. Adlercreutz, H., Fotsis, T., Bannwart, C., Wahala, K., Makela, T., Brunow, G., & Hase, T. (1986). Determination of urinary lignans and phytoestrogen metabolites, potential antiestrogens and anticarcinogens, in urine of women on various habitual diets. *Journal of Steroid Biochemistry, 25*(5B), 791–797.
20. Landete, J. M., Arques, J., Medina, M., Gaya, P., de Las Rivas, B., & Munoz, R. (2016). Bioactivation of phytoestrogens: Intestinal bacteria and health. *Critical Reviews in Food Science and Nutrition, 56*(11), 1826–1843.
21. McCann, S. E., Thompson, L. U., Nie, J., Dorn, J., Trevisan, M., Shields, P. G., Ambrosone, C. B., Edge, S. B., Li, H. F., Kasprzak, C., et al. (2010). Dietary lignan intakes in relation to survival among women with breast cancer: The Western New York Exposures and Breast Cancer (WEB) Study. *Breast Cancer Research and Treatment, 122*(1), 229–235.
22. Guglielmini, P., Rubagotti, A., & Boccardo, F. (2012). Serum enterolactone levels and mortality outcome in women with early breast cancer: A retrospective cohort study. *Breast Cancer Research and Treatment, 132*(2), 661–668.
23. Buck, K., Zaineddin, A. K., Vrieling, A., Heinz, J., Linseisen, J., Flesch-Janys, D., & Chang-Claude, J. (2011). Estimated enterolignans, lignan-rich foods, and fibre in relation to survival after postmenopausal breast cancer. *British Journal of Cancer, 105*(8), 1151–1157.
24. Hutchins, A. M., Martini, M. C., Olson, B. A., Thomas, W., & Slavin, J. L. (2001). Flaxseed consumption influences endogenous hormone concentrations in postmenopausal women. *Nutrition and Cancer, 39*(1), 58–65.
25. Mali, A. V., Wagh, U. V., Hegde, M. V., Chandorkar, S. S., Surve, S. V., & Patole, M. V. (2012). In vitro anti-metastatic activity of enterolactone, a mammalian lignan derived from flax lignan, and down-regulation of matrix metalloproteinases in MCF-7 and MDA MB 231 cell lines. *Indian Journal of Cancer, 49*(1), 181–187.
26. Adlercreutz, H. (1984). Does fiber-rich food containing animal lignan precursors protect against both colon and breast cancer? An extension of the "fiber hypothesis". *Gastroenterology, 86*(4), 761–764.
27. Serraino, M., & Thompson, L. U. (1992). The effect of flaxseed supplementation on the initiation and promotional stages of mammary tumorigenesis. *Nutrition and Cancer, 17*(2), 153–159.
28. Bommareddy, A., Zhang, X. Y., Kaushik, R. S., & Dwivedi, C. (2010). Effects of components present in flaxseed on human colon adenocarcinoma Caco-2 cells: Possible mechanisms of flaxseed on colon cancer development in animals. *Drug Discov Ther, 4*(3), 184–189.

29. Lindahl, G., Saarinen, N., Abrahamsson, A., & Dabrosin, C. (2011). Tamoxifen, flaxseed, and the lignan enterolactone increase stroma- and cancer cell-derived IL-1Ra and decrease tumor angiogenesis in estrogen-dependent breast cancer. *Cancer Research, 71*(1), 51–60.
30. Saarinen, N. M., Abrahamsson, A., & Dabrosin, C. (2010). Estrogen-induced angiogenic factors derived from stromal and cancer cells are differently regulated by enterolactone and genistein in human breast cancer in vivo. *International Journal of Cancer, 127*(3), 737–745.
31. Danbara, N., Yuri, T., Tsujita-Kyutoku, M., Tsukamoto, R., Uehara, N., & Tsubura, A. (2005). Enterolactone induces apoptosis and inhibits growth of Colo 201 human colon cancer cells both in vitro and in vivo. *Anticancer Research, 25*(3B), 2269–2276.
32. Dikshit, A., Gao, C., Small, C., Hales, K., & Hales, D. B. (2016). Flaxseed and its components differentially affect estrogen targets in pre-neoplastic hen ovaries. *The Journal of Steroid Biochemistry and Molecular Biology, 159*, 73–85.
33. Dikshit, A., Gomes Filho, M. A., Eilati, E., McGee, S., Small, C., Gao, C., Klug, T., & Hales, D. B. (2015). Flaxseed reduces the pro-carcinogenic microenvironment in the ovaries of normal hens by altering the PG and oestrogen pathways in a dose-dependent manner. *The British Journal of Nutrition, 113*(9), 1384–1395.
34. Liu, H., Liu, J., Wang, S., Zeng, Z., Li, T., Liu, Y., Mastriani, E., Li, Q. H., Bao, H. X., Zhou, Y. J., et al. (2017). Enterolactone has stronger effects than enterodiol on ovarian cancer. *J Ovarian Res, 10*(1), 49.
35. Spilmont, M., Leotoing, L., Davicco, M. J., Lebecque, P., Mercier, S., Miot-Noirault, E., Pilet, P., Rios, L., Wittrant, Y., & Coxam, V. (2014). Pomegranate and its derivatives can improve bone health through decreased inflammation and oxidative stress in an animal model of postmenopausal osteoporosis. *European Journal of Nutrition, 53*(5), 1155–1164.
36. Costantini, S., Rusolo, F., De Vito, V., Moccia, S., Picariello, G., Capone, F., Guerriero, E., Castello, G., & Volpe, M. G. (2014). Potential anti-inflammatory effects of the hydrophilic fraction of pomegranate (Punica granatum L.) seed oil on breast cancer cell lines. *Molecules, 19*(6), 8644–8660.
37. Faria, A., & Calhau, C. (2011). The bioactivity of pomegranate: Impact on health and disease. *Critical Reviews in Food Science & Nutrition, 51*(51), 626–634.
38. Amin, A. R., Kucuk, O., Khuri, F. R., & Shin, D. M. (2009). Perspectives for cancer prevention with natural compounds. *Journal of Clinical Oncology, 27*(16), 2712–2725.
39. Syed, D. N., Afaq, F., & Mukhtar, H. (2007). Pomegranate derived products for cancer chemoprevention. *Seminars in Cancer Biology, 17*(5), 377–385.
40. Jurenka, J. S. (2008). Therapeutic applications of pomegranate (Punica granatum L.): A review. *Alternative Medicine Review: A Journal of Clinical Therapeutic, 13*(2), 128–144.
41. Seeram, N. P., Adams, L. S., Henning, S. M., Niu, Y., Zhang, Y., Nair, M. G., & Heber, D. (2005). In vitro antiproliferative, apoptotic and antioxidant activities of punicalagin, ellagic acid and a total pomegranate tannin extract are enhanced in combination with other polyphenols as found in pomegranate juice. *Journal of Nutritional Biochemistry, 16*(6), 360–367.
42. Turrini, E., Ferruzzi, L., & Fimognari, C. (2015). Potential effects of pomegranate polyphenols in cancer prevention and therapy. *Oxidative Medicine & Cellular Longevity, 2015*, 1–19.
43. Lópezlázaro, M. (2009). Distribution and biological activities of the flavonoid luteolin. *Mini Reviews in Medicinal Chemistry, 9*(1), 31–59.
44. Amrutha, K., Nanjan, P., Shaji, S. K., Sunilkumar, D., Subhalakshmi, K., Rajakrishna, L., & Banerji, A. (2014). Discovery of lesser known flavones as inhibitors of NF-κB signaling in MDA-MB-231 breast cancer cells—A SAR study. *Bioorganic & Medicinal Chemistry Letters, 24*(19), 4735–4742.
45. Pandurangan, A. K., Dharmalingam, P., Sadagopan, S. K., & Ganapasam, S. (2014). Luteolin inhibits matrix metalloproteinase 9 and 2 in azoxymethane-induced colon carcinogenesis. *Human & Experimental Toxicology, 33*(11), 1176–1185.
46. Seeram, N. P., Henning, S. M., Zhang, Y., Suchard, M., Li, Z., & Heber, D. (2006). Pomegranate juice ellagitannin metabolites are present in human plasma and some persist in urine for up to 48 hours. *Journal of Nutrition, 136*(10), 2481–2485.
47. Gil, M. I., Tomásbarberán, F. A., Hesspierce, B., Holcroft, D. M., & Kader, A. A. (2000). Antioxidant activity of pomegranate juice and its relationship with phenolic composition and processing. *Journal of Agricultural & Food Chemistry, 48*(10), 4581–4589.
48. Mehta, R., & Lansky, E. P. (2004). Breast cancer chemopreventive properties of pomegranate (Punica granatum) fruit extracts in a mouse mammary organ culture. *European Journal of Cancer Prevention, 13*(4), 345–348.
49. Tsuda, H., Uehara, N., Iwahori, Y., Asamoto, M., Iigo, M., Nagao, M., Matsumoto, K., Ito, M., & Hirono, I. (1994). Chemopreventive effects of beta-carotene, alpha-tocopherol and five naturally occurring antioxidants on initiation of hepatocarcinogenesis by 2-amino-3-methylimidazo[4,5-f]quinoline in the rat. *Japanese Journal of Cancer Research, 85*(12), 1214–1219.
50. Tharappel, J. C., Lehmler, H. J., Srinivasan, C., Robertson, L. W., Spear, B. T., & Glauert, H. P. (2008). Effect of antioxidant phytochemicals on the hepatic tumor promoting activity of 3,3′,4,4′-tetrachlorobiphenyl (PCB-77). *Food and Chemical Toxicology, 46*(11), 3467–3474.
51. Shirode, A. B., Kovvuru, P., Chittur, S. V., Henning, S. M., Heber, D., & Reliene, R. (2014). Antiproliferative effects of pomegranate extract in MCF-7 breast cancer cells are associated with

reduced DNA repair gene expression and induction of double strand breaks. *Molecular Carcinogenesis, 53*(6), 458–470.
52. Wang, L., & Martins-Green, M. (2014). Pomegranate and its components as alternative treatment for prostate cancer. *International Journal of Molecular Sciences, 15*(9), 14949–14966.
53. Jaganathan, S. K., Vellayappan, M. V., Narasimhan, G., & Supriyanto, E. (2014). Role of pomegranate and citrus fruit juices in colon cancer prevention. *World Journal of Gastroenterology, 20*(16), 4618–4625.
54. Liu, H., Zeng, Z., Wang, S., Li, T., Mastriani, E., Li, Q. H., Bao, H. X., Zhou, Y. J., Wang, X., Liu, Y., et al. (2017). Main components of pomegranate, ellagic acid and luteolin, inhibit metastasis of ovarian cancer by down-regulating MMP2 and MMP9. *Cancer Biology & Therapy, 18*(12), 990–999.
55. Yoshimi, N., Matsunaga, K., Katayama, M., Yamada, Y., Kuno, T., Qiao, Z., Hara, A., Yamahara, J., & Mori, H. (2001). The inhibitory effects of mangiferin, a naturally occurring glucosylxanthone, in bowel carcinogenesis of male F344 rats. *Cancer Letters, 163*(2), 163–170.
56. Barreto, J. C., Trevisan, M. T. S., Hull, W. E., Gerhard, E., Brito, E. S., De Beate, P., Gerd, W., Bertold, S., & Owen, R. W. (2008). Characterization and quantitation of polyphenolic compounds in bark, kernel, leaves, and peel of mango (Mangifera indica L.). *Journal of Agricultural & Food Chemistry, 56*(14), 5599.
57. Chavan, J. J., Ghadage, D. M., Kshirsagar, P. R., & Kudale, S. S. (2015). Optimization of extraction techniques and RP-HPLC analysis of antidiabetic and anticancer drug mangiferin from roots of 'Saptarangi' (Salacia chinensisL.). *Journal of Liquid Chromatography & Related Technologies, 38*(9), 963–969.
58. Burton-Freeman, B. M., Sandhu, A. K., & Edirisinghe, I. (2017). Mangos and their bioactive components: Adding variety to the fruit plate for health. *Food & Function, 8*(9), 3010.
59. Pinto, M. M., Sousa, M. E., & Nascimento, M. S. (2005). Xanthone derivatives: New insights in biological activities. *Current Medicinal Chemistry, 12*(21), 2517–2538.
60. Li, H., Huang, J., Yang, B., Xiang, T., Yin, X., Peng, W., Cheng, W., Wan, J., Luo, F., & Li, H. (2013). Mangiferin exerts antitumor activity in breast cancer cells by regulating matrix metalloproteinases, epithelial to mesenchymal transition, and β-catenin signaling pathway. *Toxicology & Applied Pharmacology, 272*(1), 180–190.
61. Rajendran, P., Rengarajan, T., Nishigaki, I., Ekambaram, G., & Sakthisekaran, D. (2014). Potent chemopreventive effect of mangiferin on lung carcinogenesis in experimental Swiss albino mice. *Journal of Cancer Research & Therapeutics, 10*(4), 1033–1039.
62. Zhang, B. P., Zhao, J., Li, S. S., Yang, L. J., Zeng, L. L., Chen, Y., & Fang, J. (2014). Mangiferin activates Nrf2-antioxidant response element signaling without reducing the sensitivity to etoposide of human myeloid leukemia cells in vitro. *Acta Pharmacologica Sinica, 35*(2), 257–266.
63. Peng, Z. G., Yao, Y. B., Yang, J., Tang, Y. L., & Huang, X. (2015). Mangiferin induces cell cycle arrest at G2/M phase through ATR-Chk1 pathway in HL-60 leukemia cells. *Genetics & Molecular Research, 14*(2), 4989–5002.
64. Pan, L.-L., Wang, A.-Y., Huang, Y.-Q., Luo, Y., & Ling, M. (2014). Mangiferin induces apoptosis by regulating Bcl-2 and Bax expression in the CNE2 nasopharyngeal carcinoma cell line. *Asian Pacific Journal of Cancer Prevention, 15*(17), 7065–7068.
65. Das, S., Rao, B. N., & Rao, B. S. S. (2011). Mangiferin attenuates methylmercury induced cytotoxicity against IMR-32, human neuroblastoma cells by the inhibition of oxidative stress and free radical scavenging potential. *Chemico-Biological Interactions, 193*(2), 129–140.
66. Davydov, M., & Krikorian, A. (2000). Eleutherococcus senticosus (Rupr. & Maxim.) Maxim. (Araliaceae) as an adaptogen: A closer look. *Journal of Ethnopharmacology, 72*(3), 345–393.
67. Ho-Shan, N., I-Min, L., Juei-Tang, C., Che-Ling, L., & Feng-Lin, H. (2008). Hypoglycemic effect of syringin from Eleutherococcus senticosus in Streptozotocin-induced diabetic rats. *Planta Medica, 74*(2), 109–113.
68. Bahrke, M. S., Morgan, W. P., & Stegner, A. (2009). Is ginseng an ergogenic aid? *International Journal of Sport Nutrition & Exercise Metabolism, 19*(3), 298–322.
69. Eschbach, L. F., Webster, M. J., Boyd, J. C., Mcarthur, P. D., & Evetovich, T. K. (2000). The effect of Siberian ginseng (Eleutherococcus senticosus) on substrate utilization and performance. *International Journal of Sport Nutrition & Exercise Metabolism, 10*(4), 444–451.
70. Arouca, A., & Grassi-Kassisse, D. M. (2013). Eleutherococcus senticosus: Studies and effects. *Health, 5*(9), 1509–1515.
71. Nishibe, S., Kinoshita, H., Takeda, H., & Okano, G. (1990). Phenolic compounds from stem bark of Acanthopanax senticosus and their pharmacological effect in chronic swimming stressed rats. *Chemical & Pharmaceutical Bulletin, 38*(6), 1763–1765.
72. Linzhang, H., Hongfang, Z., Baokang, H., Chengjian, Z., Wei, P., & Luping, Q. (2011). Acanthopanax senticosus: Review of botany, chemistry and pharmacology. *Die Pharmazie, 66*(2), 83–97.
73. Chen, L. I., Wang, X. Y., Xu-Wei, H. U., Fang, H. T., & Qiao, S. Y. (2008). [Determination of eleutheroside B in antifatigue fraction of Acanthopanax senticosus by HPLC]. *Zhongguo Zhong yao za zhi, 33*(23):2800–2802.
74. Verma, R. P., & Hansch, C. (2007). Matrix metalloproteinases (MMPs): Chemical–biological functions and (Q)SARs. *Bioorganic & Medicinal Chemistry, 15*(6), 2223–2268.

75. Coussens, L. M., & Werb, Z. (1996). Matrix metalloproteinases and the development of cancer. *Chemistry & Biology, 3*(11), 895–904.
76. Cheung, L. W. T., Leung, P. C. K., & Wong, A. S. T. (2006). Gonadotropin-releasing hormone promotes ovarian cancer cell invasiveness through c-Jun NH2-terminal kinase-mediated activation of matrix metalloproteinase (MMP)-2 and MMP-9. *Cancer Research, 66*(22), 10902–10910.
77. Lu, Y. M., Rong, M. L., Shang, C., Wang, N., Li, X., Zhao, Y. Y., & Zhang, S. L. (2012). Suppression of HER-2 via siRNA interference promotes apoptosis and decreases metastatic potential of SKOV-3 human ovarian carcinoma cells. *Oncology Reports, 29*(3), 1133–1139.
78. Yu, Y., Li, H., Xue, B., Jiang, X., Huang, K., Ge, J., Zhang, H., & Chen, B. (2014). SDF-1/CXCR7 axis enhances ovarian cancer cell invasion by MMP-9 expression through p38 MAPK pathway. *DNA & Cell Biology, 33*(8), 543–549.
79. Langers, A. M., Verspaget, H. W., Hawinkels, L. J., Kubben, F. J., van Duijn, W., van der Reijden, J. J., Hardwick, J. C., Hommes, D. W., & Sier, C. F. (2012). MMP-2 and MMP-9 in normal mucosa are independently associated with outcome of colorectal cancer patients. *British Journal of Cancer, 106*(9), 1495–1498.
80. Mieszalo, K., Lawicki, S., & Szmitkowski, M. (2016). [The utility of metalloproteinases (MMPs) and their inhibitors (TIMPs) in diagnostics of gynecological malignancies]. *Polski Merkuriusz Lekarski, 40*(237):193–197.
81. Turrini, E., Ferruzzi, L., & Fimognari, C. (2015). Potential effects of pomegranate polyphenols in cancer prevention and therapy. *Oxidative Medicine & Cellular Longevity, 2014*, 1–19.
82. Nair, V., Dai, Z., Khan, M., & Ciolino, H. P. (2011). Pomegranate extract induces cell cycle arrest and alters cellular phenotype of human pancreatic cancer cells. *Anticancer Research, 31*(9), 2699–2704.
83. Huang, S. T., Wang, C. Y., Yang, R. C., Wu, H. T., Yang, S. H., Cheng, Y. C., & Pang, J. H. S. (2011). Ellagic acid, the active compound of Phyllanthus urinaria, exerts in vivo anti-angiogenic effect and inhibits MMP-2 activity. *Evidence-Based Complementary and Alternative Medicine, 2011*(5), 296–297.
84. Yuan-Chiang, C., Li-Cheng, L., Ming-Hsiu, T., Yu-Jen, C., Yi-Ying, C., Shih-Ping, Y., & Chih-Ping, H. (2013). The inhibitory effect of ellagic acid on cell growth of ovarian carcinoma cells. *Evidence-Based Complementary and Alternative Medicine, 2013*(2), 386.
85. Amakura, Y., Mai, O., Tsuji, S., & Tonogai, Y. (2000). High-performance liquid chromatographic determination with photodiode array detection of ellagic acid in fresh and processed fruits. *Journal of Chromatography A, 896*(1–2), 87–93.
86. Miean, K. H., & Mohamed, S. (2001). Flavonoid (myricetin, quercetin, kaempferol, luteolin, and apigenin) content of edible tropical plants. *Journal of Agricultural & Food Chemistry, 49*(6), 3106–3112.

Modeling the Early Steps of Ovarian Cancer Dissemination in an Organotypic Culture of the Human Peritoneal Cavity

5

Peter C. Hart, Preety Bajwa, and Hilary A. Kenny

5.1 Introduction

Ovarian cancer (OvCa) is a disease whose progression to lethality is intimately linked to the existing stromal components that comprise the tumor microenvironment (TME) of metastatic sites. During disease progression, tumor cells disseminate within the abdominal cavity and implant onto the peritoneal lining and to specific visceral adipose depots such as the omentum. These environments have been well described over decades of intensive research, and the robust presence of stromal cells, namely mesothelial cells, fibroblasts, adipocytes, immune cells, and vascular endothelial cells, have been evaluated in their contribution to metastasis of OvCa cells (Fig. 5.1). Our understanding of these interactions has often been limited by technical challenges in accurately modeling this complex TME; however, from simple 2D coculture methods to more intricate 3D organotypic and organoid cultures, there is an increasing appreciation of the multidirectional signaling mechanisms that facilitate metastatic processes, such as tumor cell adhesion, migration, invasion, and colonization of the TME. Here, we describe some of the notable advancements in delineating these interactions using these systems, as well as note some of the limitations posed by current experimental approaches that can be overcome through improvements on existing techniques.

P. C. Hart
College of Pharmacy, Roosevelt University, Chicago, IL, USA

P. Bajwa · H. A. Kenny (✉)
Department of Obstetrics and Gynecology, Gordon Center of Integrative Sciences, W116, University of Chicago, Chicago, IL, USA
e-mail: hkenny@uchicago.edu

5.2 Mesothelial Cells

As one of the primary sites for metastasis of ovarian cancer, the omentum and peritoneum are made up of a variety of stromal cell types, including mesothelial cells, fibroblasts, adipocytes, vascular endothelial cells, and immune cells. Mesothelial cells (MCs) may be of particular interest in the early stages of metastatic development, as they are the most superficial cell type forming a monolayer that lines the peritoneum and the surface of the omentum [1]. They are notable for their ability to aid in wound healing and coordinate immune responses under physiological conditions in response to mechanical injury (e.g., from surgery or peritoneal dialysis) or xenobiotic stress [2–7]. In addition to antigen presentation to mount adaptive immunity, MCs secrete a number of cytokines and growth factors to bolster leukocyte infiltration and proliferation, facilitate macrophage retention at the site of

H. Schatten (ed.), *Ovarian Cancer: Molecular & Diagnostic Imaging and Treatment Strategies*, Advances in Experimental Medicine and Biology 1330, https://doi.org/10.1007/978-3-030-73359-9_5

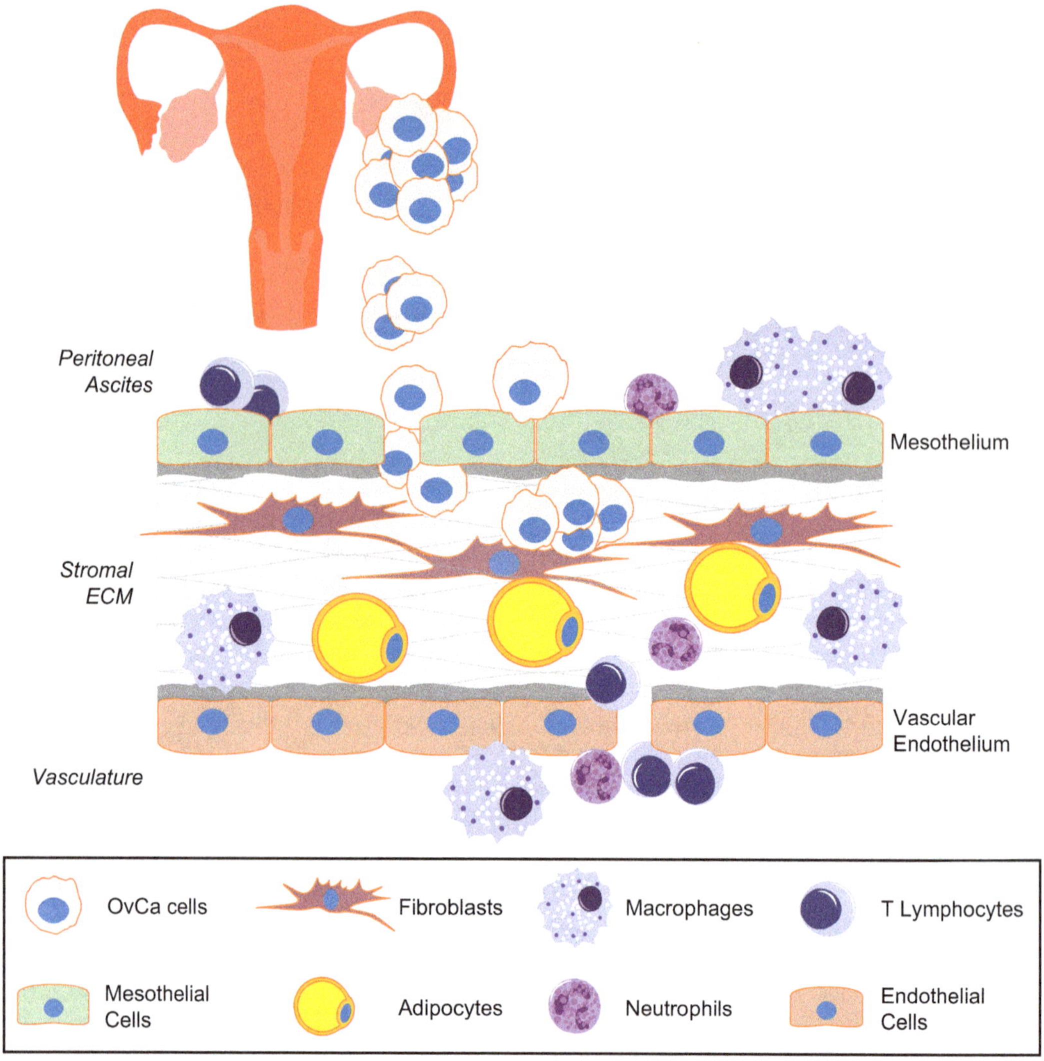

Fig. 5.1 Overview of the role of the ovarian tumor microenvironment in metastasis. The progression of ovarian cancer (OvCa) is a multistep process that includes budding and dissemination from the primary site of origin to distant sites within the abdomen, including the omentum and the peritoneal lining. Adhesion to the mesothelial surface of these tissues is followed by invasion into the endogenous stroma. Interactions between OvCa cells and stromal cells (e.g., fibroblasts, adipocytes, immune cells) promote signaling involved in successful invasion of the tissue, and secretion of metabolites and growth factors from these cells facilitate rapid proliferation. Inflammatory signaling generated in this microenvironment aids in the recruitment and infiltration of additional immune cells found in the peritoneal ascites fluid and vasculature, which in turn further propagate cancer growth

inflammation through integrin-mediated cell–cell adhesions, and are involved in regulating the composition of the extracellular matrix (ECM) [8]. Signaling mechanisms involved in these processes are often associated with promoting tumor cell adhesion and invasion of these peritoneal tissues and may regulate the initial stages of metastasis within the abdomen [9]. The ability of MCs to modulate tumor cell behavior in the TME is due in part to their contribution to the composition of ECM [10–12], as well as to produce soluble signaling proteins such as cytokines [13, 14]. Our appreciation of the role of MCs in advancing metastatic progression of OvCa in the peritoneal cavity has been facilitated by the implementation of 2D coculture systems, 3D organotypic modeling, and ex vivo approaches to identify MC-dependent factors of OvCa adhesion and invasion of the TME.

Studies utilizing 2D coculture methods had identified that MCs regulate several components of the ECM that are required for successful adhesion of OvCa cells to a mesothelial cell monolayer (e.g., fibronectin [12]) as well as transmembrane proteins that mediate cell–cell contacts such as integrins and other glycoproteins [11, 15–19]. Using similar methods, activity of several growth factor receptors have also been implicated in mediating adhesion of OvCa cells to MCs, including c-Met [20] and TGFβR2 [12]. Further, these approaches have identified vast proteomic changes of MCs in response to OvCa exposure [19], supporting a so-called mesothelial-to-mesenchymal transition (MMT) in which MCs undergo transformation to a fibroblast-like phenotype that enhance tumor cell adhesion and invasion in MC coculture systems [21]. Together, these data suggest that reprogramming of MCs result in cell–cell interactions that promote an environment conducive to successful metastasis of OvCa tumor cells.

These findings are complemented by a growing body of literature utilizing 3D organotypic modeling of the omental TME (as shown in Fig. 5.2, and described below), and ex vivo studies utilizing mouse and human omenta. To address the limitations of a less dynamic 2D coculture system, our group had established a 3D organotypic model of the omentum including MCs as well as omental fibroblasts to more faithfully recapitulate the stroma and ECM inherent to this tissue [22]. Use of this model revealed a number of complex interactions that facilitate OvCa cell adhesion and invasion in the TME. We had identified that TGFβ1 signaling from OvCa cells could induce secretion of fibronectin (FN1) from MCs and that FN1 was necessary to promote OvCa invasion in 3D culture in vitro, adhesion ex vivo, and metastases in vivo [12]. Subsequently, we observed that OvCa-derived TGFβ1 promoted metabolic reprogramming of MCs that induced a MMT-like phenotype in part through HIF1α signaling, which reciprocally promoted MC secretion of IL-8 and CCL2 to promote invasion of OvCa cells in the 3D organotypic model and colonization of the human omentum ex vivo [14]. A similar study by Natarajan and colleagues further demonstrated that MC HIF1α and HIF2α expression was necessary for lysyl oxidase-dependent collagen

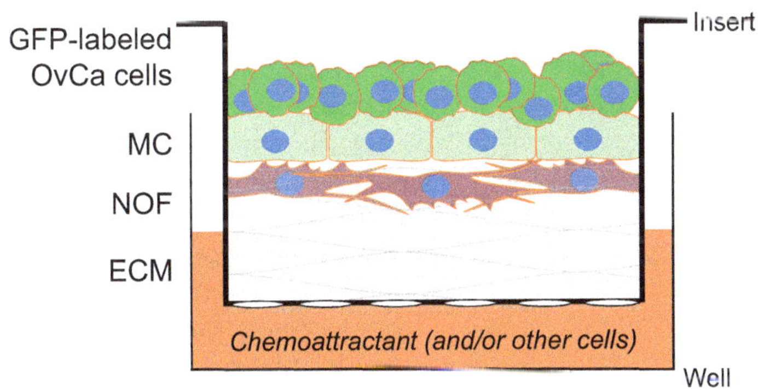

Fig. 5.2 Schematic representation of a 3D organotypic model of tumor cell invasion through the peritoneal/omental surface. Transwell inserts coated in collagen I and/or other extracellular matrix proteins (ECM) are seeded with human normal omental fibroblasts (NOF) and, following adhesion of this fibroblast layer, human primary mesothelial cells (MC). After adhesion and growth of these stromal cells (24–36 h), fluorescently tagged OvCa tumor cells are seeded and allowed to adhere and invade through the stromal culture and ECM towards a chemoattractant (e.g., 10–20% FBS, chemokines, etc.). Alternatively, plating of additional cell types (e.g., adipocytes, leukocytes) in the compartment underneath the transwell insert in place of a chemoattractant is also possible using this system to better represent the metastatic microenvironment

remodeling, which was required for invasion of OvCa cells in vitro and metastasis within the peritoneum in vivo [23]. Dissection of these complex interactions will be useful especially in the context of de novo drug development, as these models can be utilized in high-throughput screening assays to identify novel targets and signaling processes required for adhesion to and invasion of this stromal layer and its unique ECM [24, 25].

While promising, these studies would benefit from inclusion of other primary cell types including adipocytes and immune cells, as the impact of these cell types on the TME and OvCa cells are critical (discussed in detail further). Controlled assessments to account for other additional stromal cell types in 3D organotypic systems or by using mice with mesothelial-specific genetic manipulations (such as the mesothelin mouse model described in [26]) would further delineate the importance of MC-dependent interactions and of MMT on OvCa metastatic progression.

5.3 Fibroblasts

Fibroblasts are a canonical mesenchymal cell type that is present in the tissue stroma and are multifaceted in nature. They are responsible for providing scaffolding and ECM in normal tissues [27], assisting in signaling to promote wound healing [28], and can differentiate into other cell types in a context-dependent manner [29]. Fibroblasts that encounter the nascent or metastatic tumor are "activated" (referred to as cancer-associated fibroblasts, or CAF) and in turn alter the microenvironment, modifying the ECM composition as well as secreting factors that are associated with tumor progression [30]. In several cancers, use of 3D organotypic cultures has indicated that tumor cells invade through matrices resembling basement membrane more efficiently in the presence of fibroblasts [31, 32]. Thus, it is critical to interrogate CAF behavior in the microenvironment to evaluate efficacy of novel drugs aimed at inhibiting tumor progression.

Fibroblasts have been shown to participate in a multitude of oncogenic processes in most prevalent cancers. In ovarian cancer, it has been observed that coculture with CAFs promote tumor cell proliferation [33], angiogenesis [34], adhesion [35], invasion [36], and chemoresistance [37]. Recent studies have evaluated the bidirectional communication between CAFs and tumor cells in an attempt to potentially exploit these interactions therapeutically. For example, TGFβ-dependent versican expression in CAFs resulted in upregulation of NFκB driven MMP9 and CD44 expression in OvCa cells thus enhancing tumor cell migration and invasion [36]. The ability of CAFs to modulate tumor cell metabolism has also been of growing interest in understanding their impact on tumor development. While CAF-derived factors have been shown to promote glycolysis in squamous cell carcinoma [38], it was shown by Curtis and colleagues that p38-induced IL6, CXCL10, and CCL5 emanating from CAFs promoted the mobilization of glycogen in OvCa cells to activate glycolysis, leading to enhancements in proliferation and invasion [39]. In addition to signaling via secreted factors, a recent report supported a role for direct cell–cell communication between CAFs and OvCa tumor cells. In suspended/anchorage-independent coculture, CAFs promoted the ability of ascitic tumor cells to form larger spheroids in vitro, as well as improved xenograft efficiency and decreased survival in vivo [40]. Taken together, using coculture and 3D systems have revealed the numerous paracrine signaling events, both through secreted factors as well as cell–cell contacts, stemming from CAFs that drive tumor development and progression.

Interestingly, intraperitoneal xenograft of tumor spheroids prior to injection of fibroblasts resulted in their recruitment specifically to established micro-metastases rather than unaffected tissue [41]. Although the authors assert the potential for this method to visualize tumors intraoperatively, this study incidentally supports the remarkable propensity for fibroblasts to interact with tumor cells and thus strengthens the argument that this interaction is vital to target tumor development. The pervasive secretome of CAFs does in fact have profound effects on ovarian tumor progression, from cytokines and other soluble ligands [37] to exosomes [42] and metabolites [43]. Evaluating how these secreted factors act in tandem to manipulate the microenvironment, and how responses in the dynamic TME may, in turn, shape the signaling milieu, will give further insight into how to best target tumor–stromal interactions to prevent tumor growth and metastasis.

5.4 Adipocytes

Adipocytes comprise the bulk of the omentum as well as mesenteric and peritoneal adipose depots found within the visceral cavity in the abdomen. Under both normal and pathophysiologic circumstances, they secrete lipids as well as a number of signaling molecules termed adipokines, including leptin, adiponectin, and resistin, which play a role in lipid and energy homeostasis of surrounding cells and tissues (reviewed in [44]). While there is some evidence that excess adipose associated with obesity may increase the risk of several cancers [45], meta-analyses of epidemiological studies demonstrate inconsistencies on whether there is actually a strong association between obesity and ovarian cancer risk [46]. Due to these conflicting clinical results, it is required to establish causal links between adipocytes, tumor development, and metastasis using physiologically relevant approaches in order to delineate the contribution of adipocytes to disease initiation and progression.

Several approaches utilizing coculture systems and ex vivo modeling have provided insight on potential mechanisms by which adipocytes may promote OvCa metastasis. Early studies by our group had demonstrated that primary human adipocytes or their conditioned medium (CM) could markedly enhance invasion of OvCa cells through extracellular matrix; moreover, we had observed that adipocyte FABP4-dependent β-oxidation of free fatty acids in cancer cells promotes proliferation and invasion of OvCa cells, which was associated with tumor burden and metastasis in vivo [47]. In line with this, the ability for adipocyte-derived FABP4 to drive OvCa metastasis has been demonstrated to occur through activating numerous oncogenic pathways involved in proliferation, migration, and metastasis (recently reviewed in [48]). It was also observed that induction of OvCa tumor cell CD36 expression by exposure to adipocytes was a requirement for migration, invasion, and clonogenicity of OvCa cells in vitro and tumor metastasis in vivo [49], consistent with the role of adipocyte-induced CD36 in the progression of other cancers [50]. More recent work by John and colleagues demonstrated that crosstalk between adipocytes and OvCa cells induced proinflammatory signaling pathways (e.g., NFκB, AP-1) and cytokine production (e.g., IL-6, TNFα) in both cell types during coculture, and that this interaction was suppressed by osteonectin, a soluble extracellular matrix glycoprotein [51]. Furthermore, the authors had also shown that osteonectin could inhibit FABP4 and CD36 expression in OvCa cells, as well as repress the adipocyte-induced fatty acid uptake and lipid droplet formation of OvCa cells in coculture, supporting the importance of adipocytes in regulating tumor cell metabolism.

From the aforementioned studies, as well as numerous others across several cancer types, it has become clear that adipocyte-derived lipids and adipokines can promote metastasis in in vitro and in vivo models of metastasis by supporting the high energy demands of key processes such as adhesion, invasion, colonization, and rapid proliferation. However, given the complexity of the omental TME as a whole, more sophisticated approaches could delineate adipocyte-dependent signaling mechanisms required for OvCa metastasis. For example, it was recently shown that OvCa tumor xenograft efficiency was markedly reduced in $CCL2^{-/-}$ mice, but not in mice with a knockdown of its receptor CCR2, which the authors attributed to adipocyte-derived CCL2 as essential for metastasis [52]. While elegant, these studies do not unequivocally demonstrate that signaling stems from the adipocyte in particular, but rather the TME as a whole. To address this, current technological advances allowing for precise genetic manipulation of cell types in the murine model organism, such as the aP2- or adiponectin-Cre mice [53], could allow for more direct in vivo and ex vivo approaches to understand the role of the adipocyte in facilitating a conducive environment for implantation of the omentum. These types of approaches may help in overcoming the technical challenges faced in developing in vitro 3D model systems of the omental TME incorporating adipocytes, as faithfully mimicking the infrastructure of adipose tissue as it occurs in situ has yet to be resolved in current in vitro systems.

5.5 Macrophages

Tumor-associated macrophages (TAM) constitute a majority of the cells present in ovarian cancer ascites. While there is a mixed population of M1 and M2 macrophages, which are generally thought to be anti- and pro-tumorigenic, respectively, M2 TAMs are typically the prevalent phenotype observed in ascites fluid [54]. Similarly, infiltration of M2 TAMs are observed in peritoneal and omental implants [55], and prevalence of M2 over M1 phenotypes was recently associated with more aggressive ovarian cancer [56]. While there has long been interest in macrophage function in the disease's progression [57], an understanding of the impact of signals emanating from TAMs on tumor cell metastasis remains a field of intensive investigation.

The bulk of our understanding of TAM signaling in tumor progression arises from coculture and conditioned media systems. Exposure of cancer cells to TAMs in this setting has revealed that macrophages can induce proliferation [58, 59], stemness [60], migration [61], and invasion [62] in ovarian tumor cells, lending to their involvement in progression of the disease. Beyond modeling signaling between TAMs and tumor cells by traditional coculture methods, using organoid culture, Yin and colleagues [63] recently demonstrated bidirectional communication involving not only secreted factors but also cell surface interactions. In their recent report, they observed that injection of TAM/tumor cell spheroids caused tumor cells to envelop a core of activated-macrophages in vivo and that depletion of the TAMs strikingly inhibited metastasis. Mechanistically, EGF emanating from TAMs induced autocrine VEGF-driven ICAM-1 expression in tumor cells, which in turn was a requirement for their binding to integrin $\alpha_M\beta_2$ expressed on the TAMs. Using these sophisticated methods enhances our appreciation of the complex multi-directional signaling that occurs within the TME, and thus indicate equally complicated mechanisms that may warrant multiple approaches when considering therapeutic strategies.

While evidence suggesting the importance of TAM signaling on tumor progression in OvCa is rapidly growing, the breadth of the TME necessitates consideration for the multitude of other potential interactions that include other cell types of the ovarian TME. Using a custom PDMS ring apparatus to control the time and extent of contact between cultured cells [59], Pamela Kreeger's group showed that macrophage-driven P-selectin expression in mesothelial cells resulted in elevated adhesion of ovarian tumor cells via CD24 [64]. Additionally, the authors mimicked peritoneal shear stress using a parallel-plate flow chamber to demonstrate that this mesothelial cell P-selectin influenced tumor cell adhesion and rolling across the peritoneal surface. Similarly complex, Wang and colleagues identified that the milieu resulting from TAM–tumor signaling promoted angiogenesis in endothelial cells [65]. Indeed, these promising results indicate the complex interactions possible upon immune recruitment to the nascent metastasis. Further studies of TAM-influenced signaling on other components of the TME in systems, such as 3D organotypic cultures, will likely elucidate new molecular targets for drug development.

5.6 Neutrophils

Neutrophils are responsible for adaptive immunity initiated in response to infectious agents in the body and mediate the immune response through interactions with macrophages as well as T and B lymphocytes [66]. In both preclinical and clinical settings, neutrophils have been associated with several mechanisms mediating the development and progression of multiple cancers (reviewed in [67, 68]). Ascites from patients with ovarian carcinoma have been shown to contain an abundance of multiple chemokines, including the cytokines IL-6 and IL-8 [69] as well as the leukotriene LTB_4 [70], which are key for neutrophil recruitment [71]. Further, activated neutrophils have been shown to infiltrate in advanced ovarian tumors [72, 73], as well as in the early stages of omental tumor development [74], lending to the potential importance of neutrophil signaling in OvCa.

While the majority of studies on the contribution of immune cells to the ovarian tumor microenvironment largely focus on macrophages and T lymphocytes, few studies have investigated a potential role for neutrophils in processes involved in tumorigenesis and metastasis of OvCa. In a study by Mayer and colleagues [75], coculture of OvCa cells with polymorphonuclear neutrophils (PMN), as well as treatment with PMN lysate or recombinant elastase, resulted in the detachment of OvCa cells within 6 h. While the data could indicate that PMN-derived elastase promotes anoikis of OvCa cells, further studies that follow and characterize these cells over time would be required to assess the adhesion-free survival of these cells. However, the authors also observed a potential reduction of E-Cadherin in OvCa cells treated with PMN lysate, consistent with more recent reports suggesting the capacity for neutrophils to induce epithelial-to-mesenchymal transition (EMT) in gastric and breast cancer cells ([76] and [77], respectively). More recently it was observed that exposure of peripheral blood neutrophils to OvCa cells induced the formation of neutrophil extracellular traps (NETs) [74], a process which was associated with metastatic progression of colon cancer cells within the peritoneum [78] and may have a role in metastasis of other cancers [79, 80]. Importantly, adhesion of OvCa tumor cells to NETs in vitro was associated with NET-dependent omental tumor burden in vivo, together indicating an important role of activated neutrophils in the developing omental metastatic niche.

Despite that there have been advances in modeling neutrophil transendothelial migration in vitro and in vivo [81], and that there is a clear relevance to the participation of neutrophils in the tumor microenvironment of OvCa [74, 75], a mechanistic understanding of how neutrophils alter the ovarian TME requires extensive investigation moving forward. Indeed, the early observation that recruitment of neutrophils into the peritoneum in vivo is stimulated by GROα secretion from MCs activated by T cell-derived IL-17 [82] indicates a complex interactome between these stromal cells native to the metastatic niche exploited by ovarian cancer. The observation that neutrophil infiltration in excess of lymphocytes (referred to as the neutrophil to lymphocyte ratio, NLR) has been attributed to elevated proinflammatory cytokines [83] and is associated with worse overall and progression-free survival of OvCa [84] further suggests the importance of neutrophils in the ovarian TME. Overall, the current literature indicates a critical gap in knowledge that necessitates the development of novel approaches incorporating neutrophils in 3D organotypic modeling to dissect their contribution to the cell–cell signaling mechanisms that promote omental and peritoneal metastasis leading to mortality of ovarian cancer.

5.7 T Lymphocytes

Tumor-infiltrating lymphocytes (TILs) have long been of interest in solid tumors across cancer types, as cytotoxic $CD8^+$ T cells tend to indicate a better prognosis [85], whereas $CD4^+$/$Foxp3^+$ regulatory T cells (referred to as $CD4^+$ T_{reg} cells herein) are generally associated with poorer survival [86]. Together, these may indicate an overall indication of a TME being drive towards an immunosuppressive state that promotes tumor growth and dissemination. In line with observations in other cancers, OvCa with high infiltration of regulatory $CD4^+$ T_{reg} cells and less activity of cytotoxic $CD8^+$ T cells was associated with increased mortality [87]. Interestingly, while $CD8^+$ T cell infiltration in primary tumors has been consistently associated with improved progression-free survival across study populations [88], there have been conflicting reports as to whether T cell presence in ascites is a significant factor relating to patient outcome [69, 89].

These data could possibly suggest that T cell differentiation and function are dynamic throughout the progression of the disease and that they adopt unique phenotypes dependent on the niche.

Exploitation of T lymphocyte behavior in the TME is of particular interest in treating multiple cancers, as the suppression of cytotoxic $CD8^+$ T cells, in part through immunosuppression by prevailing $CD4^+$ T_{reg} populations, is linked to a permissive environment that facilitates tumor growth (reviewed in [90] and [91]). In ovarian cancer, multiple studies utilizing ascites and coculture systems have demonstrated some of the influence of TME on T cell behavior. Several reports had indicated abundant immunosuppressive $CD4^+$ T_{reg} cells present in ascites from OvCa patients [92, 93], and chemotaxis towards IL-8 (an abundant cytokine in ascites of OvCa patients) was associated with CXCR2 activity in $CD4^+$ T_{reg} cells [94]. Coculture systems using more complex approaches revealed that ascites-derived $CD14^+$ macrophages could induce the transendothelial migration of $CD4^+$ T_{reg} cells in a CCL22-dependent manner [95]. Zhu and colleagues later demonstrated that coculture of $CD4^+$ T_{reg} cells with M2-polarized macrophages induced Foxp3 expression [96], which is critical to regulating suppressive transcriptional programs associated with the immunosuppressive function of these cells (reviewed in [97]). More recently it was also observed that conditioned media (CM) from γδ-T cells, which were found to be elevated in OvCa tissue, could inhibit the proliferation and cytotoxic capacity of naïve T cells [98]. Together, these data suggest intricate communication among immune cells in the TME that may mediate T cell differentiation, recruitment, and behavior in the ovarian TME thus contributing to immune evasion and the development of metastatic tumors. Similar assessment of the influence of OvCa tumor cells on T cell behavior has utilized OvCa-derived conditioned media (CM) and coculture approaches. OvCa CM was observed to increase TGFβ-dependent Foxp3 expression in naïve $CD4^+$ [99] and $CD8^+$ T cells [100], indicating that OvCa cells can also manipulate T cell phenotypes to create an immunosuppressive environment.

Aside from a few studies which utilize a 3D coculture system to evaluate the utility of novel therapeutic strategies for specific OvCa tumor cell cytotoxicity in the TME, such as antibody-directed photoimmunotherapy or CAR-T therapy ([101] and [102], respectively), there has been little progress in utilizing physiologically relevant in vitro or ex vivo systems to evaluate the communication between T cells and their microenvironment. While the role of T cells is becoming less obscure in general, our understanding of how best to prevent an immunosuppressive environment will be markedly improved by incorporating more articulate coculture systems and 3D modeling that better represent the complex interactions between these and other cell types in the ovarian TME.

5.8 Peritoneal Model for Functional Analysis of Adhesion, Invasion, and Proliferation

OvCa metastasizes to the mesothelium-lining of the peritoneum and specific visceral adipose depots such as the omentum [103]. The mesothelium consists of a single layer of mesothelial cells with an underlying basement membrane composed of collagen I, fibronectin, vitronectin, collagen IV, and laminin [104, 105]. The tissue stroma, which consists of fibronectin, vitronectin, collagen I, fibroblasts, macrophages, T cells, neutrophils, and endothelial cells, sits just beneath the basement membrane of the mesothelial cells. The omentum is unique from other peritoneal surfaces in that it is an adipocyte-rich tissue. The main functions of the omentum include wound healing, infection prevention, and regulation of fluid exchange within the peritoneal cavity [104].

A 3D organotypic model of the human mesothelium was developed by Kenny et al. after histological analysis of the cellular and extracellular composition of the human peritoneum and omentum [22]. This model is composed of primary human mesothelial cells and fibroblasts isolated from non-diseased human omentum, and the extracellular matrices, fibronectin, and collagen I. To construct the model, primary human non-passaged fibroblasts are mixed with purified fibronectin and collagen I and plated on tissue culture plates. Primary human non-passaged mesothelial cells are layered on top of the stroma culture using a ratio of 1 fibroblast to 5–8 mesothelial cells. The cells are cultured to allow for extracellular matrix deposition and remodeling. Finally, fluorescently labeled OvCa cells were added to the 3D organotypic peritoneal model. This model allows the investigation of the differential role of extracellular matrix, mesothelial cells, and fibroblasts have during the early steps of OvCa metastasis. Furthermore, the model enables researchers to analyze the functional role of different genes, proteins, and treatments (preventative and therapeutic) in OvCa cell adhesion, migration, invasion, and proliferation in cancer and/or mesothelium microenvironment.

The 3D organotypic model of the human mesothelium was optimized to investigate OvCa cell adhesion, invasion, and proliferation. This model was publicized as a video in JoVE [106]. To investigate adhesion and proliferation, thin layer of collagen I alone (0.5 μg/0.3 cm^2) or a thin layer of collagen I (0.25 μg/0.3 cm^2) plus fibronectin (0.25 μg/0.3 cm^2) was used, while a thicker layer of collagen (10–20 mg/0.3 cm^2) was applied before the thin layer of collagen I with or without fibronectin in Boyden chambers for invasion assays (Fig. 5.2). Initially, two OvCa cell lines and primary human OvCa cells were added to the 3D organotypic model to evaluate the contribution of mesothelial cells, fibroblasts, and extracellular matrices. Modification of the model with different extracellular matrices uncovered that maximum OvCa cell adhesion and invasion occurs with collagen I when compared to vitronectin, fibronectin, or laminin [22]. Direct contact of the cancer cells with the mesothelial cells inhibited OvCa cell adhesion, while conditioned media of the mesothelial cells only minimally perturbed OvCa adhesion and invasion [22]. This finding suggests that the mesothelial cells covering the peritoneum and omentum serve as a protective barrier to the initial adhesion of OvCa cells. However, the crosstalk between mesothelial cells and OvCa cells were found to induce a pro-metastatic phenotype in mesothelial cells. These cancer-associated mesothelial cells induced OvCa invasion and proliferation through the secretion of fibronectin [12]. Direct contact of cancer cells with the fibroblasts enhanced OvCa cell adhesion, invasion, and proliferation [22].

The 3D organotypic model of the mesothelium has been used in numerous publications to unravel the mechanisms involved in early OvCa metastasis. With this modular system, one can treat or genetically modify and/or recover the individual cellular components of the 3D organotypic model to evaluate the role of certain genes and pathways, as well as, the molecular and biochemical changes in OvCa, mesothelial and/or fibroblast cells upon interaction with each other [12, 107]. Kenny and colleagues proceeded to show that OvCa cells secrete matrix-metalloproteinase 2, which cleaves fibronectin into smaller fragments, accelerates OvCa invasion [108]. Additional functional assays with antibodies, small molecule inhibitors, and gene regulation were performed with the 3D organotypic model. An antibody targeting the urokinase plasminogen activator receptor enhanced OvCa cell apoptosis [109]. Mitra et al., built on these results and found that the mesothelium induced downregulation of miR-193b levels in OvCa cells, which in turn, increased urokinase plasmin ogen activator levels using the 3D organotypic model [110]. Henry et al. found that ROR1 and ROR2, the Wnt receptors, have a synergistic role in OvCa cell adhesion to the 3D organotypic model [111]. Hart et al. explored the preventative effect of metformin on OvCa metastasis using this model [14]. Metformin abrogates OvCa cell-induced stability of HIF1α, which requires both succinate and SUGCLG2, and ultimately leads to a decrease in OvCa cell invasion. The 3D organotypic model has also contributed to our understanding of the role of c-MET [20], E-cadherin [112], α5-integrin [12, 112], β1-integrin [12], and β3-integrin [113].

The 3D organotypic model of the human mesothelium represents all the surfaces that OvCa metastasizes to in the human peritoneal cavity. However, this model is not a complete replica of human peritoneal surfaces. The main limitations with this model include access to primary human peritoneal and/or omental tissue, patient variability, the viability of the 3D model (<2 weeks), and the absence of microfluidics mimicking peritoneal fluid exchange, vasculature, and immune cells. Indeed, the 3D organotypic model of the mesothelium can be modified to include adipocytes, macrophages, T cells, neutrophils, and microfluidics, which are all principal components of the human omentum microenvironment. For example, Carroll et al. examined OvCa sphere adhesion to mesothelial cells, activated-macrophages, and extracellular matrix under continuous fluidic flow to mimic the flow of peritoneal fluid in OvCa patients [64]. Mechanistically, activated-macrophage-derived inflammatory protein-1 stimulates P-selectin protein expression in mesothelial cells, which in turn, induces rolling of OvCa spheres under flow conditions.

5.9 Peritoneal Model for High-Throughput Screening

The majority of high-grade serous OvCa patients present with dissemination throughout the peritoneal cavity, and the lethality of this disease is associated with complications from the metastatic disease. Therefore, Kenny et al., Lal-Nag et al., and colleagues reconfigured the 3D organotypic model of the mesothelium for high-throughput screening (HTS) to identify small molecule inhibitors of OvCa cell adhesion/invasion and proliferation to the peritoneal microenvironment [24, 25, 114]. Automation, incubation time, cell numbers, plating sequence, washing, and fixation parameters were optimized for a robust and reproducible 384- to 1536-well HTS primary screening assay. Oncology drug libraries, including the National Center for Advancing Translational Sciences (NCATS) Mechanism Interrogation PlatE oncology collection, the Prestwick library, and the Library of Pharmacologically Active Compounds ($LOPAC^{1280}$), were used to test the effect of these drugs on OvCa adhesion/invasion or proliferation [24, 114]. In addition, the effect of diverse small molecule compounds (44,420 compounds) from the NCATS library of compounds on OvCa cell adhesion/invasion was investigated [25].

The hit compounds from the primary quantitative-adhesion/invasion HTS assays were validated using confirmatory, counter, and in vitro and in vivo secondary biological screens [24, 25]. Ultimately, drugs that inhibited OvCa cell adhesion/invasion in a dose-dependent manner to the 3D HTS model were not cytotoxic to the 3D HTS model within 16 h (time of the assay); inhibited two of three cell lines in two of three in vitro biological function assays including adhesion, invasion, and proliferation on the 3D organotypic model of the mesothelium; and inhibited in vivo adhesion/invasion, metastasis, and survival in xenograft and syngeneic mouse models of OvCa were considered active lead compounds. Interestingly, many of these compounds are active in cancer cells on plastic, but <1% were active in the 3D HTS model. Collectively, these data show that a complex 3D culture of the omental microenvironment is effective in quantitative-HTS and is predictive of in vivo response of mouse models. Furthermore, the compounds identified (β-escin, tomatine, Milciclib, PP-121, and NCGC-117362) have promise as therapeutics for the prevention and treatment of OvCa metastasis.

The proliferation HTS assay identified compounds that inhibited the proliferation of OvCa cells grown as monolayers or forming spheroids, on plastic or the 3D HTS model of the mesothelium [114]. Target-based analysis of the pharmacological responses revealed that both the cellular context of the tumor microenvironment and cell adhesion mode have an essential role in cancer cell drug resistance. Therefore, it is important to perform screens for new drugs using model systems that more faithfully recapitulate the tissue architecture and composition.

5.10 Omental Model Ex Vivo

With the exception of the contralateral ovary, the omentum is the most frequent site of metastatic disease observed in ovarian cancer [115]. The omentum is made up of a variety of cell types that modulate the success of metastatic implantation to this niche, including mesothelial cells, fibroblasts, adipocytes, and immune cells. While many studies have utilized a 3D organotypic modeling approach to focus on pro-tumorigenic signaling from one or multiple of these cell types, evaluating the role of the omentum in disease progression may require the use of this tissue in in vivo and ex vivo assays to comprehensively assess the characteristics unique to the tissue as a whole.

Excision of the greater omentum from a human patient (e.g., omentectomy) is relatively straightforward for a trained surgeon and comes with little associated risk. Since access to human samples can be a limiting resource, transgenic mice are often used in in vivo and ex vivo assays, especially given the phenotypic similarities between murine and human omentum [116]. Multiple approaches have been used to excise this abdominal fat pad in mice. While identification and direct isolation can be mastered with practice [117], removal of the omentum with the spleen and pancreas attached with subsequent excision based on the buoyancy of the adipose can also be used to ensure reproducibility [118]. Following surgical removal, the omentum can be maintained ex vivo in growth conditions typical to primary culture (e.g., RPMI1640, 10–20% fetal bovine serum, 1% penicillin/streptomycin) for 7 or more days [118]. During this time, OvCa tumor cells can be added and allowed to adhere, colonize, and invade omental tissue, with the amount of time post-seeding indicating which process is being assessed experimentally (e.g., <6 h for adhesion, >24 h for colonization).

This useful translational approach has been successfully implemented in a number of studies to evaluate in vitro findings and evaluate multiple mechanisms by which tumor cells metastasize to this particular microenvironment. Similar to transwell migration assays using a chemoattractant to drive cell movement, omental tissue has been used ex vivo to demonstrate that secreted factors from immune competent mice enhance motility in comparison to control counterparts, suggesting that the T cell population of the omentum may modulate migration towards this specific site [119]. Other processes involved in the early stages of metastasis have also been assessed using omentum ex vivo. For example, our group had shown that the blockade of OvCa cell IL-6 and IL-8 receptors using neutralizing antibodies markedly repressed their ability to adhere to omentum [47]. Consistently, this result was reproduced in a different approach by Yung and colleagues who observed that IL-8Rβ knockdown in OvCa cells inhibited colonization of the omentum ex vivo [120]. Together, these two approaches support that IL-8 signaling within OvCa cells is required for successful implantation of the omentum; however, due to the complexity of this microenvironment, the exact mechanism concerning communication between the various cell types present in the omentum requires further delineation before it can be exploited therapeutically. The human omentum ex vivo culture has also contributed to our understanding of the therapeutic effects of metformin [14] and a fascin inhibitor [121].

Combining ex vivo techniques with transgenic animal models, especially in sophisticated systems such as Cre/Lox and CRISPR approaches, may provide direct evidence on mechanisms regulated by specific cell types and interactions within the microenvironment of the omentum. In line with this, we had recently used floxed fibronectin mice ($Fn1^{fl/fl}$) and performed intraperitoneal injection of adenoviral Cre recombinase, thereby inhibiting surface (mesothelial cell) expression of fibronectin prior to in vivo metastasis assays [12]. Using this approach revealed that stromal fibronectin expression was essential in promoting colonization of the omentum. Further extending this strategy to utilize promoter regulated Cre-driven systems using adiponectin [53, 122], collagen 1a2 [123], or mesothelin [26] would allow for direct genetic manipulation of adipocytes, fibroblasts, or mesothelial cells, respectively. In combination with loxP induced expression or floxed knockout, transgenic systems could be developed to assess which specific genes and pathways, and of which cell type(s) within the omentum, are required for tumor progression in in vivo and ex vivo metastasis assays.

5.11 Organoids

Pathophysiological modeling of different aspects of ovarian cancer requires systems that could robustly recapitulate genomic landscape of alterations observed in most primary and metastatic tumor isolates [124]. Organoids are in vitro cellular clusters derived from primary tissue grown in 3D using ECM hydrogels that resemble the original tissue in terms of architecture, histology, and genetic features [125]. 3D organoid cultures have recently emerged as a powerful modeling approach for several tumors, offering the ability to represent the original in vivo complexity of a particular tissue type and also provides mechanistic insight into human tumor biology [124, 126–128].

The cell-of-origin of OvCa remains controversial as it is diagnosed at a late stage as a large ovarian mass accompanied by widespread peritoneal metastasis [129, 130]. Historically, OvCa was thought to arise from ovarian surface epithelium (OSE) which undergoes rupture and consequent repair as a result of normal ovulatory process and any aberrations in the process leads to accumulation of mutations with advancing age [131, 132]. Lineage tracing has identified a stem cell niche in the mouse ovarian hilum consisting of $Lgr5^+$ cells that can give rise to the entire OSE and are likely to be transformed in OvCa pathogenesis [133]. More recently, with the identification of precursor lesions termed serous tubular intraepithelial carcinomas (STICS) in fallopian tube fimbria of ovarian cancer patients and women with BRCA1/2 mutations, the fallopian tube epithelium (FTE) is being considered as a likely site of origin of OvCa [134, 135]. However, the TME of FTE and OSE has not been fully explored in context of pathogenesis of ovarian cancer.

Kessler et al. reported the establishment of human FTE organoids from fallopian tube epithelial stem cells in a 3D Matrigel matrix supplemented with a cocktail of growth factors that support paracrine signaling pathways, in particular Wnt and Notch. The 3D organoids that were formed had long-term stability and faithfully mimic the phenotype of in vivo tissue with appearance of mucosal folds and invaginations. In order to study the etiology of ovarian cancer, this model can be established from FTE samples from OvCa patients which would be amenable to hormonal stimulation as well [136]. FTE organoids have also been established by using induced pluripotent stem cells (iPSCs) generated by reprogramming somatic cells which can be further differentiated into various cell types [137, 138].

Yucer et al. created a powerful human-derived FTE organoid model using this technology where iPSC cell lines were allowed to differentiate into FTE precursor cells via intermediate mesoderm-like cells through sequential exposure to BMP4, WNT4 followed by follistatin that bio-neutralizes members of the TGF-β superfamily. The spheroids from the differentiated FTE cells when placed in Matrigel containing phenol red formed a well-organized organoid structure as compared to branched and unorganized organoids formed when no phenol red was used. Since phenol red is a weak estrogen mimic, supplementation with steroid hormones such as estrogen (E2) and progesterone (P4) or treatment with conditioned media from primary FTE cells increased the architectural complexity of FTE organoids with the formation of luminal structures. Additionally, these fallopian tube structures contained ciliated (FOXJ1 and TUBB4A) and secretory (PAX8) components as demonstrated by immunocytochemistry for specific markers [139].

A recent study by Neel et al. supports the dualistic origin for OvCa suggesting that it can originate from either cell type and the cell-of-origin plays a role in modulating the therapeutic response. Using lineage-specific Cre recombinase (Cre) lines and novel organoid systems, they compared the tumor-forming capacity and behavior of FTE and OSE, harboring the same oncogenic abnormalities. They also developed serum-free defined media conditions for indefinite and long-term culture of FTE organoids and supplemented the same FTE medium with additional factors, such as hydrocortisone, estrogen, LIF, and forskolin for culturing OSE organoids. To model FT-derived OvCa, they established organoid cultures from wild type, Pax8rtTA;TetOcre;Tp53R172H/fl (PTP mice), Pax8rtTA;TetOcre;T121 (PTT) and Pax8rtTA;TetOcre;Tp53R172H/fl;T121 (PTPT) mice. Since Pax8rtTA mice reportedly enable doxycycline (Dox)-inducible selective gene expression in secretory FTE, Dox-treated PTP, PTT, and PTPT organoids activated Cre which enabled deletion of one allele of Tp53 (PTP and PTPT) and expression of T121 for inactivation of RB family (PTT and PTPT). Consistent with their genetically engineered mouse models (GEMM), the mutant organoids were larger and proliferated more as compared to wild type organoids. Also, when injected into the ovarian fat pads of nu/nu mice, mutant organoids formed tumors with PTPT organoids showing the highest tumor-forming ability as evident by hemorrhagic ascites development and widespread peritoneal studding similar to human disease. Immunohistochemical analysis of metastases showed expression of OvCa specific markers including PAX8, P53, Ki67, γH2AX, Stathmin 1, and p16 [140]. Lineage tracing using Lgr5CreERT2 mice showed that $Lgr5^+$ cells can give rise to entire OSE and $Lgr5^+$ expression is concentrated in the ovarian hilum in adult mice [133]. To mimic OSE-derived OvCa, organoids were generated from mice (wild type, LP, LT, LPT) harboring similar mutations (Tp53 and RB) but with ovarian-specific $Lgr5^+$ Cre recombinase. Orthotopic implantation of LPT organoids resulted in metastatic tumors and displayed similar histology and marker expression as seen in OvCa [140]. Although ovarian organoid cultures contain little or minimal stromal component, OSE or FTE-derived organoids can be incorporated into 3D organotypic model to study their bidirectional interaction with different cell types present in the metastatic ovarian TME [141].

Genomic characterization of OvCa presents defects in DNA repair pathway in 50% of cases. However, this might not necessarily correspond with the functional defects until clarified by functional assays. To study this, Hill et al. devised a platform for functional profiling of DNA repair in short-term (7–10 days) cultured organoids derived from 22 patients with OvCa for defects in homologous recombination (HR) and replication fork protection. Their data suggested that a stalled fork protection defect was present in 61% of organoid lines tested, irrespective of the genetic status, and was associated with therapeutic sensitivity to carboplatin, prexasertib, and VE-822 whereas only 6% of organoid lines had a functional HR defect and PARPi sensitivity [142].

One of the hallmarks of OvCa is the peritoneal dissemination accompanied by the production of peritoneal fluid (ascites) which can serve as a meaningful source of tumor cells for modeling the disease. To this end, Testa and colleagues isolated individual tumor cells from patients' metastatic ascites and cultured them in 3D based on their ability to sustain anchorage-independent growth and avoid anoikis [124]. In order to provide a favorable niche for the growth of OvCa, they supplemented the medium of growing primary cells with cells-free ascitic fluid in different ratios and found ascitic fluid-dependent increase in cell proliferation with 12.5% (Ascitic fluid: medium) showing the highest efficiency. Ascitic fluid supplementation also enabled the growth of individual cells into single cell-derived ovarian cancer organoids (scOCOs) and their serial propagation. Finally, they used single cell RNA sequencing (scRNAseq) to define the cellular composition of OvCa metastatic ascites and traced its propagation in 2D and 3D culture conditions and found that scOCOs demonstrate key features of original metastasis in a patient-specific manner that do not appear from classical 2D culture. Collectively, these findings demonstrate the power of these organoids to serve as in vitro avatars of each patient's tumor allowing mechanistic dissection of metastatic OvCa [124].

More recently, Hans Clever's group established an organoid platform amenable to genetic manipulation, consisting of 56 organoid lines derived from 32 patients, representing all main histopathological subtypes of OvCa including mucinous borderline tumors, serous borderline tumors, mucinous, low-grade serous, clear cell carcinomas, endometrioid, and high-grade serous (HGS). A comprehensive analysis of OvCa organoids revealed that they display histological features of the corresponding lesion from which they were derived, such as nuclear and cellular atypia and biomarker expression such as PAX8 and p53. Organoids and pertinent tumors remained highly similar at the genomic level, even after long-term culture and mimicked the mutational hallmarks of OC and tumor heterogeneity. Finally, based on unsupervised hierarchial clustering of gene expression data, the authors grouped the organoids according to the tumor type and found that LGS organoids are more similar to normal samples than are HGS. Also, most HGS organoids were sensitive to platinum-based treatments, whereas non-HGS organoids were more resistant. Similarly in one case, recurrent tumor organoids showed increased resistance to platinum agents as compared to matched primary tumor organoids derived from the same patient substantiating the relevance of this OvCa platform for drug screening [143].

Table 5.1 Reference of laboratories utilizing 2D coculture and 3D modeling systems of the ovarian tumor microenvironment mentioned in this chapter

	Principal investigator/laboratory
Cell type	
Mesothelial cell	B. Abendstein, E. Diamandis, K. Fujimori, E. Lengyel, M. Lopez-Cabrera, S. Mutsaers, G. Nunez, C. Ottensmeier, E. Rankin, H. Shi, A. Skubitz, S. Stack
Fibroblasts	R. Coleman, Q. Gao, W. Hongqing, E. Lengyel, S. Mok, M. Neeman, Z. Wang, C. Zou
Adipocytes	S. Benitah, J. Borras, G. Chen, E. Lengyel, H. Ginsberg, N. Said
Macrophages	F. Balkwill, T. Duan, T. Hagemann, P. Kreeger, X. Luo, W. Min, G. Mor, X. Wang
Neutrophils	M. Egeblad, M. Gaida, T. Hagemann, A. Jorres, H. Naora, S. Su, Z. Sulowska, H. Thorlacius
T lymphocytes	M. Cannon, G. Mor, S. Pan, K. Shen, B. Spring, P. Thor Straten, X. Wang
Model system	
3D organotypic model of peritoneum	C. Ford, H. Kenny, P. Kreeger, E. Lengyel, A. Mitra, I. Romero
Ex vivo omentum	H. Kenny, E. Lengyel, C. Rinker-Schaeffer, I. Romero

5.12 Conclusion

The metastatic TME of ovarian cancer is a complex composite of multiple cell types that facilitate implantation and growth of tumor cells (Fig. 5.1). Use of 2D coculture and more complex 3D modeling systems, coupled with in vivo and ex vivo validation, has greatly enhanced our understanding of the interactions that occur that are required for successful metastasis of OvCa cells, and some of the laboratories utilizing the methods described here are identified below for reference (Table 5.1). The intricate multidirectional signaling between a rather diverse population of cell types has only recently been broached by using more complicated approaches, and advancement in these techniques by focusing on mimicking the richness of this environment is necessary if we are to fully appreciate the robust signaling mechanisms involved in promoting tumor cell adhesion, invasion, and colonization of metastatic sites. Only by establishing models that fully recapitulate the tissue's complexity, from the frequency of cell types to their spatial distribution, will we be able to accurately assess actionable signaling processes to target in order to prevent metastatic disease. As indicated in Fig. 5.2, the opportunity exists to expand on the 3D organotypic modeling by systematic approaches incorporating additional cell types discussed above into the lower compartment of the existing model to elaborate on the effects of their signaling. While improving miniaturization of the TME for the implementation of HTS techniques is invaluable for testing a battery of novel or existing compounds, by also using approaches in larger platforms (e.g., 24- and 6-well inserts) it could be conceivable to examine the interaction between multiple cell types directly on the mesothelium, or similarly on the vascular endothelium. Given the overwhelming intricacy of this metastatic microenvironment, development of more articulated systems across these strategies will prove invariably useful in enhancing our knowledge of the signaling milieu that permits tumor implantation and expansion, and will ultimately provide novel actionable targets for prevention and treatment of ovarian cancer.

References

1. Lengyel, E. (2010). Ovarian cancer development and metastasis. *American Journal of Pathology, 177*(3), 1053–1064.
2. Mutsaers, S. E., Whitaker, D., & Papadimitriou, J. M. (2002). Stimulation of mesothelial cell proliferation by exudate macrophages enhances serosal wound healing in a murine model. *The American Journal of Pathology, 160*(2), 681–692.
3. Mutsaers, S. E. (2002). Mesothelial cells: Their structure, function and role in serosal repair. *Respirology, 7*, 171–191.
4. diZerega, G. S., & Campeau, J. D. (2001). Peritoneal repair and post-surgical adhesion formation. *Human Reproduction Update, 7*(6), 547–555.
5. Ye, Z. J., Yuan, M. L., Zhou, Q., Du, R. H., Yang, W. B., Xiong, X. Z., et al. (2012). Differentiation and recruitment of Th9 cells stimulated by pleural mesothelial cells in human Mycobacterium tuberculosis infection. *PLoS One, 7*(2), e31710.
6. Park, J. H., Kim, Y. G., Shaw, M., Kanneganti, T. D., Fujimoto, Y., Fukase, K., et al. (2007). Nod1/RICK and TLR signaling regulate chemokine and antimicrobial innate immune responses in mesothelial cells. *Journal of Immunology, 179*(1), 514–521.
7. Chen, Y. T., Chang, Y. T., Pan, S. Y., Chou, Y. H., Chang, F. C., Yeh, P. Y., et al. (2014). Lineage tracing reveals distinctive fates for mesothelial cells and submesothelial fibroblasts during peritoneal injury. *Journal of American Society of Nephrology, 25*(12), 2847–2858.
8. Mutsaers, S. E. (2004). The mesothelial cell. *The International Journal of Biochemistry & Cell Biology, 36*, 9–16.
9. Kenny, H. A., Nieman, K. M., Mitra, A. K., & Lengyel, E. (2011). The first line of intra-abdominal metastatic attack: Breaching the mesothelial cell layer. *Cancer Discovery, 1*(2), 100–102.
10. Wilson, A. P. (1989). Mesothelial cells stimulate the anchorage-independent growth of human ovarian tumour cells. *British Journal of Cancer, 59*(6), 876–882.
11. Moser, T. L., Pizzo, S. V., Bafetti, L., Fishman, D. A., & Stack, M. S. (1996). Evidence for preferential adhesion of ovarian epithelial carcinoma cells to type I collagen mediated by the α2ß1 integrin. *International Journal of Cancer, 67*, 695–701.
12. Kenny, H. A., Chiang, C. Y., White, E. A., Schryver, E. M., Habis, M., Romero, I. L., et al. (2014). Mesothelial cells promote early ovarian cancer metastasis through fibronectin secretion. *Journal of Clinical Investigation, 124*(10), 4614–4628.
13. Offner, F. A., Obrist, P., Stadlmann, S., Feichtinger, H., Klingler, P., Herold, M., et al. (1995). IL-6 secretion by human peritoneal mesothelial and ovarian cancer cells. *Cytokine, 7*(6), 542–547.
14. Hart, P. C., Kenny, H. A., Grassl, N., Watters, K. M., Litchfield, L. M., Coscia, F., et al. (2019). Mesothelial cell HIF1alpha expression is metabolically downregulated by metformin to prevent oncogenic tumor-stromal crosstalk. *Cell Reports, 29*(12), 4086–98.e6.
15. Cannistra, S. A., Kansas, G. S., Niloff, J., DeFranzo, B., Kim, Y., & Ottensmeier, C. (1993). Binding of ovarian cancer cells to peritoneal mesothelium in vitro is partly mediated by CD44H. *Cancer Research, 53*(August 15), 3830–3838.
16. Lee, J. G., Ahn, J. H., Jin Kim, T., Ho Lee, J., & Choi, J. H. (2015). Mutant p53 promotes ovarian cancer cell adhesion to mesothelial cells via integrin beta4 and Akt signals. *Scientific Reports, 5*, 12642.
17. Lessan, K., Aguiar, D., Oegema, T. R., Siebenson, L., & Skubitz, A. P. (1999). CD44 and β1 integrin mediate ovarian carcinoma cell adhesion to peritoneal mesothelial cells. *American Journal of Pathology, 154*(5), 1525–1537.
18. Watanabe, T., Hashimoto, T., Sugino, T., Soeda, S., Nishiyama, H., Morimura, Y., et al. (2012). Production of IL1-beta by ovarian cancer cells induces mesothelial cell beta1-integrin expression facilitating peritoneal dissemination. *Journal of Ovarian Research, 5*(1), 7.
19. Musrap, N., Karagiannis, G. S., Saraon, P., Batruch, I., Smith, C., & Diamandis, E. P. (2014). Proteomic analysis of cancer and mesothelial cells reveals an increase in Mucin 5AC during ovarian cancer and peritoneal interaction. *Journal of Proteomics, 103*, 204–215.
20. Sawada, K., Radjabi, A. R., Shinomiya, N., Kistner, E., Kenny, H., Becker, A. R., et al. (2007). c-Met overexpression is a prognostic factor in ovarian cancer and an effective target for inhibition of peritoneal dissemination and invasion. *Cancer Research, 67*(4), 1670–1680.
21. Sandoval, P., Jiménez-Heffernan, J. A., Rynne-Vidal, A., Pérez-Lozano, M. L., Gilsanz, A., Ruiz-Carpio, V., et al. (2013). Carcinoma-associated fibroblasts derive from mesothelial cells via mesothelial to mesenchymal transition in peritoneal metastasis. *Journal of Pathology, 231*(4), 517–531.
22. Kenny, H. A., Krausz, T., Yamada, S. D., & Lengyel, E. (2007). Use of a novel 3D culture model to

elucidate the role of mesothelial cells, fibroblasts and extra-cellular matrices on adhesion and invasion of ovarian cancer cells to the omentum. *International Journal of Cancer, 121*(7), 1463–1472.

23. Natarajan, S., Foreman, K. M., Soriano, M. I., Rossen, N. S., Shehade, H., Fregoso, D. R., et al. (2019). Collagen remodeling in the hypoxic tumor-mesothelial niche promotes ovarian cancer metastasis. *Cancer Research, 79*(9), 2271–2284.
24. Kenny, H. A., Lal-Nag, M., White, E. A., Shen, M., Chiang, C. Y., Mitra, A. K., et al. (2015). Quantitative high throughput screening using a primary human three-dimensional organotypic culture predicts in vivo efficacy. *Nature Communications, 6*, 6220.
25. Kenny, H. A., Lal-Nag, M., Shen, M., Kara, B., Nahotko, D. A., Wroblewski, K., et al. (2020). Quantitative high-throughput screening using an organotypic model identifies compounds that inhibit ovarian cancer metastasis. *Molecular Cancer Therapeutics, 19*(1), 52–62.
26. Rinkevich, Y., Mori, T., Sahoo, D., Xu, P. X., Bermingham, J. R., Jr., & Weissman, I. L. (2012). Identification and prospective isolation of a mesothelial precursor lineage giving rise to smooth muscle cells and fibroblasts for mammalian internal organs, and their vasculature. *Nature Cell Biology, 14*(12), 1251–1260.
27. Frantz, C., Stewart, K. M., & Weaver, V. M. (2010). The extracellular matrix at a glance. *Journal of Cell Science, 123*(Pt 24), 4195–4200.
28. des Jardins-Park, H. E., Foster, D. S., & Longaker, M. T. (2018). Fibroblasts and wound healing: An update. *Regenerative Medicine, 13*(5), 491–495.
29. Driskell, R. R., Lichtenberger, B. M., Hoste, E., Kretzschmar, K., Simons, B. D., Charalambous, M., et al. (2013). Distinct fibroblast lineages determine dermal architecture in skin development and repair. *Nature, 504*(7479), 277–281.
30. Otranto, M., Sarrazy, V., Bonte, F., Hinz, B., Gabbiani, G., & Desmouliere, A. (2012). The role of the myofibroblast in tumor stroma remodeling. *Cell Adhesion & Migration, 6*(3), 203–219.
31. Ridky, T. W., Chow, J. M., Wong, D. J., & Khavari, P. A. (2010). Invasive three-dimensional organotypic neoplasia from multiple normal human epithelia. *Nature Medicine, 16*(12), 1450–1455.
32. Attieh, Y., Clark, A. G., Grass, C., Richon, S., Pocard, M., Mariani, P., et al. (2017). Cancer-associated fibroblasts lead tumor invasion through integrin-beta3-dependent fibronectin assembly. *The Journal of Cell Biology, 216*(11), 3509–3520.
33. Sun, Y., Fan, X., Zhang, Q., Shi, X., Xu, G., & Zou, C. (2017). Cancer-associated fibroblasts secrete FGF-1 to promote ovarian proliferation, migration, and invasion through the activation of FGF-1/FGFR4 signaling. *Tumour Biology, 39*(7), 1010428317712592.
34. Xu, L. N., Xu, B. N., Cai, J., Yang, J. B., & Lin, N. (2013). Tumor-associated fibroblast-conditioned medium promotes tumor cell proliferation and angiogenesis. *Genetics and Molecular Research, 12*(4), 5863–5871.
35. Cai, J., Tang, H., Xu, L., Wang, X., Yang, C., Ruan, S., et al. (2012). Fibroblasts in omentum activated by tumor cells promote ovarian cancer growth, adhesion and invasiveness. *Carcinogenesis, 33*(1), 20–29.
36. Yeung, T. L., Leung, C. S., Wong, K. K., Samimi, G., Thompson, M. S., Liu, J., et al. (2013). TGF-beta modulates ovarian cancer invasion by upregulating CAF-derived versican in the tumor microenvironment. *Cancer Research, 73*(16), 5016–5028.
37. Deying, W., Feng, G., Shumei, L., Hui, Z., Ming, L., & Hongqing, W. (2017). CAF-derived HGF promotes cell proliferation and drug resistance by up-regulating the c-Met/PI3K/Akt and GRP78 signalling in ovarian cancer cells. *Bioscience Reports, 37*(2), BSR20160470.
38. Kumar, D., New, J., Vishwakarma, V., Joshi, R., Enders, J., Lin, F., et al. (2018). Cancer-associated fibroblasts drive glycolysis in a targetable signaling loop implicated in head and neck squamous cell carcinoma progression. *Cancer Research, 78*(14), 3769–3782.
39. Curtis, M., Kenny, H. A., Ashcroft, B., Mukherjee, A., Johnson, A., Zhang, Y., et al. (2019). Fibroblasts mobilize tumor cell glycogen to promote proliferation and metastasis. *Cell Metabolism, 29*(1), 141–55. e9.
40. Gao, Q., Yang, Z., Xu, S., Li, X., Yang, X., Jin, P., et al. (2019). Heterotypic CAF-tumor spheroids promote early peritoneal metastasis of ovarian cancer. *The Journal of Experimental Medicine, 216*(3), 688–703.
41. Oren, R., Addadi, Y., Narunsky Haziza, L., Dafni, H., Rotkopf, R., Meir, G., et al. (2016). Fibroblast recruitment as a tool for ovarian cancer detection and targeted therapy. *International Journal of Cancer, 139*(8), 1788–1798.
42. Li, W., Zhang, X., Wang, J., Li, M., Cao, C., Tan, J., et al. (2017). TGFbeta1 in fibroblasts-derived exosomes promotes epithelial-mesenchymal transition of ovarian cancer cells. *Oncotarget, 8*(56), 96035–96047.
43. Romero, I. L., Mukherjee, A., Kenny, H. A., Litchfield, L. M., & Lengyel, E. (2015). Molecular pathways: Trafficking of metabolic resources in the tumor microenvironment. *Clinical Cancer Research, 21*(4), 680–686.
44. Yu, Y. H., & Ginsberg, H. N. (2005). Adipocyte signaling and lipid homeostasis: Sequelae of insulin-resistant adipose tissue. *Circulation Research, 96*(10), 1042–1052.
45. Lauby-Secretan, B., Scoccianti, C., Loomis, D., Grosse, Y., Bianchini, F., Straif, K., et al. (2016). Body fatness and cancer—Viewpoint of the IARC working group. *The New England Journal of Medicine, 375*(8), 794–798.

46. Foong, K. W., & Bolton, H. (2017). Obesity and ovarian cancer risk: A systematic review. *Post Reproductive Health, 23*(4), 183–198.
47. Nieman, K. M., Kenny, H. A., Penicka, C. V., Ladanyi, A., Buell-Gutbrod, R., Zillhardt, M. R., et al. (2011). Adipocytes promote ovarian cancer metastasis and provide energy for rapid tumor growth. *Nature Medicine, 17*(11), 1498–1503.
48. Guaita-Esteruelas, S., Guma, J., Masana, L., & Borras, J. (2018). The peritumoural adipose tissue microenvironment and cancer. The roles of fatty acid binding protein 4 and fatty acid binding protein 5. *Molcular and Cellular Endocrinology, 462*(Pt B), 107–118.
49. Ladanyi, A., Mukherjee, A., Kenny, H. A., Johnson, A., Mitra, A. K., Sundaresan, S., et al. (2018). Adipocyte-induced CD36 expression drives ovarian cancer progression and metastasis. *Oncogene, 37*, 2285–2301.
50. Pascual, G., Avgustinova, A., Mejetta, S., Martin, M., Castellanos, A., Attolini, C. S., et al. (2017). Targeting metastasis-initiating cells through the fatty acid receptor CD36. *Nature, 541*(7635), 41–45.
51. John, B., Naczki, C., Patel, C., Ghoneum, A., Qasem, S., Salih, Z., et al. (2019). Regulation of the bi-directional cross-talk between ovarian cancer cells and adipocytes by SPARC. *Oncogene, 38*(22), 4366–4383.
52. Sun, C., Li, X., Guo, E., Li, N., Zhou, B., Lu, H., et al. (2020). MCP-1/CCR-2 axis in adipocytes and cancer cell respectively facilitates ovarian cancer peritoneal metastasis. *Oncogene, 39*(8), 1681–1695.
53. Jeffery, E., Berry, R., Church, C. D., Yu, S., Shook, B. A., Horsley, V., et al. (2014). Characterization of Cre recombinase models for the study of adipose tissue. *Adipocytes, 3*(3), 206–211.
54. Wang, X., Deavers, M., Patenia, R., Bassett, R. L., Jr., Mueller, P., Ma, Q., et al. (2006). Monocyte/macrophage and T-cell infiltrates in peritoneum of patients with ovarian cancer or benign pelvic disease. *Journal of Translational Medicine, 4*, 30.
55. Liu, J., Geng, X., & Li, Y. (2016). Milky spots: Omental functional units and hotbeds for peritoneal cancer metastasis. *Tumour Biology, 37*(5), 5715–5726.
56. Zhang, M., He, Y., Sun, X., Li, Q., Wang, W., Zhao, A., et al. (2014). A high M1/M2 ratio of tumor-associated macrophages is associated with extended survival in ovarian cancer patients. *Journal of Ovarian Research, 7*, 19.
57. Kaneko, M., Nishida, M., & Iwasaki, H. (1985). A macrophage activating factor is present and active in the ascitic fluid of patients with ovarian cancer. *Journal of Cancer Research and Clinical Oncology, 110*(2), 131–135.
58. Liu, L., Wang, X., Li, X., Wu, X., Tang, M., & Wang, X. (2018). Upregulation of IGF1 by tumor-associated macrophages promotes the proliferation and migration of epithelial ovarian cancer cells. *Oncology Reports, 39*(2), 818–826.
59. Carroll, M. J., Kapur, A., Felder, M., Patankar, M. S., & Kreeger, P. K. (2016). M2 macrophages induce ovarian cancer cell proliferation via a heparin binding epidermal growth factor/matrix metalloproteinase 9 intercellular feedback loop. *Oncotarget, 7*(52), 86608–86620.
60. Ning, Y., Cui, Y., Li, X., Cao, X., Chen, A., Xu, C., et al. (2018). Co-culture of ovarian cancer stem-like cells with macrophages induced SKOV3 cells stemness via IL-8/STAT3 signaling. *Biomedicine & Pharmacotherapy, 103*, 262–271.
61. Neyen, C., Pluddemann, A., Mukhopadhyay, S., Maniati, E., Bossard, M., Gordon, S., et al. (2013). Macrophage scavenger receptor a promotes tumor progression in murine models of ovarian and pancreatic cancer. *Journal of Immunology, 190*(7), 3798–3805.
62. Hagemann, T., Wilson, J., Kulbe, H., Li, N. F., Leinster, D. A., Charles, K., et al. (2005). Macrophages induce invasiveness of epithelial cancer cells via NF-kappa B and JNK. *Journal of Immunology, 175*(2), 1197–1205.
63. Yin, M., Li, X., Tan, S., Zhou, H. J., Ji, W., Bellone, S., et al. (2016). Tumor-associated macrophages drive spheroid formation during early transcoelomic metastasis of ovarian cancer. *The Journal of Clinical Investigation, 126*(11), 4157–4173.
64. Carroll, M. J., Fogg, K. C., Patel, H. A., Krause, H. B., Mancha, A. S., Patankar, M. S., et al. (2018). Alternatively-activated macrophages upregulate mesothelial expression of P-selectin to enhance adhesion of ovarian cancer cells. *Cancer Research, 78*(13), 3560–3573.
65. Wang, X., Zhao, X., Wang, K., Wu, L., & Duan, T. (2013). Interaction of monocytes/macrophages with ovarian cancer cells promotes angiogenesis in vitro. *Cancer Science, 104*(4), 516–523.
66. Mantovani, A., Cassatella, M. A., Costantini, C., & Jaillon, S. (2011). Neutrophils in the activation and regulation of innate and adaptive immunity. *Nature Reviews. Immunology, 11*(8), 519–531.
67. Coffelt, S. B., Wellenstein, M. D., & de Visser, K. E. (2016). Neutrophils in cancer: Neutral no more. *Nature Reviews. Cancer, 16*(7), 431–446.
68. Shaul, M. E., & Fridlender, Z. G. (2019). Tumour-associated neutrophils in patients with cancer. *Nature Reviews. Clinical Oncology, 16*(10), 601–620.
69. Giuntoli, R. L., II, Webb, T. J., Zoso, A., Rogers, O., Diaz-Montes, T. P., Bristow, R. E., et al. (2009). Ovarian cancer-associated ascites demonstrates altered immune environment: Implications for antitumor immunity. *Anticancer Research, 29*(8), 2875–2884.
70. Reinartz, S., Finkernagel, F., Adhikary, T., Rohnalter, V., Schumann, T., Schober, Y., et al. (2016). A transcriptome-based global map of signaling pathways in the ovarian cancer microenvironment associated with clinical outcome. *Genome Biology, 17*(1), 108.

71. Sokol, C. L., & Luster, A. D. (2015). The chemokine system in innate immunity. *Cold Spring Harbor Perspectives in Biology, 7*(5), a016303.
72. Klink, M., Jastrzembska, K., Nowak, M., Bednarska, K., Szpakowski, M., Szyllo, K., et al. (2008). Ovarian cancer cells modulate human blood neutrophils response to activation in vitro. *Scandinavian Journal of Immunology, 68*(3), 328–336.
73. Charles, K. A., Kulbe, H., Soper, R., Escorcio-Correia, M., Lawrence, T., Schultheis, A., et al. (2009). The tumor-promoting actions of TNF-alpha involve TNFR1 and IL-17 in ovarian cancer in mice and humans. *The Journal of Clinical Investigation, 119*(10), 3011–3023.
74. Lee, W., Ko, S. Y., Mohamed, M. S., Kenny, H. A., Lengyel, E., & Naora, H. (2018). Neutrophils facilitate ovarian cancer premetastatic niche formation in the omentum. *Journal of Experimental Medicine, 216*(1), 176–194.
75. Mayer, C., Darb-Esfahani, S., Meyer, A. S., Hubner, K., Rom, J., Sohn, C., et al. (2016). Neutrophil granulocytes in ovarian cancer—Induction of epithelial-to-mesenchymal-transition and tumor cell migration. *Journal of Cancer, 7*(5), 546–554.
76. Li, S., Cong, X., Gao, H., Lan, X., Li, Z., Wang, W., et al. (2019). Tumor-associated neutrophils induce EMT by IL-17a to promote migration and invasion in gastric cancer cells. *Journal of Experimental & Clinical Cancer Research, 38*(1), 6.
77. Wang, Y., Chen, J., Yang, L., Li, J., Wu, W., Huang, M., et al. (2019). Tumor-contacted neutrophils promote metastasis by a CD90-TIMP-1 juxtacrine-paracrine loop. *Clinical Cancer Research, 25*(6), 1957–1969.
78. Al-Haidari, A. A., Algethami, N., Lepsenyi, M., Rahman, M., Syk, I., & Thorlacius, H. (2019). Neutrophil extracellular traps promote peritoneal metastasis of colon cancer cells. *Oncotarget, 10*(12), 1238–1249.
79. Park, J., Wysocki, R. W., Amoozgar, Z., Maiorino, L., Fein, M. R., Jorns, J., et al. (2016). Cancer cells induce metastasis-supporting neutrophil extracellular DNA traps. *Science Translational Medicine, 8*(361), 361ra138.
80. Jung, H. S., Gu, J., Kim, J. E., Nam, Y., Song, J. W., & Kim, H. K. (2019). Cancer cell-induced neutrophil extracellular traps promote both hypercoagulability and cancer progression. *PLoS One, 14*(4), e0216055.
81. Akk, A., Springer, L. E., Yang, L., Hamilton-Burdess, S., Lambris, J. D., Yan, H., et al. (2019). Complement activation on neutrophils initiates endothelial adhesion and extravasation. *Molecular Immunology, 114*, 629–642.
82. Witowski, J., Pawlaczyk, K., Breborowicz, A., Scheuren, A., Kuzlan-Pawlaczyk, M., Wisniewska, J., et al. (2000). IL-17 stimulates intraperitoneal neutrophil infiltration through the release of GRO alpha chemokine from mesothelial cells. *Journal of Immunology, 165*(10), 5814–5821.
83. Rodrigues, I. S. S., Martins-Filho, A., Micheli, D. C., Lima, C. A., Tavares-Murta, B. M., Murta, E. F. C., et al. (2019). IL-6 and IL-8 as prognostic factors in peritoneal fluid of ovarian cancer. *Immunological Investigations, 49*, 510–521.
84. Yin, X., Wu, L., Yang, H., & Yang, H. (2019). Prognostic significance of neutrophil-lymphocyte ratio (NLR) in patients with ovarian cancer: A systematic review and meta-analysis. *Medicine (Baltimore), 98*(45), e17475.
85. Barnes, T. A., & Amir, E. (2017). HYPE or HOPE: The prognostic value of infiltrating immune cells in cancer. *British Journal of Cancer, 117*(4), 451–460.
86. Shang, B., Liu, Y., Jiang, S. J., & Liu, Y. (2015). Prognostic value of tumor-infiltrating FoxP3+ regulatory T cells in cancers: A systematic review and meta-analysis. *Scientific Reports, 5*, 15179.
87. Zhang, G., Lu, J., Yang, M., Wang, Y., Liu, H., & Xu, C. (2020). Elevated GALNT10 expression identifies immunosuppressive microenvironment and dismal prognosis of patients with high grade serous ovarian cancer. *Cancer Immunology, Immunotherapy, 69*(2), 175–187.
88. Li, J., Wang, J., Chen, R., Bai, Y., & Lu, X. (2017). The prognostic value of tumor-infiltrating T lymphocytes in ovarian cancer. *Oncotarget, 8*(9), 15621–15631.
89. Wefers, C., Duiveman-de Boer, T., Yigit, R., Zusterzeel, P. L. M., van Altena, A. M., Massuger, L., et al. (2018). Survival of ovarian cancer patients is independent of the presence of DC and T cell subsets in ascites. *Frontiers in Immunology, 9*, 3156.
90. Farhood, B., Najafi, M., & Mortezaee, K. (2019). CD8(+) cytotoxic T lymphocytes in cancer immunotherapy: A review. *Journal of Cellular Physiology, 234*(6), 8509–8521.
91. Togashi, Y., Shitara, K., & Nishikawa, H. (2019). Regulatory T cells in cancer immunosuppression—Implications for anticancer therapy. *Nature Reviews. Clinical Oncology, 16*(6), 356–371.
92. Wertel, I., Surowka, J., Polak, G., Barczynski, B., Bednarek, W., Jakubowicz-Gil, J., et al. (2015). Macrophage-derived chemokine CCL22 and regulatory T cells in ovarian cancer patients. *Tumour Biology, 36*(6), 4811–4817.
93. Bu, M., Shen, Y., Seeger, W. L., An, S., Qi, R., Sanderson, J. A., et al. (2016). Ovarian carcinoma-infiltrating regulatory T cells were more potent suppressors of CD8(+) T cell inflammation than their peripheral counterparts, a function dependent on TIM3 expression. *Tumour Biology, 37*(3), 3949–3956.
94. Idorn, M., Olsen, M., Halldorsdottir, H. R., Skadborg, S. K., Pedersen, M., Hogdall, C., et al. (2018). Improved migration of tumor ascites lymphocytes to ovarian cancer microenvironment by CXCR2 transduction. *Oncoimmunology, 7*(4), e1412029.
95. Goyne, H. E., Stone, P. J., Burnett, A. F., & Cannon, M. J. (2014). Ovarian tumor ascites CD14+ cells suppress dendritic cell-activated CD4+ T-cell

responses through IL-10 secretion and indoleamine 2,3-dioxygenase. *Journal of Immunotherapy, 37*(3), 163–169.
96. Zhu, Q., Wu, X., Wu, Y., & Wang, X. (2016). Interaction between Treg cells and tumor-associated macrophages in the tumor microenvironment of epithelial ovarian cancer. *Oncology Reports, 36*(6), 3472–3478.
97. Zhao, H., Liao, X., & Kang, Y. (2017). Tregs: Where we are and what comes next? *Frontiers in Immunology, 8*, 1578.
98. Chen, X., Shang, W., Xu, R., Wu, M., Zhang, X., Huang, P., et al. (2019). Distribution and functions of gammadelta T cells infiltrated in the ovarian cancer microenvironment. *Journal of Translational Medicine, 17*(1), 144.
99. Alvero, A. B., Montagna, M. K., Craveiro, V., Liu, L., & Mor, G. (2012). Distinct subpopulations of epithelial ovarian cancer cells can differentially induce macrophages and T regulatory cells toward a protumor phenotype. *American Journal of Reproductive Immunology, 67*(3), 256–265.
100. Wu, M., Chen, X., Lou, J., Zhang, S., Zhang, X., Huang, L., et al. (2016). TGF-beta1 contributes to CD8+ Treg induction through p38 MAPK signaling in ovarian cancer microenvironment. *Oncotarget, 7*(28), 44534–44544.
101. Kercher, E. M., Nath, S., Rizvi, I., & Spring, B. Q. (2020). Cancer cell-targeted and activatable photoimmunotherapy spares T cells in a 3D coculture model. *Photochemistry and Photobiology, 96*(2), 295–300.
102. Ando, Y., Siegler, E. L., Ta, H. P., Cinay, G. E., Zhou, H., Gorrell, K. A., et al. (2019). Evaluating CAR-T cell therapy in a hypoxic 3D tumor model. *Advanced Healthcare Materials, 8*(5), e1900001.
103. Kenny, H. A., Dogan, S., Zillhardt, M., Mitra, A. B., Yamada, S. D., Krausz, T., et al. (2009). Organotypic models of metastasis: A three dimensional culture mimicking the human peritoneum and omentum for the study of the early steps of ovarian cancer metastasis. *Cancer Treatment and Research, 149*, 335–351.
104. Liebermann-Meffert, D. (2000). The greater omentum, anatomy, embryology, and surgical applications. *The Surgical Clinics of North America, 80*(1), 275–293.
105. Witz, C. A., Montoya-Rodriguez, I. A., Cho, S., Centonze, V. E., Bonewald, L., & Schenken, R. S. (2001). Composition of the extracellular matrix of the peritoneum. *Journal of the Society for Gynecologic Investigation, 8*(5), 299–304.
106. Peters, P. N., Schryver, E. M., Lengyel, E., & Kenny, H. (2015). Modeling the early steps of ovarian cancer dissemination in an organotypic culture of the human peritoneal cavity. *Journal of Visualized Experiments, 106*, e53541.
107. Kenny, H. A., & Lengyel, E. (2009). MMP-2 functions as an early response protein in ovarian cancer metastasis. *Cell Cycle, 8*(5), 683–688.
108. Kenny, H. A., Kaur, S., Coussens, L. M., & Lengyel, E. (2008). The initial steps of ovarian cancer cell metastasis are mediated by MMP-2 cleavage of vitronectin and fibronectin. *The Journal of Clinical Investigation, 118*(4), 1367–1379.
109. Kenny, H. A., Leonhardt, P., Ladanyi, A., Yamada, S. D., Montag, A., Im, H. K., et al. (2011). Targeting the urokinase plasminogen activator receptor inhibits ovarian cancer metastasis. *Clinical Cancer Research, 17*(3), 459–471.
110. Mitra, A. K., Chiang, C. Y., Tiwari, P., Tomar, S., Watters, K. M., Peter, M. E., et al. (2015). Microenvironment-induced downregulation of miR-193b drives ovarian cancer metastasis. *Oncogene, 34*(48), 5923–5932.
111. Henry, C., Hacker, N., & Ford, C. (2017). Silencing ROR1 and ROR2 inhibits invasion and adhesion in an organotypic model of ovarian cancer metastasis. *Oncotarget, 8*(68), 112727–112738.
112. Sawada, K., Mitra, A. K., Radjabi, A. R., Bhaskar, V., Kistner, E., Tretiakova, M. S., et al. (2008). Loss of E-cadherin promotes ovarian cancer metastasis via alpha 5-integrin, which is a therapeutic target. *Cancer Research, 68*(7), 2329–2339.
113. Kaur, S., Kenny, H. A., Jagadeeswaran, S., Zillhardt, M., Montag, A. G., Kistner, E., et al. (2009). β3-integrin expression on tumor cells inhibits tumor progression, reduces metastasis, and is associated with a favorable prognosis in patients with ovarian cancer. *The American Journal of Pathology, 175*(5), 2184–2196.
114. Lal-Nag, M., McGee, L., Guha, R., Lengyel, E., Kenny, H. A., & Ferrer, M. (2017). A high-throughput screening model of the tumor microenvironment for ovarian cancer cell growth. *SLAS Discovery, 22*(5), 494–506.
115. Sehouli, J., Senyuva, F., Fotopoulou, C., Neumann, U., Denkert, C., Werner, L., et al. (2009). Intra-abdominal tumor dissemination pattern and surgical outcome in 214 patients with primary ovarian cancer. *Journal of Surgical Oncology, 99*(7), 424–427.
116. Wilkosz, S., Ireland, G., Khwaja, N., Walker, M., Butt, R., de Giorgio-Miller, A., et al. (2005). A comparative study of the structure of human and murine greater omentum. *Anatomy and Embryology, 209*(3), 251–261.
117. Krishnan, V., Clark, R., Chekmareva, M., Johnson, A., George, S., Shaw, P., et al. (2015). In vivo and ex vivo approaches to study ovarian cancer metastatic colonization of milky spot structures in peritoneal adipose. *Journal of Visualized Experiments, 105*, e52721.
118. Khan, S. M., Funk, H. M., Thiolloy, S., Lotan, T. L., Hickson, J., Prins, G. S., et al. (2010). In vitro metastatic colonization of human ovarian cancer cells to the omentum. *Clinical & Experimental Metastasis, 27*(3), 185–196.
119. Clark, R., Krishnan, V., Schoof, M., Rodriguez, I., Theriault, B., Chekmareva, M., et al. (2013). Milky spots promote ovarian cancer metastatic colonization

of peritoneal adipose in experimental models. *The American Journal of Pathology, 183*(2), 576–591.

120. Yung, M. M., Tang, H. W., Cai, P. C., Leung, T. H., Ngu, S. F., Chan, K. K., et al. (2018). GRO-alpha and IL-8 enhance ovarian cancer metastatic potential via the CXCR2-mediated TAK1/NFkappaB signaling cascade. *Theranostics, 8*(5), 1270–1285.
121. McGuire, S., Kara, B., Hart, P. C., Montag, A., Wroblewski, K., Fazal, S., et al. (2019). Inhibition of fascin in cancer and stromal cells blocks ovarian cancer metastasis. *Gynecologic Oncology, 153*(2), 405–415.
122. Berry, R., & Rodeheffer, M. S. (2013). Characterization of the adipocyte cellular lineage in vivo. *Nature Cell Biology, 15*(3), 302–308.
123. Ieda, M., Fu, J. D., Delgado-Olguin, P., Vedantham, V., Hayashi, Y., Bruneau, B. G., et al. (2010). Direct reprogramming of fibroblasts into functional cardiomyocytes by defined factors. *Cell, 142*(3), 375–386.
124. Velletri, T., Villa, C. E., Lupia, M., Lo Riso, P., Luongo, R., Lopez Tobon, A., et al. (2018). Single cell derived organoids capture the self-renewing subpopulations of metastatic ovarian cancer. *bioRXiv, 2018*, 484121.
125. Fatehullah, A., Tan, S. H., & Barker, N. (2016). Organoids as an in vitro model of human development and disease. *Nature Cell Biology, 18*(3), 246–254.
126. Bartfeld, S., Bayram, T., van de Wetering, M., Huch, M., Begthel, H., Kujala, P., et al. (2015). In vitro expansion of human gastric epithelial stem cells and their responses to bacterial infection. *Gastroenterology, 148*(1), 126–36.e6.
127. Sato, T., Stange, D. E., Ferrante, M., Vries, R. G., Van Es, J. H., Van den Brink, S., et al. (2011). Long-term expansion of epithelial organoids from human colon, adenoma, adenocarcinoma, and Barrett's epithelium. *Gastroenterology, 141*(5), 1762–1772.
128. Roerink, S. F., Sasaki, N., Lee-Six, H., Young, M. D., Alexandrov, L. B., Behjati, S., et al. (2018). Intra-tumour diversification in colorectal cancer at the single-cell level. *Nature, 556*(7702), 457–462.
129. Klotz, D. M., & Wimberger, P. (2017). Cells of origin of ovarian cancer: Ovarian surface epithelium or fallopian tube? *Archives of Gynecology and Obstetrics, 296*(6), 1055–1062.
130. Kurman, R. J. (2013). Origin and molecular pathogenesis of ovarian high-grade serous carcinoma. *Annals of Oncology, 24*(Suppl 10), x16–x21.
131. Auersperg, N. (2013). Ovarian surface epithelium as a source of ovarian cancers: Unwarranted speculation or evidence-based hypothesis? *Gynecologic Oncology, 130*(1), 246–251.
132. Auersperg, N. (2011). The origin of ovarian carcinomas: A unifying hypothesis. *International Journal of Gynecological Pathology, 30*(1), 12–21.
133. Flesken-Nikitin, A., Hwang, C. I., Cheng, C. Y., Michurina, T. V., Enikolopov, G., & Nikitin, A. Y. (2013). Ovarian surface epithelium at the junction area contains a cancer-prone stem cell niche. *Nature, 495*(7440), 241–245.
134. Zweemer, R. P., van Diest, P. J., Verheijen, R. H., Ryan, A., Gille, J. J., Sijmons, R. H., et al. (2000). Molecular evidence linking primary cancer of the fallopian tube to BRCA1 germline mutations. *Gynecologic Oncology, 76*(1), 45–50.
135. Rebbeck, T. R., Lynch, H. T., Neuhausen, S. L., Narod, S. A., Van't Veer, L., Garber, J. E., et al. (2002). Prophylactic oophorectomy in carriers of BRCA1 or BRCA2 mutations. *The New England Journal of Medicine, 346*(21), 1616–1622.
136. Kessler, M., Hoffmann, K., Brinkmann, V., Thieck, O., Jackisch, S., Toelle, B., et al. (2015). The Notch and Wnt pathways regulate stemness and differentiation in human fallopian tube organoids. *Nature Communications, 6*, 8989.
137. Takahashi, K., Tanabe, K., Ohnuki, M., Narita, M., Ichisaka, T., Tomoda, K., et al. (2007). Induction of pluripotent stem cells from adult human fibroblasts by defined factors. *Cell, 131*(5), 861–872.
138. Takahashi, K., & Yamanaka, S. (2006). Induction of pluripotent stem cells from mouse embryonic and adult fibroblast cultures by defined factors. *Cell, 126*(4), 663–676.
139. Yucer, N., Holzapfel, M., Jenkins Vogel, T., Lenaeus, L., Ornelas, L., Laury, A., et al. (2017). Directed differentiation of human induced pluripotent stem cells into fallopian tube epithelium. *Scientific Reports, 7*(1), 10741.
140. Neel, B. G., Zhang, S., Zhang, T., Dolgalev, I., Ran, H., & Levine, D. A. (2018). Both fallopian tube and ovarian surface epithelium can act as cell-of-origin for high grade serous ovarian carcinoma. *Nature Coomunications, 2018*, 481200.
141. Watters, K. M., Bajwa, P., & Kenny, H. A. (2018). Organotypic 3D models of the ovarian cancer tumor microenvironment. *Cancers (Basel), 10*(8), 265.
142. Hill, S. J., Decker, B., Roberts, E. A., Horowitz, N. S., Muto, M. G., Worley, M. J., Jr., et al. (2018). Prediction of DNA repair inhibitor response in short-term patient-derived ovarian cancer organoids. *Cancer Discovery, 8*(11), 1404–1421.
143. Kopper, O., de Witte, C. J., Lohmussaar, K., Valle-Inclan, J. E., Hami, N., Kester, L., et al. (2019). An organoid platform for ovarian cancer captures intra- and interpatient heterogeneity. *Nature Medicine, 25*, 838–849.

PAX8, an Emerging Player in Ovarian Cancer

6

Priyanka Gokulnath, Amata Amy Soriano, Tiziana de Cristofaro, Tina Di Palma, and Mariastella Zannini

6.1 Ovarian Cancer

Ovarian Cancers (OC) are named after a heterogeneous group of cancers that involve the ovary. The tumors may have different sites of origin and vary in their histotype but are included as ovarian cancer simply because of their association with ovary. It is also the most prevalent cancer, having around 240,000 new cases diagnosed each year and very lethal with around 152,000 succumbing to the disease every year [1]. Thus, making it the seventh most common cancer and the leading cause for all gynecological deaths worldwide. It is more rampant in the developed countries and is the fifth leading cause of death in women [2].

OC can be divided into epithelial (95%), germ cell (3%), and sex-cord stroma (2%) tumors. The epithelial tumors can be further subdivided into five main histotypes on the basis of their pattern of differentiation and tumor cell morphology: high-grade serous carcinoma (HGSC, 68%), endometrioid carcinoma (EMC, 20%), clear-cell carcinoma (CCC, 4%), mucinous carcinoma (MC, 3%), and low-grade serous carcinoma (LGSC, prevalence less than 5%) [3].

Epithelial ovarian cancers are the most commonly occurring and have the most fatalities among all ovarian malignancies. Out of this, about 80% is attributed to HGSC [4], widely regarded to follow the Type II pathway of ovarian carcinogenesis, and its development is thought to be de novo or abrupt. This pathway is in contrast with the development of LGSC that are thought to follow the Type I pathway of tumorigenesis. In the Type I pathway, there is a gradual progression of the disease with the marked precursor that is indicative of the stage. These tumors are slow-growing and are usually diagnosed when confined to the ovaries. They are genetically stable and have characteristic gene signatures such as mutations in KRAS, PTEN, ARID1A, CTNBB1 and have a good prognosis [5].

In contrast, the Type II pathway characteristic of HGSC proposes a drastic development of the disease to a rapidly growing, highly proliferative, metastatic cancer. In most of the cases the tumors are detected at an advanced stage usually showing an intra-abdominal spread. Though very few cases, there is a possibility of the LGSC to accumulate further mutations to progress into HGSC. However, HGSC and LGSC have very different mechanisms of carcinogenesis and have distinctive features of histopathology and molecular gene signatures. HGSC are genetically unstable with several characteristic gene signa-

P. Gokulnath · A. A. Soriano · T. de Cristofaro · T. Di Palma · M. Zannini (✉)
Institute of Experimental Endocrinology and Oncology 'G. Salvatore' (IEOS) – CNR, National Research Council, Naples, Italy
e-mail: s.zannini@ieos.cnr.it

H. Schatten (ed.), *Ovarian Cancer: Molecular & Diagnostic Imaging and Treatment Strategies*, Advances in Experimental Medicine and Biology 1330, https://doi.org/10.1007/978-3-030-73359-9_6

tures such as TP53 mutations, BRCA1 loss, and PIK3CA mutations. They are very aggressive with a high mitotic index and have a very poor prognosis. The high mortality observed in HGSC is because 75% of women are diagnosed in the advanced stage of the disease (FIGO stage III–IV) and have a very poor 5-year survival rate. The late diagnosis is mainly attributed to the absence of disease symptoms and a lack of proper screening methodology during the early stages of the disease [5].

6.1.1 Origin of HGSC

For the better management of HGSC, it is necessary to understand the origin and pathogenesis of this disease. Since High-Grade Serous Carcinoma is only detected during the latter stages after the disease onset, the exact process of pathogenesis is still obscure. However, there have been several theories with supporting evidences that have been proposed to explain the nature and development of the disease.

Though the tumorigenic pathway of HGSC is termed as de novo, it is more than likely that the preceding events occur in a step-wise manner and the transition probably occurs rapidly as there are no precursor lesions detected. Principally, there are two main schools of thought to better explain the pathogenesis of HGSC—one supporting the ovarian origin and the other advocating the extra-ovarian origin of HGSC.

Until the 1990s, it was ovary that was regarded as the principal site for initiating the tumorigenic events that precede HGSC. One of the most popular theory to advocate ovarian origin was the "incessant ovulation theory" wherein the ovarian surface epithelium is transformed [6]. Ovarian Surface Epithelium (OSE) or ovarian mesothelium is the mesodermally derived pelvic epithelium that lines the external surface of the ovary. According to this theory, the cells of the ovarian surface epithelium accumulate mutations and DNA damage upon repeated breach of the epithelial layer caused by continuous ovulation in every menstrual cycle. The risk for the transformation process increases with increasing number of ovulations. The best model system for this theory is the domestic fowl that shows similar ovarian neoplasm due to repetitive egg laying throughout its reproductive life. Since the injury is hormone-induced, oral contraceptives to contain ovulation were associated with lesser risk of ovarian cancer [6].

The other theory supporting ovarian origin of HGSC is the "OSE or coelomic metaplasia." This theory also advocates the hormonal influence on ovarian epithelium where upon each cycle of repair the OSE undergoes serous metaplasia and progresses to a full-fledged carcinoma. To undergo metaplasia, the cells should retain their primitive stem cell like pluripotent nature and

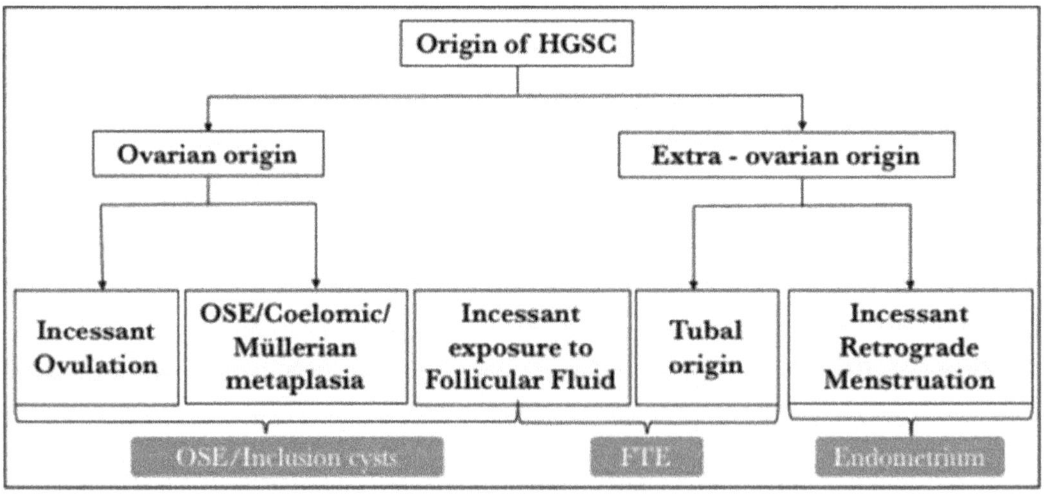

this was demonstrated by the presence of stem cell markers like NANOG in the OSE. These cells also had the ability to become fibroblast-like upon changing the microenvironment. Apart from these, mouse models that generate spontaneous serous ovarian tumors by inducing mutations in Rb1 and p53 inactivation have supported the OSE metaplasia theory to explain the carcinogenesis of HGSC [7].

However, the theories that support the ovarian origin of HGSC do not convincingly explain why Serous Tubal Intraepithelial Carcinoma (STIC) is found along with HGSC in 67% of cases [8]. BRCA1 and BRCA2 mutations were associated with an increased risk for ovarian cancer. Therefore, in the decade of 1990 and 2000 women with BRCA1 and BRCA2 germline mutations were suggested to undergo the prophylactic tubectomy and oophorectomy (removal of ovaries and Fallopian tubes). In many of these women, there were no ovarian lesions but had lesions in the Fallopian tube that resembled carcinoma (invasive and noninvasive) [9]. Subsequently, gene expression profiles comparing HGSC and Fallopian tube secretory epithelial cells showed more similarity than HGSC and OSE. Moreover, the STICs and HGSC shared similar histomorphological and expression profiles because 92% of the STICs express TP53 mutations [10]. It is proposed that the transformation process occurs in the secretory epithelial cells of the Fallopian tube that undergo mutations in TP53 and attain a p53 signature [11]. These cells further accumulate hormone-induced mutations that promoted the development of noninvasive STIC in the Fallopian tube. The proximity to ovary and possibly the paracrine factors secreted by it induces further neoplastic changes. These STIC, on reaching the adjoining OSE become fully metastatic and aggressive HGSC. Figure 6.1 demonstrates the classical demarcating junctions during the development of the highly invasive HGSC from the normal Fallopian Tube epithelium (FTE) through intermediary accumulation of p53 signature and development of STIC. This development has been demonstrated in mouse models that have oviduct epithelium-specific (Anti-Mullerian hormone receptor type 2 directed Cre) mutations in DICER and PTEN from the PI3K pathway are able to generate spontaneous tumors that resemble HGSC [12]. These evidences along with others [13–16] seem to suggest that majority of the HGSC possibly originate as STICs in the FTE.

However, the role of the ovary cannot be ruled out as it is unlikely that STICs that are inherently not aggressive become highly invasive HGSC. This transition could be possibly attributed to the microenvironment provided by the ovarian stroma. There are a couple of theories that have been recently developed to explain the link between ovary, fallopian tube, and HGSC. "Incessant exposure to Follicular Fluid theory" is a theory that is comparable to the "incessant ovulation theory" and the "tubal origin," and advocates that the continuous exposure of follicular fluid that contains inflammatory signals, DNA damage, and oxidative stress triggers the process of transformation in the ovarian and fallopian tube epithelial tissues [17].

Another theory "Incessant Retrograde Menstruation Hypothesis" bridges retrograde menstruation wherein the endometrium flows the fallopian tube to the pelvic cavity as a causative for the development of HGSC. The theory suggests that the menstruation-associated cytokines from the endometrium accumulate in the fallopian tube and its exposure induces tubal inflammation which when persistent could initiate tumorigenesis and develop into HGSC [18].

There are indeed some HGSC without any lesions in the Fallopian tube highlighting the multimodal mechanisms for the development of the HGSC. Though not discussed here, it is possible that other pathways of carcinogenesis leading to HGSC exist. Nevertheless, as evidenced by numerous case reports and research studies, FTE seems to be the prime site for the preceding events of a majority of HGSC.

It would be pertinent to mention that PAX8 protein that is expressed in the secretory cells of Fallopian tube epithelium and a lineage marker of the Mullerian duct [19] is also an immunohistological marker for HGSC as its expression is observed in most HGSC patients [20]. Infact, PAX8 histological staining is able to efficiently

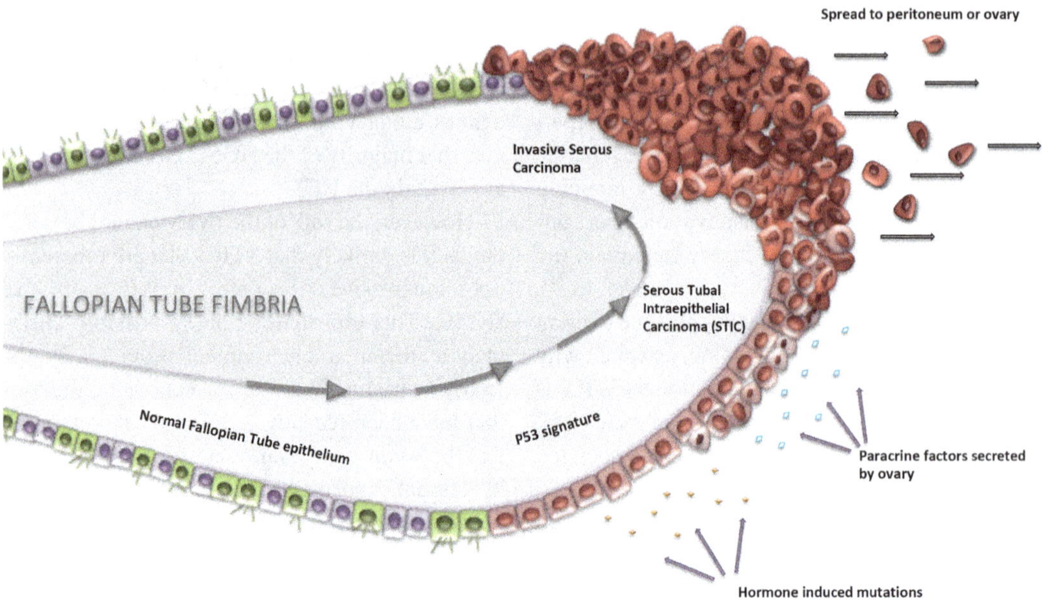

Fig. 6.1 Transformation process of secretory epithelial cells of Fallopian tube fimbria. Following TP53 mutations, the cells further accumulate hormone-induced mutations that promote the development of noninvasive STIC. The proximity to the ovary and possibly the paracrine factors secreted by it induces further neoplastic changes becoming fully metastatic and aggressive HGSC

distinguish and profile HGSC, even though PAX8 is not expressed in the normal ovarian epithelium [21, 22]. This strengthens the tubal origin of HGSC wherein PAX8 primarily expressed only in Fallopian tube secretory cells is further retained in HGSC cells [23], suggesting the site of origin of the cancer and the role of PAX8 in progression to ovarian cancer.

6.2 Overview of PAX Genes

PAX8 is a member of the Paired-boX (PAX) gene family transcription factors that are well known for their role in embryogenesis. The PAX gene family is composed of nine members (PAX1–9) and their expression is tightly regulated temporally and spatially as their expression is critical for normal embryonic development [24]. The PAX genes first described in Drosophila are evolutionarily well-conserved with expression in many species including humans, mice, zebrafish, birds, frog, flies, and worms [25, 26].

The PAX gene family is named after the paired-box DNA binding domain that is common for all nine members. As indicated in Fig. 6.2, the PAX genes are further structurally divided into four subgroups (I–IV) based on the differences in their three typical regions: Paired-box (common in all), Octapeptide (presence or absence), and Homeodomain (presence, absence or truncation) [27].

The Paired-box domain is 128 amino acids long and is located at the amino terminal end of the protein and is highly sequence-specific in its DNA binding region. The principle DNA binding paired-box domain is made up of two subdomains, namely, PAI and RED, each of which recognize distinct half-sites in the adjacent major grooves of the DNA helix. The N-terminal PAI sub-domain interacts directly with the DNA and the C-terminal RED sub-domain though does not bind directly to DNA is involved in the interaction that causes the entire PAIRED domain to bind to the DNA. The Paired (Pd) domain was first observed in Drosophila and it is highly conserved across the animal kingdom

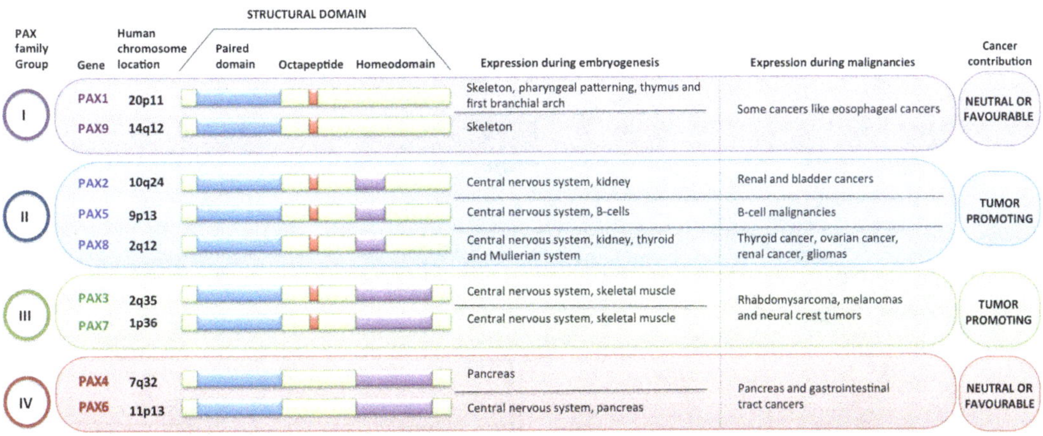

Fig. 6.2 Structural domains, expression, and cancer contribution of PAX family members subgroups

with a specific consensus sequence that is required to bind to the DNA. The Paired domain can also interact and bind to other proteins to exert its regulatory role [27].

The additional or partial homeodomain is also capable of interacting with DNA. PAX3, PAX4, PAX6, and PAX7 have a three-helix homeodomain while PAX2, PAX5, and PAX8 have a partial one-helix homeodomain. The homeodomain present in several groups of transcription factors like the Hox genes, while structurally conserved helix-turn-helix secondary structure, vary in their amino acid sequences. Generally, homeodomains assist in sequence-specific binding to the DNA, and the homeodomain of the PAX proteins recognize the palindromic sequence TAAT(N)2–3ATTA [28, 29].

The octapeptide is an eight amino acid long domain found in all PAX proteins except PAX4 and PAX6. It is highly conserved and functions as a transcriptionally inhibitory motif. The direct interaction of the octapeptide domain of the PAX protein is functionally important as this repressive activity is associated with dysregulation of the Wnt pathway that is attributed to several cancers [27].

Along with the octapeptide and the homeodomain, the PAX transcription factor also binds to other protein motifs that are a part of cofactors important for transcriptional regulation. The PAX proteins also have a transactivating domain towards the C-terminal region. Almost all PAX genes, except for PAX4 and PAX9, produce alternatively spliced variants of the respective PAX genes [30].

PAX proteins are expressed dynamically during embryogenic development and their roles are well-conserved across species. Their expression is generally observed only until the organogenesis and may persist in few organs in the adult. PAX1 and PAX9 form are well-characterized for their role in the organogenesis of the skeleton, pharyngeal patterning, development of thymus and first branchial arch; PAX2, PAX3, PAX5, PAX6, PAX7, and PAX8 in central nervous system; PAX2 and PAX8 in kidney; PAX5 in B-cells, PAX8 in thyroid and Mullerian system; PAX4 and PAX6 in pancreas; and PAX3 and PAX7 in skeletal muscle. Mutations in PAX genes result in severe growth abnormalities and are highly correlated with their expression pattern during development [31, 32].

The PAX gene expression is critical during development as their presence is attributed to the regulation of cell fate decisions and has also been associated with enhanced cell proliferation, repression of apoptosis, inhibition of terminal differentiation, and promotion of stem cell features [33]. However, their aberrant expression is now increasingly associated with several malignancies and other pathogenesis. Though PAX protein expression is seen in several cancers, their precise role in tumor progression is being explored and is still unclear. However, their over-

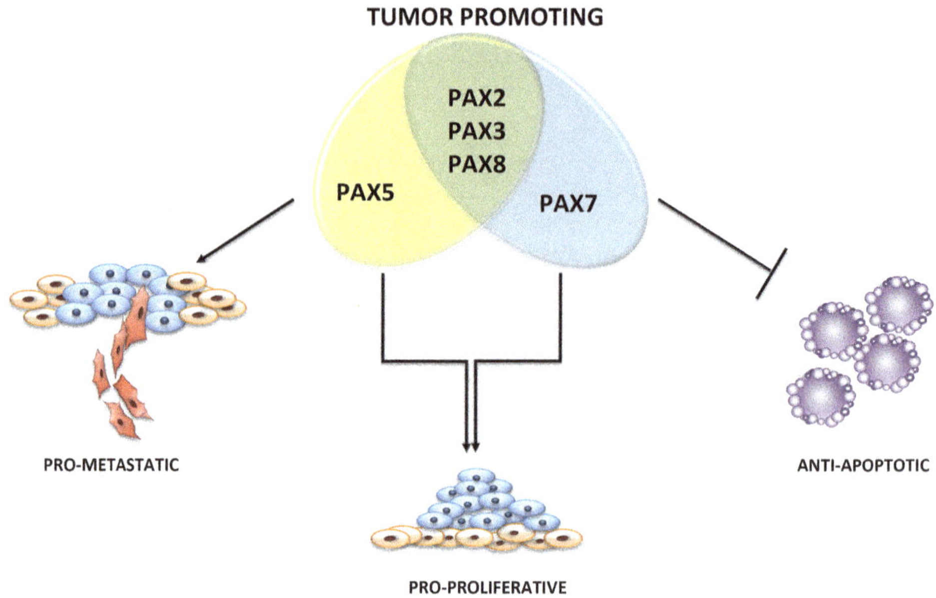

Fig. 6.3 Overview of the roles of individual PAX genes in tumor-promoting processes

expression or aberrant expression does not seem sufficient to cause malignancy [33].

There seems to be a definite correlation between the developmental roles of PAX genes and the tumors that they are involved. Figure 6.2 summarizes the role of PAX genes in normal tissue and in pathogenesis. PAX2, PAX3, PAX5, PAX7, and PAX8 are known for their tumor-promoting functions. PAX3 and PAX7 are known for their role in sarcoma specifically, Rhabdomyosarcomas, melanomas, and neural crest tumors [34]. PAX5 is expressed in several B-cell malignancies [35]. PAX2 in several renal and bladder cancers [33, 36]. PAX4 and PAX6 are implicated in the cancers of the pancreas and gastrointestinal tract [37, 38]. PAX8 is involved in thyroid cancers, ovarian cancers, renal cancers, and gliomas [39]. However, PAX1 and PAX9 are correlated with favorable prognoses of some cancers like esophageal cancers [40]. The PAX genes also are involved through chromosomal translocations in cancers by producing a fusion oncopeptide that is constitutively expressed. The best examples are PAX3-FKHR, PAX7-FKHR in alveolar rhabdomyosarcoma [41, 42], PAX5–IGH in Non-Hodgkins Lymphoma [43], PAX8-PPAR in Follicular thyroid carcinoma [44].

In summary, PAX genes are necessary for the development and differentiation during and after embryogenesis. In malignancies, their exact mechanism is still being explored. However, as shown in Fig. 6.3 PAX genes such as PAX2, PAX3, PAX5, PAX7, and PAX8 with role that are indicative of anti-apoptotic, pro-proliferative, and pro-metastatic are collectively categorized as tumor-promoting. The other PAX genes such as PAX1, PAX4, PAX6, and PAX9 have not been reported to have tumor enhancing functions. It is important to note that PAX8 exhibits role in major hallmarks for metastasis as in inhibiting cell death, propagating self-renewal, and in Epithelial Mesenchymal Transition [45].

6.3 The Transcription Factor PAX8

The PAX8 transcription factor was first isolated in 1990 in mice and in 1992 in humans [46]. It is now well known for its important role in the cell fate determination and development of several organs like thyroid, kidney, eyes, inner ear, brain, and Müllerian tract [40]. The gene that codes for the human PAX8 protein is present in Chromosome 2 at position 2q12-14 [47]. It is composed of 12 exons and the translation begins from exon 2 which has the start codon. As in other PAX subgroup II, PAX8 is composed of Paired domain, an octapeptide, and a truncated homeodomain. Exons 3 and 4 code for the DNA binding Paired-box domain, which is present in the amino terminal region of the protein. Exon 5 codes for the octapeptide while exons 7 form the truncated transactivating homeodomain [48].

PAX8 has five different isoforms, namely, PAX8A, B, C, D, and E, from different RNA transcripts produced as a result of alternative splicing of 8–10 exons. PAX8A with 450aa is the longest and most common isoform that includes all the codons from exon 2 to 12. PAX8B does not have exon 9. The C-terminal region contains proline, serine, and threonine-rich regions that contribute to the transcriptional activating function of the PAX8 protein. PAX8A and PAX8B which have exon 10–11 with unique serine, threonine, and tyrosine-rich transcriptional activation domain unlike other isoforms. PAX8C has a shorted exon 9 as it utilizes an internal exon 9 5′-splice site and due to this has an altered reading frame producing the stop codon in exon 11 thereby having a shorter proline-rich carboxyl-terminal. PAX8D has lacks exon 8 and 9, while PAX8E has exons 8–10 deleted. Both PAX8D and PAX8E have the reading frame identical to PAX8C and produce truncated proteins. Of all these the transcriptional activity is higher in PAX8A and PAX8B as compared to PAX8C [49].

Regarding the post-transcriptional modifications of the PAX8 protein, very little is known. Since there are serine and threonine sites in the domains of the PAX8 protein, they are phosphorylated with the possible involvement of PKA (Protein Kinase A) [50]. However, the exact sites for phosphorylation are yet to be defined. In thyroid, PAX8 has been known to undergo sumoylation by the conjugation of SUMO at lysin residue 309 thereby stabilizing the protein by preventing its degradation [51].

Since PAX8 is well documented and analyzed in the development and functioning of thyroid, most of the transcriptional role is known only in this context. However, it is important to note that PAX8 is involved in several pathways that contribute to carcinogenesis. Retinoblastoma (RB), a tumor suppressor, is a known positive transcriptional coactivator of PAX8 and interacts with its partial homeodomain [52]. PAX8 also is involved with RB in a reciprocal relation wherein PAX8 regulates E2F1 promotor and stabilizes RB helping in tumor cell growth [53]. PAX8 is also known to promote tumor cell survival by suppressing the expression of another well-known tumor suppressor, TP53 through TP53inp1 [54]. TP53 repression is very common in several malignancies. Wilms Tumor 1 (WT1), another tumor suppressor is reported to be transcriptionally activated by PAX8 [55]. PAX8 is also reported to be involved in the activation of Bcl2, an important anti-apoptotic protein [56]. Transforming Growth Factor-β1 (TGFB1), whose role is implicated in favoring the tumor microenvironment, is also reported to control PAX8 transcription [57].

In addition, PAX8 is implicated in an oncogenic rearrangement in thyroid carcinomas caused by a translocation between chromosomal regions 2q13 and 3p25. This rearrangement results in a fusion transcript wherein most of the coding sequence of PAX8 (2q13) is fused in frame with the entire coding exons of PPARγ1 (3p25). The PAX8 promoter is highly active in thyroid follicular cells and drives the expression of the fusion transcript, resulting in high-level expression of the fusion transcript and protein PPFP (PAX8-PPARγ fusion protein). Although the specific mechanism of PPFP action is yet to be defined, it is known that PPFP has the DNA binding domains of both PAX8 and PPARγ. Therefore, a plausible mechanism of oncogenesis is the modulation of the downstream pathways of PAX8 or PPARγ [58].

Thus, PAX8 though primarily known as a differentiation promoting transcription factor has been in different contexts reported to aid in tumor progression and maintenance.

6.3.1 PAX8 Expression in Normal Tissues

PAX8 is an important transcription factor involved in the development of various organs such as thyroid, kidney, Müllerian tract, vertebral column, hindbrain, eye, and inner ear [31]. It is also necessary for the maintenance and has been shown to be a lineage-specific marker of organs like thyroid, kidney, and the Müllerian tract [39].

Due to the high sequence homology between PAX subgroup II, certain reports of PAX8 staining in lymph nodes, pancreas, and neuroendocrine cells of stomach and colon cannot be considered as completely true. This is attributed to the cross-reactivity of PAX8 polyclonal antibody used and not because of PAX8 expression in these tissues. To resolve this issue, a monoclonal PAX8 antibody with high specificity was used to identify PAX8 positive tissues [59].

Consistent with the role of PAX8 as a lineage-specific marker, it has been shown to be expressed in normal adult tissues of thyroid, kidney, and the Müllerian tract. Its expression has been demonstrated in the developing thyroid gland [60] and it efficiently determines the differentiated phenotype typically seen in the adult follicular thyroid cells [61].

PAX8 is involved in the organogenesis of the kidney and its staining was detected in the normal kidney with focal segmental staining of glomerular parietal epithelial cells and diffuse staining of collecting duct epithelial cells [46]. In the male genital tract, seminal vesicles and epididymis were diffusely positive, but not germ cells like, Leydig cells or Sertoli cells [62].

With respect to the Müllerian duct, PAX8 is detected in the embryo during the organogenesis and development of the Müllerian duct. Moreover, PAX8 is said to be involved in the formation of the epithelial layer [63]. It is retained throughout the formation of the fallopian tube and uterine epithelium. In the female genital tract, strong and diffuse PAX8 staining was observed in the epithelial cells of the endocervix and the endometrium. When the epithelial layer of the fallopian tube differentiates into ciliated and secretory cells, PAX8 stained only basal and secretory cells where its expression is retained and absent in ciliated cells [64].

6.3.2 PAX8 Expression in Cancer Tissues

The expression of PAX8 in neoplastic tissues is well studied in several cancers by analyzing the tissue expression profiles of cancer patients. PAX8 is observed in carcinomas of ovary, uterus, kidney, prostate, gliomas, Wilms' tumor, Kaposi sarcoma. Here, we review some of the histological studies, to underline major prevalence of PAX8 in significant number of cancers.

The first report of PAX8 in malignancy was in Wilms' tumor in 1992 [46]. The expression of PAX8 was then reported in several human thyroid neoplasms [65]. Reports of PAX8 expression in several cancers such as renal neoplasm, ovarian carcinomas, gliomas were described establishing a definite correlation between PAX8 and several epithelial malignancies [66]. Further, PAX8 is now known to be a useful immunohistological marker that helps differentiate Mullerian from non-Mullerian tumors [67].

Tacha et al. examined the immunohistochemical expression of PAX8 in multiple normal and neoplastic tissues. Renal cell carcinomas tested positive for PAX8 in 90% of the cases, ovarian cancers for 79% of the cases, and thyroid cancer for 90% of all cases [39]. The *PAX8/PPARG* gene fusion, that was previously mentioned, was found in 30–35% of follicular thyroid carcinomas and in a substantially smaller fraction of follicular variant papillary thyroid carcinomas. This rearrangement is very occasionally found in follicular adenomas [58].

In endometrial cancers, 84% of the cases were positive for PAX8 expression and in cervical cancers, PAX8 was observed in 83% of cervical adenocarcinomas whereas 98% of squamous cell carcinomas cases were negative for the protein expression. In bladder cancers, PAX8 was negative in 93% of all the cases including all bladder adenocarcinomas. PAX8 expression was observed in only one case of lung cancer (99% negative) and was 100% negative in cancers of the colon, breast, prostate, liver, testicular, stomach, esophagus, melanoma, gastrointestinal stromal tumors, leiomyosarcoma, and pheochromocytoma [39].

Laury et al. conducted a study in which PAX8 immunohistochemistry was performed on 1357 tumors (486 tumors in whole-tissue sections and 871 tumors in tissue microarrays, predominantly epithelial) from multiple organs. Nuclear PAX8 staining was present in 91% of thyroid tumors, 90% of renal cell carcinomas (RCCs), 81% of renal oncocytomas, 91% of cervical epithelial lesions, and 98% of endometrial adenocarcinomas. The remaining tumors such as those from the prostate, colon, stomach, liver, adrenal gland, head and neck, small cell carcinomas from the lung were PAX8 negative [68].

PAX8 expression and its association with ovarian epithelial cancers deserve a special focus as its presence in most of the tissues is well replicated and documented in several independent studies apart from the ones mentioned above. Laury et al. reported that PAX8 staining was present in 99% (164 of 165) of high-grade serous ovarian carcinomas, 71% (32 of 49) of nonserous ovarian epithelial neoplasms, and all (100%) low-grade ovarian carcinomas and serous borderline tumors. It is important to mention that strong PAX8 staining was highly specific for ovarian serous tumors according to both Laury and Tacha et al. [39, 68].

Recently, Hong-Juan Chai et al. demonstrated that PAX8 was highly expressed in primary epithelial ovarian cancer (PEOC) with an overall 92% positivity. In addition, their study revealed that PAX8 expression level was associated with the degree of cancer cell differentiation, FIGO stage, and survival rate, indicating that PAX8 is a potential marker for the diagnosis of PEOC [22]. This could be relevant in the therapeutic approach.

Apart from the ones mentioned above, there are various recent reports that vouch for the specificity of PAX8 expression specifically in HGSC. Due to this PAX8 has become a reliable immunohistological marker to identify and diagnose HGSC.

Figure 6.4 highlights the role of PAX8 in organogenesis, expression in adult tissue in both normal and neoplastic conditions. If one observes closely, there seems to be a definite association between expression of PAX8 in normalcy and malignancy, as in many transcriptional factors that are involved in development and cancers.

6.4 PAX8 as a Player in Ovarian Cancer

Given the fact that PAX8 is not generally expressed in the normal ovarian tissues, it would be interesting to note how PAX8 came to be accepted as a histological marker of ovarian cancer in the first place. The first indication of PAX8-Ovarian Cancer association was a result of the study by Schaner et al. in 2003 [69], who performed DNA microarray studies to understand histopathologic subtypes, grades of ovarian cancers, and genes that differentiate breast and ovarian carcinomas. In this study, PAX8 was among the genes that were highly expressed by the ovarian cancer cells. Another significantly important study that further strengthened the link between PAX8 and ovarian carcinoma was the microarray studies done by Marquez et al. in 2005 [70]. This study also associated the expression of PAX8 with the Fallopian tube and a possible tubal origin of ovarian carcinoma. Several subsequent studies continued to reinforce PAX8 expression in ovarian cancer tissues, especially in HGSC [64]. In the previous section, PAX8 is shown to be expressed in most of the HGSC patients' tissue samples with a very high frequency rate of 99% [39]. Having established the importance of PAX8 as a histological marker in HGSC, it is necessary to understand if there is a correlation between the role of PAX8 protein with the development and progression of HGSC.

6.4.1 PAX8-Dependent Tumorigenic Phenotype

The first report on the role of PAX8 in ovarian carcinogenesis dates back to 2011 when CG Li et al. [53] showed that PAX8 is capable of promoting tumor cell growth through direct

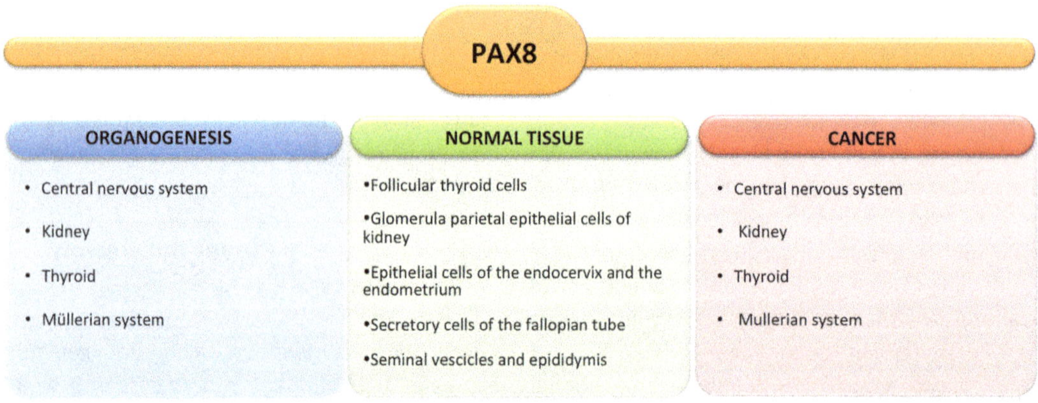

Fig. 6.4 Role of PAX8 in organogenesis and its expression in adult tissue in both normal and neoplastic conditions

regulation of the *E2F1* promoter and upregulation of *E2F1* expression. They proposed a model in which PAX8-mediated E2F1 regulation could be essential to maintain cell proliferation signal before S-phase entry.

Another important study that established and reinforced the significance PAX8 had on ovarian cancer and ovarian cancer cell lines was that of Cheung et al. in 2011 [71]. They studied 102 human cancer cell lines and checked for vulnerabilities, gene signatures, and mutations. One of the unifying aspect that they discovered was the dependency of PAX8 expression in Ovarian cancer cell lines, specifically amplified in high-grade serous cancer cell line. They also validated their study by characterizing the lineage-specific dependencies using shRNA in ovarian cancer cell line OVCAR-8. They concluded that PAX8 expression is necessary for the proliferation and survival of the cell where it is expressed, even ones such as renal and endometrial cancer cell lines.

Subsequently, in 2014, our group investigated the role of PAX8 in ovarian cancer in vitro and in vivo [45]. We demonstrated that PAX8 plays a critical role in cell cycle progression and cell survival of differentiated epithelial cells, reinforcing the crucial involvement of this transcription factor in different biological processes specifically in the context of ovarian cancer.

We analyzed the expression level of PAX8 in a series of ovarian cancer cell lines (SKOV-3, TOV-21G, OVCAR-3, TOV-112D, and A2780) and we chose SKOV-3 cell lines to assess PAX8 involvement in ovarian tumorigenesis. As shown in Fig. 6.5a, our results indicated that PAX8 knock-down elicited a dramatic effect on SKOV-3 cell growth, inhibited the invasion rate of these cells through the matrigel, and reduced the migration rate in wound-healing assay. These experiments demonstrated the tumorigenic potential of PAX8 in vitro.

To assess the ability of PAX8 in contributing towards tumor growth in vivo, we injected SKOV-3 cells constitutively silenced by shPAX8 into nude mice. The results obtained in this study showed for the first time that PAX8 was capable of inducing in vivo tumor growth. The size of palpable lesions well correlated with PAX8 expression level of single clones, confirming the role of PAX8 as oncogene in vivo Fig. 6.5b.

The fact that PAX8 was able to elicit an effect in invasion and migration raised the question of its possible role in Epithelial Mesenchymal Transition (EMT). To further understand this, expression markers of EMT such as Snail, Twist, Zeb2, Vimentin were analyzed before and after PAX8 overexpression in IOSE-80 (an ovarian epithelial cancer cell line) using RT-PCR and Western blot. Figure 6.6a, b confirms that PAX8 is indeed able to trigger EMT; however, it is necessary to probe in this direction to understand how the exact mechanism occurs.

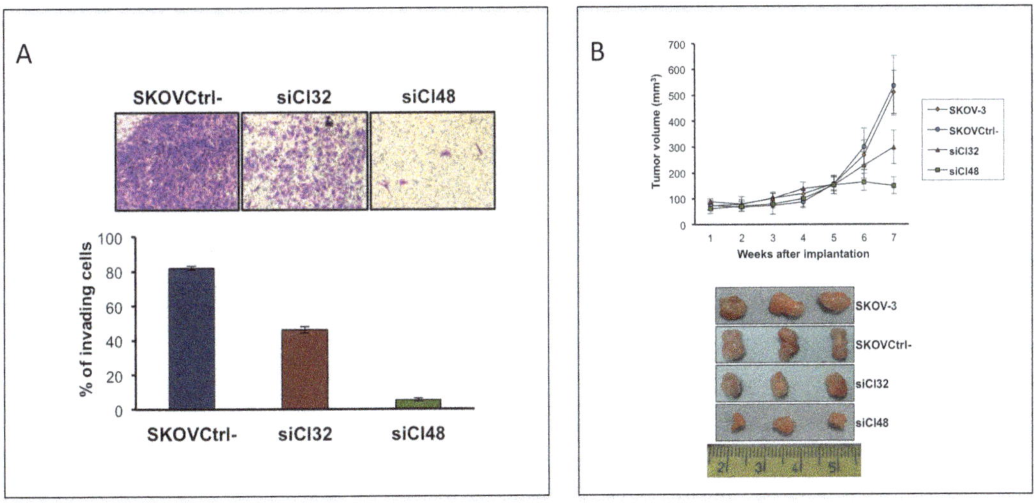

Fig. 6.5 (**a**) Matrigel invasion assay of SKOVCtrl- and siPAX8 clones and relative quantitation. (**b**) Growth curves of tumors in tumor-bearing nude mice and representative images of the tumors following surgical resection

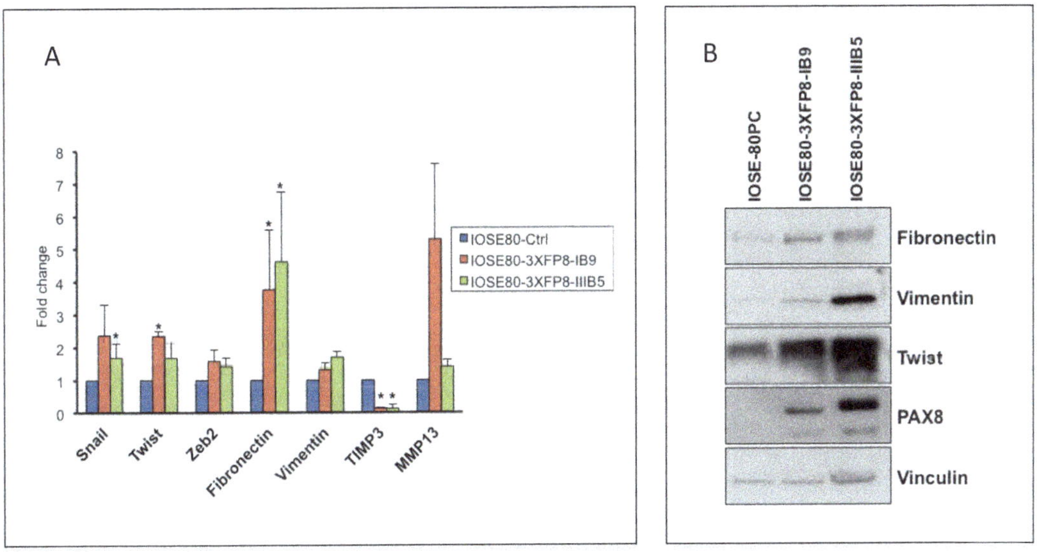

Fig. 6.6 Q-PCR (**a**) and Western blot analysis (**b**) for EMT markers in IOSE-80 cells before and after PAX8 overexpression

This was also when the theory of tubal origin of HGSC was gaining strength and PAX8 being one of the unifying factors being present in both fallopian tube and HGSC was of key interest to understand the pathogenesis of this ovarian carcinoma. Rodgers et al. [72] tried to address this and the findings presented in this study demonstrated that genes with known functional binding sites for PAX8 such as E2F1, BRCA1, and WT1 are transcriptionally upregulated by PAX8 in the fallopian tube but not in the ovary. This study also confirmed the role of PAX8 in proliferation, migration, and EMT in ovarian cancer mouse model. In addition, several human HGSC cell lines were tested (Kuramochi, OVSAHO, OVCAR4, OVKATE, OVCAR3, OVCAR5, OVCA432, and SKOV3) and they showed an uniform decrease in proliferation and increase in apoptosis after PAX8 silencing. They observed that reducing PAX8 levels resulted in decreased expression of FOXM1 and its downstream targets including AURKB. FOXM1 is a transcription factor known to increase cell proliferation, migration, and EMT. It is important to note that TCGA highlighted FOXM1 as one of the top altered genes in HGSC, with 87% of cases showing FOXM1 alteration. It has been previously shown that PAX8 can transcriptionally repress p53 and since p53 is a negative regulator of FOXM1 expression, it was hypothesized that PAX8 increases FOXM1 expression by inhibiting p53 activity [72].

It might be relevant to mention that PAX8 has been well correlated with the regulation of epithelial cell survival by altering apoptotic pathways reported by our group in the context of thyroid [54]. However, the exact mechanism should be investigated and could be context dependant.

From the abovementioned studies, one can conceive that PAX8 seems to have concrete role in progression of ovarian cancer, though the pathway for its involvement in tumorigenesis is still unclear. It would be very important to unravel and analyze the signaling that can explain the intimate relation PAX8 has with ovarian cancer.

6.4.2 Pathways Regulated by PAX8 in Ovarian Cancer

To have better insights regarding the possible underlying mechanisms by which PAX8, that has a normal expression in Fallopian tube is aberrantly retained in metastatic ovarian cancers like HGSC, several groups studied the gene expression analysis before and after PAX8 silencing in Fallopian tube cell line and ovarian cancer cell line.

In 2016, our group reported regarding the altered gene expression analysis before and after PAX8 silencing in a Fallopian tube secretory cell line (FT-194) and PAX8 expressing ovarian cancer cell line (SKOV-3) and analyzing their transcriptome [73]. There were approximately around 60% of the genes that were differentially regulated between the fallopian tube and the ovarian cancer cell lines. Upon PAX8 silencing, the pathways that were commonly downregulated in both FT-194 and SKOV-3 were p53 pathway, Estrogen response, Kras signaling, UV response; and those that were commonly upregulated were interferon response, TNFα signaling, inflammatory response, apoptosis, and epithelial mesenchymal transition. Using this RNA-seq analysis, genes and pathways that could be candidates downstream of PAX8 transcription factor were identified in both normal context and in ovarian cancer. This study was able to provide certain insights helpful in chalking out the tubal development of HGSC. Many of the genes that were altered by PAX8 were secretory factors and their relation should be further carefully examined to reveal any suitable biomarkers that could be therapeutically relevant.

In the same year, Elias et al. [74] published their findings of how the epigenetic changes that govern the PAX8 binding sites are modified between fallopian tube and ovarian cancer. They studied the PAX8 cistrome by comparing three fallopian tube cell lines (FT33, FT194, and FT246) and three high-grade serous cancer cell lines (KURAMOCHI, OVSAHO, and JHOS4) before and after PAX8 silencing. Cistrome is defined as the genome-wide map of the binding sites of a transcription factor. Their study showed that between fallopian tube and the ovarian cancer cell lines, the ovarian cancer cells had significantly reprogrammed their PAX8 cistrome. Further on analyzing by RNA-seq and ChIP-Seq, the genes that are differentially expressed between these two cell types were located and clustered around the PAX8 binding sites. Additionally, on careful scrutiny of the PAX8 cistrome alterations, there seems to be an increase in the interactions between PAX8 and TEAD. This suggests that the Hippo-YAP signaling pathway, which is known for their important role in many cancers, could interact with PAX8 and be an important mediator to regulate transformation in ovarian cancer. Their group further investigated this relation and proposed that PAX8 functions differentially upon transformation and thereby promote cancer progression through Hippo-YAP pathway.

To highlight the significance in exploring the PAX8 pathway in epithelial ovarian cancer, Kar et al. [75] in 2017 showed that of loci that are susceptible to alterations in Serous Epithelial Ovarian Cancers (SOC), the putative PAX8 target genes with binding sites for PAX8 were enriched. This is the first study that distinguishes binding sites for PAX8 as the governing change among SNP (Single Nucleotide Polymorphisms) that are enriched in epithelial ovarian cancer and increases its risk, much like Estrogen receptors in breast cancer or androgen receptors in prostate cancer. This study analyzed the 615 Transcription factor (TF)-target gene sets from the two largest and available Serous Ovarian Cancer risk Genome-wide Association Studies data sets. They concluded that the PAX8 targets were the most altered among the gene sets. Their study indicated how the SOC risk was driven by a network of PAX8 and 15 selected genes such as BNC2 and HOXB7. These genes were further validated for their binding to PAX8 by ChIP-seq analysis.

Along the same line, Adler et al. [76] also in 2017 investigated PAX8 cistrome in epithelial ovarian cancer. This report along with others demonstrated the decrease in anchorage-dependent and independent growth in ovarian cancer cell lines (HeyA8 and IGROV1) upon PAX8 silencing. PAX8 cistrome in these cell lines were characterized using ChIP-seq along with ChIP-seq for acetylated histone subunit 3 (H3K27ac) to analyze PAX8 binding sites in active chromatin. The two cell lines showed difference in the number of PAX8 binding sites corresponding with active chromatin, approximately 60% in IGROV1 and 25% in HeyA8 cells. The enormous difference could be attributed to the difference in histotype of cancer having isolated from different patients. This study also characterized the PAX8 binding motif and identified candidate PAX8 co-regulators and target genes. They concluded that PAX8 binding was enriched at superenhancers and controlled genes that encompass differentiation, development, and tumorigenesis. They reported that among the many developmental pathways that were augmented in the PAX8 cistrome, tissue morphogenesis, apoptosis, EMT, Notch signaling were notable. This study indicates the direct and indirect regulatory gene targets of PAX8 throughout the genome.

More recently, Ghannam-Shahbari et al. in a study published in 2018 [77] have reported that PAX8 can directly bind to the first intron-exon boundary of TP53 in Fallopian tube and also to mutant TP53 in HGSC. The PAX8 activated mutant TP53 further causes a cytoplasmic p21 accumulation resulting in a proliferative effect observed in HGSC. It is more than plausible that PAX8 could have other pathways to execute its pro-oncogenic roles in HGSC. However, from the above studies, one can consider PAX8 as a putative and an important regulator of HGSC. This also indicates that inhibition of the PAX8 pathway or many of its downstream regulators in a combinatorial method could be a possible therapeutic methodology to contain HGSC.

6.5 Conclusion and Future Perspectives

HGSC, despite years of research, evades the realm of effective diagnosis and therapy. It is still one of the most lethal malignancies that affect women worldwide. Thus, finding an early diagnostic marker for this cancer is very important. PAX8 is emerging as an important player in HGSC diagnosis because the currently more prevalent biomarker CA-125 can only test positive during the very late stages of the disease. Further, CA-125 is limited in its specificity and diagnosis of gynecological malignancies. Therefore, it is important to find a suitable marker that can indicate development of the disease. PAX8 not only reliably identifies HGSC but also has a certain diagnostic utility with respect to metastatic epithelial ovarian cancers [77]. Upon staining cells isolated from the fluid and fine needle aspirations of ovarian cancer patients, PAX8 has a 90% sensitivity without staining mesothelial cells as compared to WT1, another EOC marker [78]. Furthermore, Wang et al. [20] reported that PAX8 along with Calretinin staining was used to determine the administration of neoadjuvant therapy for ovarian cancer patients exhibiting PAX8+/Calretinin—in cells obtained from ascitic sample. In spite of its reliability, PAX8 as a marker is still too late for an effective therapy to be executed.

PAX8 is associated with secretory pathways in thyroid and it is plausible that it is associated with a similar role in the Müllerian tract as well. Even though the expression of PAX8 in the development of the Mullerian tract, specifically the development of Fallopian tube and its retention in the secretory cells is long known, the exact role of PAX8 in these cells is yet to be understood. Though several studies have reasonably strengethened the tubal origin of HGSC, the exact mechanism and PAX8's involvement in its development and maintenance is still a mystery.

To this end, it is necessary to study and analyze the role of PAX8 in the context of fallopian tube and the possible development of HGSC. It should be noted that PAX8 is expressed in most of the Ovarian cancer cell lines and it was reasonable to explore the specific role of PAX8 in ovarian cancer. Many gene expression profiles and RNA-seq studies were conducted to identify the specific bioprocess or signaling pathway that is regulated by PAX8 in the context of ovarian cancer. However, reports from various sources indicate that PAX8 has several approaches in contributing towards the development of tumor progression—through inhibition of apoptosis, enhancing proliferation, and inducing EMT. PAX8 could possibly be involved in stemness and angiogenesis, indicated by the pathway analysis and each of this arena has to be carefully explored and critically analyzed. Given the importance of PAX8 in secretory functions, it would be reasonable to assume that its putative targets could have an important role as secretory factors and thereby steer us in identifying a suitable biomarker for HGSC. This could have an enormous effect either with respect to early diagnosis of HGSC or could help in identifying a suitable therapeutic target against this malignancy. Therefore, not only is it very imperative to explore the role of PAX8 and its signaling pathway in the context of HGSC, it is very crucial and insightful to examine in the context of its secretory functions in fallopian tube so that we could identify a major biomarker that is indicative of the early stages of HGSC.

Acknowledgments This author's work was supported by the FLAGSHIP InterOmics project funded and supported by the Italian Ministry of Education, University and Research, and by the National Research Council organizations.

Priyanka Gokulnath is a student of the international Ph.D. program INCIPIT co-funded by the Marie Skłodowska-Curie actions; Amata Amy Soriano is a student of the Ph.D. program in *Molecular Medicine and Molecular Biotechnology* of the University of Naples Federico II.

References

1. Ferlay, J., Soerjomataram, I., Dikshit, R., Eser, S., Mathers, C., Rebelo, M., Parkin, D., Forman, D., & Bray, F. (2014). Cancer incidence and mortality worldwide: Sources, methods and major patterns in GLOBOCAN 2012. *International Journal of Cancer, 136*(5), E359–E386.
2. Siegel, R. L., Miller, K. D., & Jemal, A. (2015). Cancer statistics, 2015. *CA: A Cancer Journal for Clinicians, 65*(1), 5–29. https://doi.org/10.3322/caac.21254.
3. Kuhn, E., Meeker, A. K., Visvanathan, K., Gross, A. L., Wang, T. L., Kurman, R. J., & Shih Ie, M. (2011). Telomere length in different histologic types of ovarian carcinoma with emphasis on clear cell carcinoma. *Modern Pathology, 24*, 1139–1145. https://doi.org/10.1038/modpathol.2011.67.
4. Bowtell, D. D., Böhm, S., Ahmed, A. A., Aspuria, P.-J., Bast, R. C., Beral, V., et al. (2015). Rethinking ovarian cancer II: Reducing mortality from high-grade serous ovarian cancer. *Nature Reviews Cancer, 15*(11), 668–679. https://doi.org/10.1038/nrc4019.
5. Vang, R., Shih, I.-M., & Kurman, R. J. (2009). Ovarian low-grade and high-grade serous carcinoma. *Advances in Anatomic Pathology, 16*(5), 267–282. https://doi.org/10.1097/pap.0b013e3181b4fffa.
6. Fathalla, M. F. (2013). Incessant ovulation and ovarian cancer—A hypothesis re-visited. *Facts, Views and Vision in ObGyn, 5*(4), 292–297.
7. Auersperg, N. (2013). Ovarian surface epithelium as a source of ovarian cancers: Unwarranted speculation or evidence-based hypothesis? *Gynecologic Oncology, 130*(1), 246–251. https://doi.org/10.1016/j.ygyno.2013.03.021.
8. Kessler, M., Fotopoulou, C., & Meyer, T. (2013). The molecular fingerprint of high grade serous ovarian cancer reflects its fallopian tube origin. *International Journal of Molecular Sciences, 14*(4), 6571–6596. https://doi.org/10.3390/ijms14046571.

9. Olivier, R. I., van Beurden, M., Lubsen, M. A. C., Rookus, M. A., Mooij, T. M., van de Vijver, M. J., & van't Veer, L. J. (2004). Clinical outcome of prophylactic oophorectomy in BRCA1/BRCA2 mutation carriers and events during follow-up. *British Journal of Cancer, 90*(8), 1492–1497.
10. Przybycin, C. G., Kurman, R. J., Ronnett, B. M., Shih, I.-M., & Vang, R. (2010). Are all pelvic (nonuterine) serous carcinomas of tubal origin? *The American Journal of Surgical Pathology, 34*(10), 1407–1416. https://doi.org/10.1097/pas.0b013e3181ef7b16.
11. Kuhn, E., Kurman, R. J., Vang, R., et al. (2012). TP53 mutations in serous tubal intraepithelial carcinoma and concurrent pelvic high-grade serous carcinoma—Evidence supporting the clonal relationship of the two lesions. *The Journal of Pathology, 226*(3), 421–426. https://doi.org/10.1002/path.3023.
12. Kim, J., Coffey, D. M., Creighton, C. J., Yu, Z., Hawkins, S. M., & Matzuk, M. M. (2012). High-grade serous ovarian cancer arises from fallopian tube in a mouse model. *Proceedings of the National Academy of Sciences, 109*(10), 3921–3926. https://doi.org/10.1073/pnas.1117135109.
13. Carlson, J. W., Miron, A., Jarboe, E. A., Parast, M. M., Hirsch, M. S., Lee, Y., et al. (2008). Serous tubal intraepithelial carcinoma: Its potential role in primary peritoneal serous carcinoma and serous cancer prevention. *Journal of Clinical Oncology, 26*(25), 4160–4165. https://doi.org/10.1200/jco.2008.16.4814.
14. Lim, D., & Oliva, E. (2013). Precursors and pathogenesis of ovarian carcinoma. *Pathology, 45*(3), 229–242. https://doi.org/10.1097/pat.0b013e32835f2264.
15. Jarboe, E., Folkins, A., Nucci, M. R., Kindelberger, D., Drapkin, R., Miron, A., et al. (2008). Serous carcinogenesis in the fallopian tube. *International Journal of Gynecological Pathology, 27*(1), 1–9. https://doi.org/10.1097/pgp.0b013e31814b191f.
16. Diniz, P. M., Carvalho, J. P., Baracat, E. C., & Carvalho, F. M. (2011). Fallopian tube origin of supposed ovarian high-grade serous carcinomas. *Clinics (São Paulo, Brazil), 66*(1), 73–76. https://doi.org/10.1590/S1807593220110001000l3.
17. Tone, A. A. (2017). Taking the tube. *Clinical Obstetrics and Gynecology, 60*(4), 697–710. https://doi.org/10.1097/grf.0000000000000313.
18. Vercellini, P., Crosignani, P., Somigliana, E., Vigano, P., Buggio, L., Bolis, G., & Fedele, L. (2011). The "incessant menstruation" hypothesis: A mechanistic ovarian cancer model with implications for prevention. *Human Reproduction, 26*(9), 2262–2273. https://doi.org/10.1093/humrep/der211.
19. Tong, G. X., Devaraj, K., Hamele-Bena, D., Yu, W. M., Turk, A., Chen, X., Wright, J. D., & Greenebaum, E. (2011). Pax8: A marker for carcinoma of Müllerian origin in serous effusions. *Diagnostic Cytopathology, 39*(8), 567–574. https://doi.org/10.1002/dc.21426. Epub 2010 Jul 6.
20. Wang, Y., Wang, Y., Li, J., Yuan, Z., Yuan, B., Zhang, T., Cragun, J. M., Kong, B., & Zheng, W. (2013). PAX8: A sensitive and specific marker to identify cancer cells of ovarian origin for patients prior to neoadjuvant chemotherapy. *Journal of Hematology and Oncology, 6*, 60. https://doi.org/10.1186/1756-8722-6-60.
21. Xiang, L., & Kong, B. (2013). PAX8 is a novel marker for differentiating between various types of tumor, particularly ovarian epithelial carcinomas. *Oncology Letters, 5*(3), 735–738.
22. Chai, H. J., Ren, Q., Fan, Q., Ye, L., Du, G. Y., Du, H. W., Xu, W., Li, Y., Zhang, L., & Cheng, Z. P. (2017). PAX8 is a potential marker for the diagnosis of primary epithelial ovarian cancer. *Oncology Letters, 14*(5), 5871–5875. https://doi.org/10.3892/ol.2017.6949.
23. Perets, R., Wyant, G. A., Muto, K. W., Bijron, J. G., Poole, B. B., Chin, K. T., Chen, J. Y. H., Ohman, A. W., Stepule, C. D., Kwak, S., Karst, A. M., Hirsch, M. S., Setlur, S. R., Crum, C. P., Dinulescu, D. M., & Drapkin, R. (2013). Transformation of the fallopian tube secretory epithelium leads to high-grade serous ovarian cancer in Brca; Tp53; Pten models. *Cancer Cell, 24*(6), 751–765. https://doi.org/10.1016/j.ccr.2013.10.013.
24. Mansouri, A., Hallonet, M., & Gruss, P. (1996). Pax genes and their roles in cell differentiation and development. *Current Opinion in Cell Biology, 8*(6), 851–857. https://doi.org/10.1016/s0955-0674(96)80087-1.
25. Bopp, D., Burri, M., Baumgartner, S., Frigerio, G., & Noll, M. (1986). Conservation of a large protein domain in the segmentation gene paired and in functionally related genes of Drosophila. *Cell, 47*(6), 1033–1040.
26. Dahl, E., Koseki, H., & Balling, R. (1997). Pax genes and organogenesis. *BioEssays, 19*, 755–765.
27. Treisman, J., Harris, E., & Desplan, C. (1991). The paired box encodes a second DNA-binding domain in the paired homeo domain protein. *Genes & Development, 5*(4), 594–604.
28. Mayran, A., Pelletier, A., & Drouin, J. (2015). Pax factors in transcription and epigenetic remodelling. *Seminars in Cell & Developmental Biology, 44*, 135–144. https://doi.org/10.1016/j.semcdb.2015.07.007.
29. Wilson, D., Sheng, G., Lecuit, T., Dostatni, N., & Desplan, C. (1993). Cooperative dimerization of paired class homeo domains on DNA. *Genes & Development, 7*, 2120–2134. https://doi.org/10.1101/gad.7.11.21201993.
30. Kozmik, Z., Czerny, T., & Busslinger, M. (1997). Alternatively spliced insertions in the paired domain restrict the DNA sequence specificity of Pax6 and Pax8. *The EMBO Journal*. https://doi.org/10.1093/emboj/16.22.6793.
31. Blake, J. A., & Ziman, M. R. (2014). Pax genes: Regulators of lineage specification and progenitor cell maintenance. *Development, 141*, 737–751. https://doi.org/10.1242/dev.091785.
32. Lang, D., Powell, S. K., Plummer, R. S., Young, K. P., & Ruggeri, B. A. (2007). PAX genes: Roles in development, pathophysiology, and cancer. *Biochemical*

Pharmacology, 73(1), 1–14. https://doi.org/10.1016/j.bcp.2006.06.024.
33. Muratovska, A., Zhou, C., He, S., Goodyer, P., & Eccles, M. R. (2003). Paired-box genes are frequently expressed in cancer and often required for cancer cell survival. *Oncogene, 22*, 7989–7997.
34. Relaix, F., Rocancourt, D., Mansouri, A., & Buckingham, M. (2004). Divergent functions of murine Pax3 and Pax7 in limb muscle development. *Genes & Development, 18*, 1088–1105.
35. Busslinger, M. (2004). Transcriptional control of early B cell development. *Annual Review of Immunology, 22*, 55–79.
36. Daniel, L., et al. (2001). Pax-2 expression in adult renal tumors. *Human Pathology, 32*, 282–237.
37. Miyamoto, T., et al. (2001). Expression of dominant negative form of PAX4 in human insulinoma. *Biochemical and Biophysical Research Communications, 282*, 34–40.
38. Yamaoka, T., et al. (2000). Diabetes and pancreatic tumours in transgenic mice expressing Pax 6. *Diabetologia, 43*, 332–339.
39. Tacha, D., Zhou, D., & Cheng, L. (2011). Expression of PAX8 in normal and neoplastic tissues: A comprehensive immunohistochemical study. *Applied Immunohistochemistry & Molecular Morphology, 19*(4), 293–299. https://doi.org/10.1097/PAI.0b013e3182025f66.
40. Gerber, J. K., et al. (2002). Progressive loss of PAX9 expression correlates with increasing malignancy of dysplastic and cancerous epithelium of the human oesophagus. *The Journal of Pathology, 197*, 293–297.
41. Liu, L., & Chen, T. (2012). PAX3-FKHR regulates the expression of pleiotrophin to mediate motility in alveolar rhabdomyosarcoma cells. *Journal of Cancer Research Updates*. https://doi.org/10.6000/1929-2279.2012.01.01.09.
42. Du, S., Lawrence, E. J., Strzelecki, D., Rajput, P., Xia, S. J., Gottesman, D. M., & Barr, F. G. (2005). Co-expression of alternatively spliced forms of PAX3, PAX7, PAX3-FKHR and PAX7-FKHR with distinct DNA binding and transactivation properties in rhabdomyosarcoma. *International Journal of Cancer, 115*(1), 85–92.
43. Souabni, A., Jochum, W., & Busslinger, M. (2007). Oncogenic role of Pax5 in the T-lymphoid lineage upon ectopic expression from the immunoglobulin heavy-chain locus. *Blood, 109*(1), 281–289.
44. Sugg, S. L., Ezzat, S., Rosen, I. B., Freeman, J. L., & Asa, S. L. (1998). Distinct multiple RET/PTC gene rearrangements in multifocal papillary thyroid neoplasia. *The Journal of Clinical Endocrinology and Metabolism, 83*(11), 4116–4122.
45. Di Palma, T., Lucci, V., de Cristofaro, T., Filippone, M. G., & Zannini, M. (2014). A role for PAX8 in the tumorigenic phenotype of ovarian cancer cells. *BMC Cancer, 14*, 292. https://doi.org/10.1186/1471-2407-14-292.
46. Poleev, A., Fickenscher, H., Mundlos, S., Winterpacht, A., Zabel, B., Fidler, A., Gruss, P., & Plachov, D. (1992). PAX8, a human paired box gene: Isolation and expression in developing thyroid, kidney and Wilms' tumors. *Development, 116*(3), 611–623.
47. Human Protein Atlas. Retrieved from www.proteinatlas.org.
48. Yusuf, D., Butland, S. L., Swanson, M. I., Bolotin, E., Ticoll, A., Cheung, W. A., et al. (2012). The transcription factor encyclopedia. *Genome Biology, 13*(3), R24. https://doi.org/10.1186/gb-2012-13-3-r24.
49. Kozmik, Z., Kurzbauer, R., Dörfler, P., & Busslinger, M. (1993). Alternative splicing of Pax-8 gene transcripts is developmentally regulated and generates isoforms with different transactivation properties. *Molecular and Cellular Biology, 13*(10), 6024–6035.
50. Poleev, A., Okladnova, O., Musti, A. M., Schneider, S., Royer-Pokora, B., & Plachov, D. (1997). Determination of functional domains of the human transcription factor PAX8 responsible for its nuclear localization and transactivating potential. *European Journal of Biochemistry, 247*(3), 860–869.
51. de Cristofaro, T., Mascia, A., Pappalardo, A., D'Andrea, B., Nitsch, L., & Zannini, M. (2009). Pax8 protein stability is controlled by sumoylation. *Journal of Molecular Endocrinology, 42*(1), 35–46. https://doi.org/10.1677/JME-08-0100.
52. Miccadei, S., Provenzano, C., Mojzisek, M., Natali, P. G., & Civitareale, D. (2005). Retinoblastoma protein acts as Pax 8 transcriptional coactivator. *Oncogene, 24*(47), 6993–7001.
53. Li, C. G., Nyman, J. E., Braithwaite, A. W., & Eccles, M. R. (2011). PAX8 promotes tumor cell growth by transcriptionally regulating E2F1 and stabilizing RB protein. *Oncogene, 30*(48), 4824–4834. https://doi.org/10.1038/onc.2011.190.
54. Di Palma, T., Filippone, M. G., Pierantoni, G. M., Fusco, A., Soddu, S., & Zannini, M. (2013). Pax8 has a critical role in epithelial cell survival and proliferation. *Cell Death & Disease, 4*, e729. https://doi.org/10.1038/cddis.2013.262.
55. Siehl, J. M., Thiel, E., Heufelder, K., et al. (2003). Possible regulation of Wilms' tumour gene 1 (WT1) expression by the paired box genes PAX2 and PAX8 and by the haematopoietic transcription factor GATA-1 in human acute myeloid leukaemias. *British Journal of Haematology, 123*, 235–242.
56. Kang, H.-C., Ohmori, M., Harii, N., Endo, T., & Onaya, T. (2001). Pax-8 is essential for regulation of the thyroglobulin gene by transforming growth factor-β1. *Endocrinology, 142*(1), 267–275. https://doi.org/10.1210/endo.142.1.7918.
57. Hewitt, S. M., Hamada, S., Monarres, A., Kottical, L. V., Saunders, G. F., & McDonnell, T. J. (1997). Transcriptional activation of the bcl-2 apoptosis suppressor gene by the paired box transcription factor PAX8. *Anticancer Research, 17*(5A), 3211–3215.
58. Raman, P., & Koenig, R. J. (2014). PAX8-PPARγ fusion protein in thyroid carcinoma. *Nature Reviews. Endocrinology, 10*(10), 616–623.

59. Toriyama, A., Mori, T., Sekine, S., Yoshida, A., Hino, O., & Tsuta, K. (2014). Utility of PAX8 mouse monoclonal antibody in the diagnosis of thyroid, thymic, pleural and lung tumours: A comparison with polyclonal PAX8 antibody. *Histopathology, 65*(4), 465–472. https://doi.org/10.1111/his.12405.
60. Damante, G., Tell, G., & Di Lauro, R. (2001). A unique combination of transcription factors controls differentiation of thyroid cells. *Progress in Nucleic Acid Research and Molecular Biology, 66*, 307–356.
61. Pasca di Magliano, M., Di Lauro, R., & Zannini, M. (2000). Pax8 has a key role in thyroid cell differentiation. *Proceedings of the National Academy of Sciences of the United States of America, 97*(24), 13144–13149.
62. Ozcan, A., Shen, S. S., Hamilton, C., Anjana, K., Coffey, D., Krishnan, B., & Truong, L. D. (2011). PAX 8 expression in non-neoplastic tissues, primary tumors, and metastatic tumors: A comprehensive immunohistochemical study. *Modern Pathology, 24*(6), 751–764. https://doi.org/10.1038/modpathol.2011.3.
63. Kobayashi, A., Shawlot, W., Kania, A., & Behringer, R. R. (2004). Requirement of Lim1 for female reproductive tract development. *Development, 131*(3), 539–549.
64. Bowen, N. J., Logani, S., Dickerson, E. B., Kapa, L. B., Akhtar, M., Benigno, B. B., & McDonald, J. F. (2007). Emerging roles for PAX8 in ovarian cancer and endosalpingeal development. *Gynecologic Oncology, 104*(2), 331–337.
65. Fabbro, D., Di Loreto, C., Beltrami, C. A., Belfiore, A., Di Lauro, R., & Damante, G. (1994). Expression of thyroid-specific transcription factors TTF-1 and PAX-8 in human thyroid neoplasms. *Cancer Research, 54*(17), 4744–4749.
66. Hung, N., Chen, Y. J., Taha, A., Olivecrona, M., Boet, R., Wiles, A., Warr, T., Shaw, A., Eiholzer, R., Baguley, B. C., Eccles, M. R., Braithwaite, A. W., Macfarlane, M., Royds, J. A., & Slatter, T. (2014). Increased paired box transcription factor 8 has a survival function in glioma. *BMC Cancer, 14*, 159. https://doi.org/10.1186/1471-2407-14-159.
67. Heidarpour, M., & Tavanafar, Z. (2014). Diagnostic utility of PAX8 in differentiation of mullerian from non-mullerian tumors. *Advanced Biomedical Research, 3*, 96. https://doi.org/10.4103/2277-9175.129366.
68. Laury, A. R., Perets, R., Piao, H., Krane, J. F., Barletta, J. A., French, C., Chirieac, L. R., Lis, R., Loda, M., Hornick, J. L., Drapkin, R., & Hirsch, M. S. (2011). A comprehensive analysis of PAX8 expression in human epithelial tumors. *The American Journal of Surgical Pathology, 35*(6), 816–826. https://doi.org/10.1097/PAS.0b013e318216c112.
69. Schaner, M. E., Ross, D. T., Ciaravino, G., Sorlie, T., Troyanskaya, O., Diehn, M., Wang, Y. C., Duran, G. E., Sikic, T. L., Caldeira, S., Skomedal, H., Tu, I. P., Hernandez-Boussard, T., Johnson, S. W., O'Dwyer, P. J., Fero, M. J., Kristensen, G. B., Borresen-Dale, A. L., Hastie, T., Tibshirani, R., van de Rijn, M., Teng, N. N., Longacre, T. A., Botstein, D., Brown, P. O., & Sikic, B. I. (2003). Gene expression patterns in ovarian carcinomas. *Molecular Biology of the Cell, 14*(11), 4376–4386.
70. Marquez, R. T., Baggerly, K. A., Patterson, A. P., Liu, J., Broaddus, R., Frumovitz, M., et al. (2005). Patterns of gene expression in different histotypes of epithelial ovarian cancer correlate with those in normal fallopian tube, endometrium, and colon. *Clinical Cancer Research, 11*, 6116–6126.
71. Cheung, H. W., Cowley, G. S., Weir, B. A., Boehm, J. S., Rusin, S., Scott, J. A., East, A., Ali, L. D., Lizotte, P. H., Wong, T. C., Jiang, G., Hsiao, J., Mermel, C. H., Getz, G., Barretina, J., Gopal, S., Tamayo, P., Gould, J., Tsherniak, A., Stransky, N., Luo, B., Ren, Y., Drapkin, R., Bhatia, S. N., Mesirov, J. P., Garraway, L. A., Meyerson, M., Lander, E. S., Root, D. E., & Hahn, W. C. (2011). Systematic investigation of genetic vulnerabilities across cancer cell lines reveals lineage-specific dependencies in ovarian cancer. *Proceedings of Natiional Academy of Science of United States of America, 108*(30), 12372–12377. https://doi.org/10.1073/pnas.1109363108.
72. Rodgers, L. H., hAinmhire, E. Ó., Young, A. N., & Burdette, J. E. (2016). Loss of PAX8 in high-grade serous ovarian cancer reduces cell survival despite unique modes of action in the fallopian tube and ovarian surface epithelium. *Oncotarget, 7*(22), 32785–32795.
73. de Cristofaro, T., Di Palma, T., Soriano, A. A., Monticelli, A., Affinito, O., Cocozza, S., & Zannini, M. (2016). Candidate genes and pathways downstream of PAX8 involved in ovarian high-grade serous carcinoma. *Oncotarget, 7*(27), 41929–41947.
74. Elias, K. M., Emori, M. M., Westerling, T., Long, H., Budina-Kolomets, A., Li, F., MacDuffie, E., Davis, M. R., Holman, A., Lawney, B., Freedman, M. L., Quackenbush, J., Brown, M., & Drapkin, R. (2016). Epigenetic remodeling regulates transcriptional changes between ovarian cancer and benign precursors. *JCI Insight, 1*(13), e87988.
75. Kar, S. P., Adler, E., Tyrer, J., Hazelett, D., Anton-Culver, H., Bandera, E. V., Beckmann, M. W., Berchuck, A., Bogdanova, N., Brinton, L., Butzow, R., Campbell, I., Carty, K., Chang-Claude, J., Cook, L. S., Cramer, D. W., Cunningham, J. M., Dansonka-Mieszkowska, A., Doherty, J. A., Dörk, T., Dürst, M., Eccles, D., Fasching, P. A., Flanagan, J., Gentry-Maharaj, A., Glasspool, R., Goode, E. L., Goodman, M. T., Gronwald, J., Heitz, F., Hildebrandt, M. A., Høgdall, E., Høgdall, C. K., Huntsman, D. G., Jensen, A., Karlan, B. Y., Kelemen, L. E., Kiemeney, L. A., Kjaer, S. K., Kupryjanczyk, J., Lambrechts, D., Levine, D. A., Li, Q., Lissowska, J., Lu, K. H., Lubiński, J., Massuger, L. F., McGuire, V., McNeish, I., Menon, U., Modugno, F., Monteiro, A. N., Moysich, K. B., Ness, R. B., Nevanlinna, H., Paul, J.,

Pearce, C. L., Pejovic, T., Permuth, J. B., Phelan, C., Pike, M. C., Poole, E. M., Ramus, S. J., Risch, H. A., Rossing, M. A., Salvesen, H. B., Schildkraut, J. M., Sellers, T. A., Sherman, M., Siddiqui, N., Sieh, W., Song, H., Southey, M., Terry, K. L., Tworoger, S. S., Walsh, C., Wentzensen, N., Whittemore, A. S., Wu, A. H., Yang, H., Zheng, W., Ziogas, A., Freedman, M. L., Gayther, S. A., Pharoah, P. D., & Lawrenson, K. (2017). Enrichment of putative PAX8 target genes at serous epithelial ovarian cancer susceptibility loci. *British Journal of Cancer, 116*(4), 524–535. https://doi.org/10.1038/bjc.2016.426.

76. Adler, E. K., Corona, R. I., Lee, J. M., Rodriguez-Malave, N., Mhawech-Fauceglia, P., Sowter, H., Hazelett, D. J., Lawrenson, K., & Gayther, S. A. (2017). The PAX8 cistrome in epithelial ovarian cancer. *Oncotarget, 8*(65), 108316–108332.
77. Ghannam-Shahbari, D., Jacob, E., Kakun, R. R., Wasserman, T., Korsensky, L., Sternfeld, O., Kagan, J., Bublik, D. R., Aviel-Ronen, S., Levanon, K., Sabo, E., Larisch, S., Oren, M., Hershkovitz, D., & Perets, R. (2018). PAX8 activates a p53-p21-dependent pro-proliferative effect in high grade serous ovarian carcinoma. *Oncogene*. https://doi.org/10.1038/s41388-017-0040-z.
78. McKnight, R., Cohen, C., & Siddiqui, M. T. (2010). Utility of paired box gene 8 (PAX8) expression in fluid and fine-needle aspiration cytology: An immunohistochemical study of metastatic ovarian serous carcinoma. *Cancer Cytopathology, 118*(5), 298–302.

Single-Cell RNA Sequencing of Ovarian Cancer: Promises and Challenges

7

Shobhana Talukdar, Zenas Chang,
Boris Winterhoff, and Timothy K. Starr

7.1 Introduction

Ovarian cancer (OC) is the most lethal gynecologic malignancy in the developed world, accounting for more than 14,000 estimated new deaths in the United States, 2018 [1]. Despite improvement in overall survival from 37% to 46% [2] in the last three decades with the use of radical surgery and cytotoxic chemotherapy, the mortality rate for patients with advanced ovarian cancer, which is greater than 50% [3], continues to be a growing concern. The inability to cure and halt the progression of this disease has prompted researchers to explore the unique cellular and molecular characteristics of these tumors in an effort to develop tailored treatment for individual patients based on molecular phenotyping of these tumors [4–7]. Over the past several decades, with the use of -omics level technology, we have a much better understanding of the molecular complexity of ovarian cancer, including the systematic analyses of transcriptome, genome, proteome, and epigenome in hundreds of patient samples [8–12]. Based on these -omics level studies, it has been increasingly recognized that ovarian cancer represents a disease that is characterized by cellular heterogeneity caused by multiple molecular and environmental factors such as intratumor evolution, cellular plasticity, and multiple sources of stochastic variability [13]. Profound chromosomal instability, which promotes intratumor heterogeneity often portends poor prognosis [14] and patients harboring such diverse tumor clones have a higher chance of relapse [15]. A key challenge in cancer therapeutics is the detection of rare subpopulations of cells with clonal diversity and genomic instability which can lead to possible drug resistance. High-resolution microarrays and next-generation

S. Talukdar · Z. Chang
Division of Gynecologic Oncology, Department of Obstetrics, Gynecology and Women's Health, University of Minnesota School of Medicine, Minneapolis, MN, USA

B. Winterhoff
Division of Gynecologic Oncology, Department of Obstetrics, Gynecology and Women's Health, University of Minnesota School of Medicine, Minneapolis, MN, USA

Masonic Cancer Center, University of Minnesota, Minneapolis, MN, USA

T. K. Starr (✉)
Division of Gynecologic Oncology, Department of Obstetrics, Gynecology and Women's Health, University of Minnesota School of Medicine, Minneapolis, MN, USA

Masonic Cancer Center, University of Minnesota, Minneapolis, MN, USA

Institute of Health Informatics, University of Minnesota, Minneapolis, MN, USA
e-mail: star0044@umn.edu

H. Schatten (ed.), *Ovarian Cancer: Molecular & Diagnostic Imaging and Treatment Strategies*, Advances in Experimental Medicine and Biology 1330, https://doi.org/10.1007/978-3-030-73359-9_7

sequencing technologies have rapidly increased our understanding of the molecular landscape of epithelial ovarian cancers. However, the vast majority of these studies perform analysis on bulk tissue samples, consisting of thousands to millions of cells, and the resulting data represents an average of all of these cells. Rare cell subpopulations are often undetectable in these analyses. Single-cell sequencing technologies provide means for high-resolution molecular phenotyping of large numbers of individual cancer cells and also cancer-associated stromal and immune cells and enables detailed characterization of intratumor heterogeneity in ovarian cancer patients.

This chapter highlights the contributions of single-cell RNA sequencing technologies in possible clinical management of ovarian cancer we discuss the challenges and future implications.

7.2 Why Consider scRNA-Seq in Ovarian Cancer?

The cancer environment consists of an amalgamation of cell types that coordinately result in growth and metastasis leading to morbidity in the patient. Cancer cells evolve and the cancer environment causes changes in supporting stroma, vascular machinery, and recruitment of various immune cell types. Each cell in this environment is bestowed with its unique genome, transcriptome, epigenome, and proteome. Even genetically identical cells will undergo random fluctuations in the mechanisms driving and regulating transcription and translation leading to stochastic gene expression [16, 17]. This underlying heterogeneity is evident in ovarian cancer and single-cell sequencing analysis now makes it possible to study each cell individually at the transcriptome level.

Previous attempts to analyze the molecular complexity of the tumor environment were based on bulk RNA sequencing (RNA-Seq) methods. This technique measures aggregate molecular abundance of RNA levels across several thousands of cells contained in a cancer biopsy. Lost is the information about what types of cells exist in the tumor, what genetic structural diversity exists, and what functional behavior (e.g., gene expression or protein abundance/function) they exhibit at the individual cellular level, and what interactions they have with the host (e.g., with the host immune system). Some of these limitations can now be overcome using single-cell RNA sequencing (scRNA-Seq), which performs transcriptome-wide RNA analysis on several thousand individual cells present within the cancer microenvironment, with the ability to individually interrogate rare cell types from a single tumor sample. ScRNA-Seq has emerged as a powerful tool to dissect tumor tissues at the single-cell level into various cell types and/or cell states (Fig. 7.1) enabling a cell-by-cell molecular characterization of thousands of cells within a tumor specimen thus capable of providing new insights into the mechanism of platinum resistance, with the potential to discover new therapeutic targets.

7.3 Evolution of RNA Sequencing Techniques

The history of single-cell experiments can be traced back to the late 1900s when the pioneering experiments on cell staining techniques and cytological methods allowed scientists to directly visualize genetic differences on chromosomes in single cells. However, these cytogenetic and immunostaining methods were limited to measuring targeted genes and proteins and gross structural anomalies. Starting in the 1980s, quantitative microarray technologies were developed for measuring genome-wide DNA and RNA information, although these methods suffered from certain inherent drawbacks of requiring too much input material for single-cell analysis. At the same time, PCR technologies were capable of amplifying small targeted regions of the genome using small sample sizes. In an effort to overcome these limitations, whole-transcriptome amplification (WTA) [18] and whole-genome-amplification (WGA) [19, 20] methods were developed to amplify genome-wide DNA and RNA which was an important milestone in the

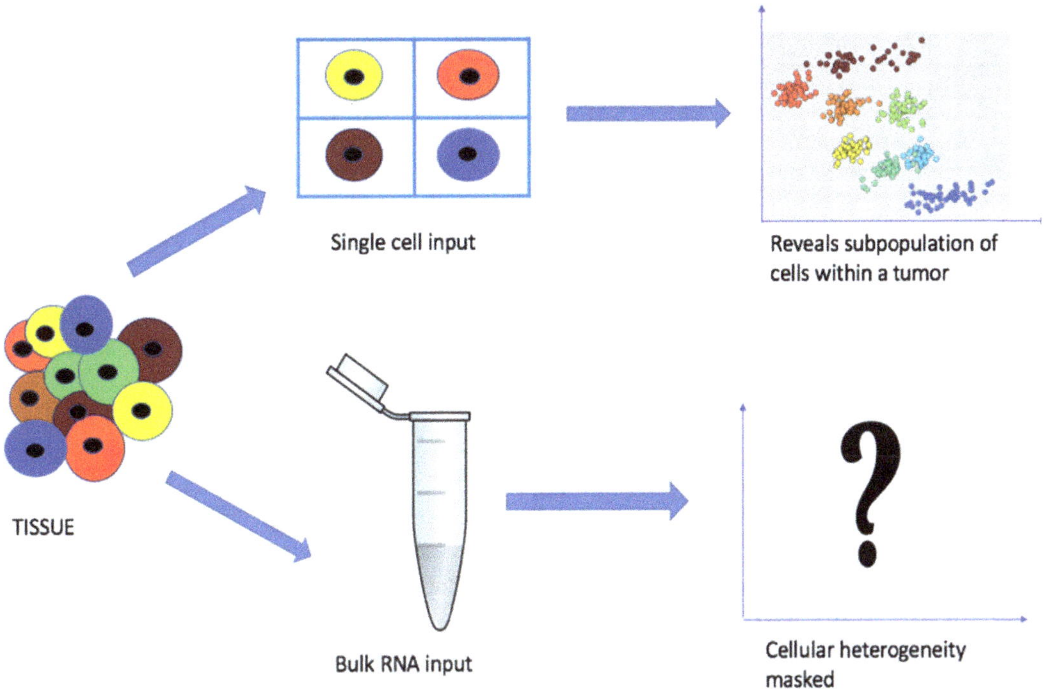

Fig. 7.1 Bulk RNA versus scRNA-seq genomic analysis: scRNA-seq provides the expression profile of individual cells whereas conventional bulk sequencing can provide only the average expression signal for an ensemble of cells, while masking the biologic heterogeneity

field of genomics. The next significant milestone occurred in 2005 with the development of the first next-generation sequencing (NGS) technologies, which enabled genome-wide sequencing of DNA and RNA [21]. It was these pioneering studies that opened the gateway to the new field of single-cell genomics (Fig. 7.2). The first transcriptomes generated via single-cell RNA sequencing (scRNA-seq) were published in 2009, when Tang et al. reported an analysis of the mouse blastomere transcriptome at a single-cell resolution [22]. Two years later, Navin et al. [23] generated the first single-cell genomes from breast cancer samples. Since then, a plethora of scRNA-seq technologies has been developed, providing an unbiased measurement of expression profiles at a single-cell resolution.

7.4 Basic Framework of Single-Cell Isolation Protocols

Each scRNA-seq protocol involves several fundamental steps: (1) sample acquisition from patient; (2) creation of single-cell suspension; (3) cell lysis, mRNA capture, and reverse transcription into cDNA; (4) cDNA amplification by PCR; (5) sequencing of PCR amplicons; and (6) bioinformatic analyses of the sequence data. Capturing single cells may seem trivial but capturing these cells quickly and accurately with high efficiency is one of the main challenges of single-cell sequencing.

Several different protocols have been developed to perform these required steps. Current technologies generally require that the cells be alive immediately preceding step (3), which means that the time required for sample acquisition and creation of the single-cell suspension should be minimized to avoid cell death. Samples are normally acquired from freshly obtained

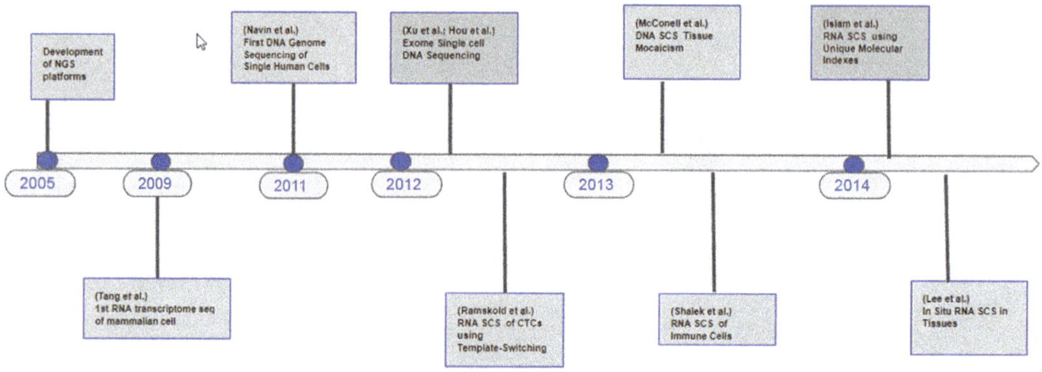

Fig. 7.2 Milestones in single cell sequencing research

biopsies or waste tissue from debulking surgery after clinicopathologic examination. It is helpful to have a trained pathologist identify sections of the tumor that are free from necrosis. The sample to be profiled is dissociated by enzymatic digestion and suspended in a buffer using numerous protocols. Cell dissociation can be very challenging in some cases, as enzymatic treatment of a tissue with trypsin or collagenase may impact cell viability and incubation time required will affect transcriptional activity, potentially adding confounding variables to gene expression. An alternative approach is to perform sequencing on single nuclei, instead of single whole cells, which can reduce artifacts introduced by single-cell dissociation [24].

Isolation of individual cells is the most critical part of the entire process as it is a primary determinant of the throughput of the method. This technique uses several different protocols in order to profile enough cells to capture infrequent or rare cells. Initial methods used manual pipetting of cells into wells, Flow Activated Cell Sorting (FACs) to isolate single cells into plates or microfluidic chips (e.g., Fluidigm C1) to capture single cells in nanoliter chambers which used to be laborious and were prone to error [25–27]. More recent technologies such as droplet-based microfluidics and nanowell-based technologies to randomly capture single cells into isolated nanoliter compartments (droplets or nanowells) are more powerful increasing throughput to tens of thousands of cells thus reducing manual labor to a significant extent [28, 29].

Once cells are isolated, they are lysed to capture as many RNA molecules as possible. A single-cell can only supply very limited starting material (about 0.1 pg of mRNA in each cell), so amplification methods are needed to produce high fidelity, high coverage, and reliable data [18]. In order to specifically analyze polyadenylated mRNA molecules, and to avoid capturing ribosomal RNAs, poly[T]-primers are commonly used. Analysis of non-polyadenylated mRNAs is typically more challenging and requires specialized protocols [30, 31]. Next, poly [T]-primed mRNA is converted to complementary DNA (cDNA) by reverse transcriptase. The poly[T]-primers usually contain one or more barcodes (short unique sequences of DNA), which serve as molecular tags for the amplified mRNA, and can be used to assign sequences to individual cells and individual transcripts. For example, droplet-based sequencing uses a bead coated with millions of poly[T]-primers. All of the poly[T]-primers attached to a single bead will have a cell barcode that is the same on all primers, and then a second barcode called a Unique Molecular Identifier (UMI) which is unique to each primer on the bead. The first barcode is used to assign the sequence to the cell and the UMI barcode is used to assign the sequence to a specific mRNA transcript. This technique allows for digital molecule quantification. Incorporation of UMI also reduces amplification bias such as overrepresentation or underrepresentation of certain regions of cDNA [32]. The minute amounts of cDNA are then amplified either by

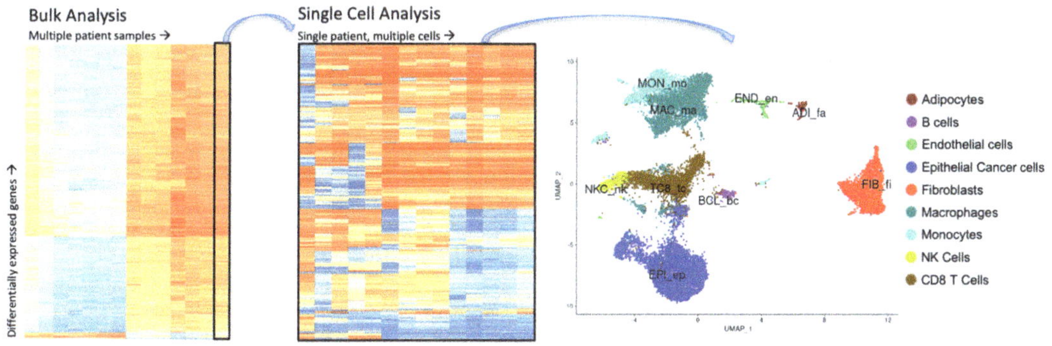

Fig. 7.3 Real-life single cell analyses data from a viable omental metastases obtained during primary surgery from an ovarian cancer patient treated at the University of Minnesota. The detailed information about the number, gene function, pathway function, of individual cells and cell types, as well as their interactions is preserved in scRNA-seq, compared to the bulk sample

PCR or, in some instances, by in vitro transcription followed by another round of reverse transcription. After barcode tagging and PCR amplification, the PCR amplicons are sequenced using Next-Generation Sequencing (NGS) platforms. Single-cell transcriptome data are then analyzed using statistical techniques for clustering cells and identifying cell types, cell states, and signaling pathways (Fig. 7.3).

7.4.1 Single-Cell RNA Sequencing Data Analysis

Numerous specialized software packages are available for single-cell data analysis: this data analysis consists of several tasks which can be completed using various bioinformatic approaches [33–37]. There has been an explosion of published methods for analyzing scRNA-Seq data. The software development website, Github, hosts a useful database of software tools developed for scRNA-seq (https://www.scrna-tools.org). The basic analyses tasks can be grouped into four broad phases: data acquisition, data cleaning, cell assignment, and gene identification. In Phase 1 (data acquisition) raw nucleotide sequencing reads are de-multiplexed and mapped to the transcriptome. Quantification based on read count or UMI is then used to calculate gene expression of each gene in each cell. The data is then "cleaned" (Phase 2), which involves removal of low-quality cells and uninformative genes. The data can also be scaled, normalized and missing values imputed during this phase. Phase 3 assigns cells, either in a discrete manner to known (classification) or unknown (clustering) groups or along a continuous trajectory from one cell type to another using selected genes with a high variation of expression in the dataset. There are many methods proposed for this phase and there is currently no consensus on the best method to use. In general, clustering can be based on principle components, imputation, and/or nonnegative matrix factorization. Machine learning techniques have also been incorporated. In Phase 4, informative genes (e.g., differentially expressed, markers, specific patterns of expression) are then identified to explain these groups or trajectories. Despite the availability of numerous tools, there is little agreement on which method is the best one to use for single-cell analysis [38].

7.5 Applications of Single-Cell RNA Sequencing in Ovarian Cancer

Single-cell RNA sequencing technology has been used in various cancer types to address a range of general and specific research questions related to etiology, evolution, and treatment. Most notable work using this technology has been carried out in the field of melanoma, breast, and brain cancer [39–41]. Its applications in ovarian cancer are now emerging and will likely soon expand our knowledge in the understanding of the disease in the following key areas:

7.5.1 Characterizing the Heterogeneity of a Tumor Population and Identifying Cell Types

Intratumor heterogeneity poses a major challenge to cancer diagnosis and treatment and has been identified as a major cause of treatment failure and drug resistance in ovarian cancer [17]. Tumors evolve from normal cells. During this process, the cancer cells accumulate genomic and epigenomic alterations and evolve to form distinct lineages and subpopulations. This heterogeneity can be seen as variability between tumors across patients, wherein different stages, genetic lesions, or expression programs are associated with distinct outcomes or therapeutic responses [42–44]. In addition, cells from the same tumor are also very diverse, displaying different mutations or distinct phenotypic or epigenetic states [23, 45–47]. This heterogeneity is especially important to study as we recognize the significance of individual or subsets of clones to treatment resistance and tumor recurrence. Bulk sequencing of such tumor samples has inherent challenges, as small subpopulations of tumor cells are masked when analyzed alongside other low and high RNA expressing cells within a bulk tumor sample. Single-cell sequencing of tumor cells has been able to successfully address this issue by identifying subpopulations of cancer cells within a single patient thus facilitating the ability to characterize intratumor heterogeneity.

7.5.1.1 Identification of Cell Subtypes and Molecular Subtypes

We performed scRNA-Seq on 66 individual cells isolated from a tumor specimen obtained from a patient with high-grade serous ovarian cancer during primary debulking surgery [48]. Before sequencing, cells were sorted by FACS to remove cells with major immune markers including CD45 and CD3. After removing the immune cells, we performed scRNA-seq on the remainder using the Fluidigm C1 platform, which uses a microfluidics system to separate cells into 96 single cell wells. We were able to use the single-cell gene expression data generated in this experiment to classify cells as epithelial or stromal cells. Importantly, within each subgroup, we were able to identify individual cells with unique gene expression patterns that differed from the other cells. For example, we were able to identify a few cells that expressed genes associated with ovarian cancer stem cells.

Many groups have attempted to classify ovarian cancer patients based on bulk RNA-seq gene expression patterns [8, 11, 12, 48, 49]. Four consensus molecular subtypes have been defined based on this work. The subtypes, labeled Mesenchymal, Differentiated, Immunoreactive, and Proliferative, are defined by differential expression of a core group of ~1000 genes. Interestingly, when we used this set of genes to classify each single cell, we found that all four molecular subtypes were present within the patient's tumor sample. This finding indicates that molecular subtyping may need to be done at the single-cell level to have more reliable prognostic and predictive value.

7.5.1.2 Studying Immune Microenvironment of a Tumor

It has been increasingly recognized that ovarian cancer is an immunogenic tumor. Several immune cell types have been identified as key players which are associated positively or negatively with antitumor activity. The accumulation of tumor-infiltrating lymphocytes (TILs) in ovarian tumor is associated with antitumor activity, while several other immune cell types are correlated with evasion of immune surveillance and promotion of tumor growth, evasion, and metastases. These pro-tumor cells include tumor-associated macrophages (TAMs), regulatory T cells (Tregs), myeloid-derived suppressor cells (MDSCs), and Tumor-associated dendritic cells (tDCs) [50]. Studying the make-up of the immune cell population within ovarian tumor using scRNA-seq will be instrumental in finding more promising immunotherapeutic targets for cancers.

7.5.1.3 Evolution of Therapy Resistance

Although ovarian cancers show initial response to chemotherapy or targeted therapies, the majority eventually develop resistance which is a major clinical obstacle in the treatment of these patients. The precise mechanism by which tumors evolve into chemoresistance clones remains is difficult to ascertain in each individual patient. It is unclear whether resistant clones are rare subpopulations that are preexisting in the tumor mass and gradually evolved after therapy (adaptive resistance) or therapeutic agents directly cause new mutations that confer a resistant phenotype within a tumor (acquired resistance). Studies in other human cancers have shown that intratumor heterogeneity itself might play an important role in resistance evolution [51]. The role of epithelial-to-mesenchymal transition (EMT) and cell plasticity in conferring a resistance phenotype has also been proposed and recent studies have shown that tumor cells may switch to a mesenchymal phenotype in response to chemotherapy [52]. ScRNA-seq in ovarian cancer holds great potential for improving our understanding of resistance evolution in response to therapy.

7.5.2 Identification of Ovarian Cancer Stem Cells/Ovarian Cancer Stem like Cells

There is strong evidence that ovarian cancer is driven and sustained by cancer stem cells (CSC) [53, 54]. These cells frequently exhibit a slow cycling rate making them inherently resistant to standard chemotherapy [55–57] that, by definition, targets actively proliferating cells. Thus, the high frequency of relapse despite optimal cytoreduction and adjuvant chemotherapy in ovarian cancer might be accounted for by a subpopulation of quiescent CSC that survive treatment. These cells then become active in a later phase, playing a crucial role in cancer chemoresistance, metastasis, and tumor recurrence. Although CSCs constitute a few percent of the tumor mass, identifying them will be crucial for improving current cure rates of less than 50% for advanced stage patients. Studies have investigated these CSCs in ovarian cancer however molecular markers that can reliably identify these cells are not well-defined [53, 54, 58]. We analyzed the expression of cancer stem cell markers in our research and identified many cell subtypes, predominantly in the stromal subgroup that expressed markers associated with CSCs [48]. Future research in this area will be instrumental in reliably identifying and quantifying the frequency of CSCs in ovarian cancer. Our preliminary scRNA-seq analysis of viable primary ovarian cancer samples demonstrates the presence of small cancer stem like subpopulations (comprising 1–2% of all cells analyzed).

7.5.3 Circulating Tumor Cells and Metastatic Dissemination

Cancer metastasis is a complex biological process accounting for more than 90% of cancer-related deaths [59]. Several models of metastasis have been proposed (late dissemination, early seeding, and self-seeding) but remain unresolved. Our understanding of the lineage of metastatic cells is limited. The metastatic cascade is a multistep phenomenon that involves invasion into the bloodstream by cancer cells, their survival in the blood as circulating tumor cells (CTCs), followed by extravasation and seeding of metastatic CTCs into distant sites [59]. Ovarian cancer metastasis can also occur by disseminating tumor cells (DTCs) through the peritoneal cavity, independent of the vascular system, followed by attachment and seeding within the peritoneal cavity. In general, it is thought that the majority of CTCs and DTCs die in the bloodstream or upon arrival at a distant site, due to high shear forces and anoikis signals in circulation, immune defense, or limited capability to adapt to a foreign microenvironment [60]. It is therefore crucial to identify features that distinguish DTCs and CTCs that are able to survive and initiate metastasis.

ScRNA-seq has been used to investigate metastatic dissemination and CTCs in the blood. One study identified three distinct gene signatures in CTCs associated with metastasis in melanoma patients [61]. The dissemination of single CTCs and CTC clusters was also studied in metastatic breast cancer patients and mouse models [55] in which plakoglobin was identified as a key regulator of CTC clusters; CTC clusters were found to have increased metastatic potential relative to individual CTCs. Another study found that CTCs express their own extracellular matrix proteins in the blood in metastatic pancreatic cancer patients [62]. The use of single-cell RNA-seq technique to investigate cell lineage of metastatic cells or to study the relevance of CTCs in ovarian cancer has yet to be reported.

7.5.4 Studying Tumor-Derived Exosomes

Another emerging use of scRNA-seq is the study of tumor-derived exosomes, with the goal of gaining insight into several aspects of the tumor microenvironment. Exosomes are membrane bound vesicles, 50–100 nm in size that are secreted by tumor and immune cells for short- and long-distance intercellular communication and are also involved in mediating exchange of protein and genetic material between cells [63]. Thus, exosomes are thought to play an important role in mechanisms of therapeutic response and resistance [63, 64]. Recent studies have attempted to study several components of exosomes, including DNA, RNA, miRNA, and proteins [64]. ScRNA-seq can be adapted to single-exosome RNA sequencing and used as a powerful tool for longitudinally monitoring of RNA expression profiles in circulating exosomes to study changes in immune pathway genes during immunotherapy as well as to study differential patterns of expression between responders and nonresponders [63, 65].

7.6 Published and Ongoing Work in Ovarian Cancer Using ScRNA-Seq

There is a paucity of published literature on utilization of scRNA-seq to understand genetic complexity of high-grade ovarian cancer. Dr. Winterhoff et al. [48] published the first study on gene expression patterns in single cells from a patient with high-grade serous ovarian cancer using scRNA-seq technology. Analysis of the RNA expression patterns on 66 evaluable single cells from a primary tumor identified two major subsets of cells: epithelial and stromal cells. The epithelial group was characterized by proliferative genes including genes associated with oxidative phosphorylation and MYC activity, while the stromal group was characterized by increased expression of extracellular matrix (ECM) genes and genes associated with epithelial-to-mesenchymal transition (EMT). Neither of these groups displayed gene expression patterns associated with chemoresistance which was consistent with clinical history of the patient who showed no evidence of disease recurrence 19 months post-surgery. This study provided a first view of the application of single-cell gene expression analysis in ovarian cancer for understanding the etiology, progression, and drug resistance in ovarian cancer.

Later, Shih et al. studied primary and metastatic tumor tissue samples from women with HGSOC using high-throughput single-cell RNA-seq analysis and found that while there was considerable heterogeneity among primary tumor cells from different patients, the expression profiles of metastatic lesions from different patients were remarkably similar, and were distinct from the primary lesions [66]. The group identified 16 distinct cell populations with specific cells correlated to high-grade tumors, low-grade tumors, benign, and one population unique to a patient with a breast cancer relapse. The proportion of these populations changed from primary to metastatic in a shift from mainly epithelial cells to leukocytes with few cancer epithelial cells in the metastases. Differential gene expression showed myeloid lineage cells were the primary cell group expressing soluble factors in primary samples while fibroblasts did so in metastatic samples.

Izar et al. in their recent work analyzed ascitic fluid samples from high-grade ovarian cancer patients using single-cell RNA-seq technology and explored potential role of JAK/STAT pathway inhibition as a therapeutic option for these women [67].

A European consortium initiated the HERCULES project in 2016. The stated goals are to find solutions to drug resistance in high-grade ovarian cancer via single-cell analysis [68].

7.7 Conclusion

The field of single-cell sequencing is rapidly developing and will undoubtedly play an increasingly important role in the field of precision medicine by shedding light on intratumor heterogeneity, development of treatment resistance, and genetic drivers of metastasis and tumor evolution in ovarian cancer patients. This unprecedented understanding of tumor biology at the single-cell level will be instrumental for developing novel therapies in the future.

Key Points

- Intratumor heterogeneity is an inherent property of ovarian cancer which impacts clinical outcomes.
- Single-cell sequencing technologies enable characterization of intratumor heterogeneity in ovarian cancer.
- To date, single-cell studies on ovarian cancers are scarce, which limits the opportunity to investigate effects on clinical outcomes.
- Larger data sets are needed to establish potential associations between the unique information captured by single-cell sequencing and clinically relevant outcomes in ovarian cancer.

References

1. Siegel, R. L., Miller, K. D., & Jemal, A. (2018). Cancer statistics, 2018. *CA: A Cancer Journal for Clinicians, 68*(1), 7–30.
2. Howlader, N., Noone, A. M., Krapcho, M., Miller, D., Bishop, K., Kosary, C. L., Yu, M., Ruhl, J., Tatalovich, Z., Mariotto, A., Lewis, D. R., Chen, H. S., Feuer, E. J., & Cronin, K. A. (Eds.). *SEER cancer statistics review, 1975–2014*. Bethesda, MD: National Cancer Institute. Retrieved from https://seer.cancer.gov/csr/1975_2014/, based on November 2016 SEER data submission, posted to the SEER web site, April 2017.
3. Ng, J. S., Low, J. J., & Ilancheran, A. (2012). Epithelial ovarian cancer. *Best Practice & Research Clinical Obstetrics & Gynaecology, 26*, 337–345.
4. Garraway, L. A., Verweij, J., & Ballman, K. V. (2013). Precision oncology: An overview. *Journal of Clinical Oncology, 31*(15), 1803–1805.
5. Werner, H. M., Mills, G. B., & Ram, P. T. (2014). Cancer systems biology: A peek into the future of patient care? *Nature Reviews. Clinical Oncology, 11*(3), 167–176.
6. Rubin, E. H., et al. (2014). Developing precision medicine in a global world. *Clinical Cancer Research, 20*(6), 1419–1427.
7. Schwaederle, M., et al. (2015). Impact of precision medicine in diverse cancers: A meta-analysis of phase II clinical trials. *Journal of Clinical Oncology, 33*(32), 3817–3825.
8. Cancer Genome Atlas Research Network. (2011). Integrated genomic analyses of ovarian carcinoma. *Nature, 474*, 609–615.
9. Patch, A. M., Christie, E. L., Etemadmoghadam, D., Garsed, D. W., George, J., Fereday, S., Nones, K., Cowin, P., Alsop, K., Bailey, P. J., et al. (2015). Whole-genome characterization of chemoresistant ovarian cancer. *Nature, 521*, 489–494.
10. Verhaak, R. G., Hoadley, K. A., Purdom, E., Wang, V., Qi, Y., Wilkerson, M. D., Miller, C. R., Ding, L., Golub, T., Mesirov, J. P., et al. (2010). Integrated genomic analysis identifies clinically relevant subtypes of glioblastoma characterized by abnormalities in PDGFRA, IDH1, EGFR, and NF1. *Cancer Cell, 17*, 98–110.
11. Verhaak, R., Tamayo, P., Yang, J. Y., Hubbard, D., Zhang, H., Creighton, C. J., Fereday, S., Lawrence, M., Carter, S. L., Mermel, C. H., et al. (2013). Prognostically relevant gene signatures of high-grade serous ovarian carcinoma. *Journal of Clinical Investigation, 123*, 517–525.
12. Tothill, R. W., Tinker, A. V., George, J., et al. (2008). Novel molecular subtypes of serous and endometrioid ovarian cancer linked to clinical outcome. *Clinical Cancer Research, 14*, 5198–5208.
13. Lee, J. Y., Yoon, J. K., Kim, B., Kim, S., Kim, M. A., Lim, H., et al. (2015). Tumor evolution and intratu-

mor heterogeneity of an epithelial ovarian cancer investigated using next-generation sequencing. *BMC Cancer, 15*, 85.
14. Donati, G. (2016). The niche in single-cell technologies. *Immunology and Cell Biology, 94*, 250–255.
15. Van Dijk, E. L., Auger, H., Jaszczyszyn, Y., & Thermes, C. (2014). Ten years of next-generation sequencing technology. *Trends in Genetics, 30*, 418–426.
16. Li, G. W., & Xie, X. S. (2011). Central dogma at the single-molecule level in living cells. *Nature, 475*, 308–315.
17. Raj, A., & van Oudenaarden, A. (2008). Nature, nurture, or chance: Stochastic gene expression and its consequences. *Cell, 135*, 216–226.
18. Van Gelder, R. N., von Zastrow, M. E., Yool, A., Dement, W. C., Barchas, J. D., & Eberwine, J. H. (1990). Amplified RNA synthesized from limited quantities of heterogeneous cDNA. *Proceedings of the National Academy of Sciences of the United States of America, 87*, 1663–1667.
19. Dean, F. B., Hosono, S., Fang, L., Wu, X., Faruqi, A. F., Bray-Ward, P., Sun, Z., Zong, Q., Du, Y., Du, J., et al. (2002). Comprehensive human genome amplification using multiple displacement amplification. *Proceedings of the National Academy of Sciences of the United States of America, 99*, 5261–5266.
20. Telenius, H., Carter, N. P., Bebb, C. E., Nordenskjold, M., Ponder, B. A., & Tunnacliffe, A. (1992). Degenerate oligonucleotide-primed PCR: General amplification of target DNA by a single degenerate primer. *Genomics, 13*, 718–725.
21. Mardis, E. R. (2011). A decade's perspective on DNA sequencing technology. *Nature, 470*, 198–203.
22. Tang, F., Barbacioru, C., Wang, Y., Nordman, E., Lee, C., Xu, N., Wang, X., Bodeau, J., Tuch, B. B., Siddiqui, A., et al. (2009). mRNA-Seq whole-transcriptome analysis of a single cell. *Nature Methods, 6*, 377–382.
23. Navin, N., Kendall, J., Troge, J., Andrews, P., Rodgers, L., McIndoo, J., et al. (2011). Tumour evolution inferred by single-cell sequencing. *Nature, 472*, 90–94.
24. Lake, B. B., Codeluppi, S., Yung, Y. C., Gao, D., Chun, J., Kharchenko, P. V., et al. (2017). A comparative strategy for single-nucleus and single-cell transcriptomes confirms accuracy in predicted cell-type expression from nuclear RNA. *Scientific Reports, 7*, 6031.
25. Gawad, C., Koh, W., & Quake, S. R. (2016). Single-cell genome sequencing: Current state of the science. *Nature Reviews. Genetics, 17*, 175–188.
26. Navin, N. E. (2014). Cancer genomics: One cell at a time. *Genome Biology, 15*(8), 452.
27. Gierahn, T. M., Wadsworth, M. H., et al. (2017). Seq-Well: Portable, low cost RNA sequencing of single cells at high throughput. *Nature Methods, 14*, 395–398.
28. Vitak, S. A., Torkenczy, K. A., Rosenkrantz, J. L., et al. (2017). Sequencing thousands of single-cell genomes with combinatorial indexing. *Nature Methods, 14*, 302–308.
29. Cao, J., Packer, J. S., Ramani, V., et al. (2017). Comprehensive single-cell transcriptional profiling of a multicellular organism. *Science, 357*(6352), 661–667.
30. Sheng, K., Cao, W., Niu, Y., Deng, Q., & Zong, C. (2017). Effective detection of variation in single-cell transcriptomes using MATQ-seq. *Nature Methods, 14*, 267–270.
31. Fan, X., Zhang, X., Wu, X., Guo, H., Hu, Y., Tang, F., et al. (2015). Single-cell RNA-seq transcriptome analysis of linear and circular RNAs in mouse preimplantation embryos. *Genome Biology, 16*, 148.
32. Kivioja, T., Vähärautio, A., Karlsson, K., Bonke, M., Enge, M., Linnarsson, S., et al. (2011). Counting absolute numbers of molecules using unique molecular identifiers. *Nature Methods, 9*, 72–74.
33. Bacher, R., & Kendziorski, C. (2016). Design and computational analysis of single-cell RNA-sequencing experiments. *Genome Biology, 17*, 63.
34. Wagner, A., Regev, A., & Yosef, N. (2016). Revealing the vectors of cellular identity with single-cell genomics. *Nature Biotechnology, 34*, 1145–1160.
35. Miragaia, R. J., Teichmann, S. A., & Hagai, T. (2017). Single-cell insights into transcriptomic diversity in immunity. *Current Opinion in Systems Biology, 5*, 63–71.
36. Poirion, O. B., Zhu, X., Ching, T., & Garmire, L. (2016). Single-cell transcriptomics bioinformatics and computational challenges. *Frontiers in Genetics, 7*, 163.
37. Rostom, R., Svensson, V., Teichmann, S. A., & Kar, G. (2017). Computational approaches for interpreting scRNA-seq data. *FEBS Letters, 591*(15), 2213–2225.
38. Freytag, S., Tian, L., Lönnstedt, I., Ng, M., & Bahlo, M. (2018). Comparison of clustering tools in R for medium-sized 10x Genomics single-cell RNA-sequencing data. *F1000Res, 7*, 1297.
39. Ramskold, D., Luo, S., Wang, Y. C., Li, R., Deng, Q., Faridani, O. R., Daniels, G. A., Khrebtukova, I., Loring, J. F., Laurent, L. C., et al. (2012). Full-length mRNA-Seq from single-cell levels of RNA and individual circulating tumor cells. *Nature Biotechnology, 30*, 777–782.
40. Aceto, N., Bardia, A., Miyamoto, D. T., Donaldson, M. C., Wittner, B. S., Spencer, J. A., Yu, M., Pely, A., Engstrom, A., Zhu, H., et al. (2014). Circulating tumor cell clusters are oligoclonal precursors of breast cancer metastasis. *Cell, 158*, 1110–1122.
41. Patel, A. P., Tirosh, I., Trombetta, J. J., et al. (2014). Single-cell RNA-seq highlights intratumoral heterogeneity in primary glioblastoma. *Science, 344*(6190), 1396–1401.
42. Yachida, S., et al. (2010). Distant metastasis occurs late during the genetic evolution of pancreatic cancer. *Nature, 467*, 1114–1117.
43. Eppert, K., et al. (2011). Stem cell gene expression programs influence clinical outcome in human leukemia. *Nature Medicine, 17*, 1086–1093.

44. Parsons, D. W., et al. (2008). An integrated genomic analysis of human glioblastoma multiforme. *Science, 321*, 1807–1812.
45. Gerlinger, M., et al. (2012). Intratumor heterogeneity and branched evolution revealed by multiregion sequencing. *The New England Journal of Medicine, 366*, 883–892.
46. Driessens, G., et al. (2012). Defining the mode of tumour growth by clonal analysis. *Nature, 488*, 527–530.
47. Schepers, A. G., et al. (2012). Lineage tracing reveals Lgr5+ stem cell activity in mouse intestinal adenomas. *Science, 337*, 730–735.
48. Winterhoff, B. J., Maile, M., Mitra, A. K., Sebe, A., Bazzaro, M., Geller, M. A., et al. (2017). Single cell sequencing reveals heterogeneity within ovarian cancer epithelium and cancer associated stromal cells. *Gynecologic Oncology, 144*, 598–606.
49. Konecny, G. E., Wang, C., Hamidi, H., Winterhoff, B., Kalli, K. R., Dering, J., et al. (2014). Prognostic and therapeutic relevance of molecular subtypes in high-grade serous ovarian cancer. *Journal of the National Cancer Institute, 106*, dju249.
50. Cai, D. L., & Jin, L. P. (2017). Immune cell population in ovarian tumor microenvironment. *Journal of Cancer, 8*, 2915–2923.
51. Navin, N. E. (2014). Tumor evolution in response to chemotherapy: Phenotype versus genotype. *Cell Reports, 6*, 417–419.
52. Almendro, V., Cheng, Y. K., Randles, A., Itzkovitz, S., Marusyk, A., Ametller, E., et al. (2014). Inference of tumor evolution during chemotherapy by computational modeling and in situ analysis of genetic and phenotypic cellular diversity. *Cell Reports, 6*, 514–527.
53. Shah, M. M., & Landen, C. N. (2014). Ovarian cancer stem cells: Are they real and why are they important? *Gynecologic Oncology, 132*, 483–489.
54. Burgos-Ojeda, D., Rueda, B. R., & Buckanovich, R. J. (2012). Ovarian cancer stem cell markers: Prognostic and therapeutic implications. *Cancer Letters, 322*, 1–7.
55. Chen, W., Dong, J., Haiech, J., Kilhoffer, M. C., & Zeniou, M. (2016). Cancer stem cell quiescence and plasticity as major challenges in cancer therapy. *Stem Cells International, 2016*, 1740936.
56. Nassar, D., & Blanpain, C. (2016). Cancer stem cells: Basic concepts and therapeutic implications. *Annual Review of Pathology, 11*, 47–76.
57. Takeishi, S., & Nakayama, K. I. (2016). To wake up cancer stem cells, or to let them sleep, that is the question. *Cancer Science, 107*, 875–881.
58. Bast, R. C., Hennessy, B., & Mills, G. B. (2009). The biology of ovarian cancer: New opportunities for translation. *Nature Reviews. Cancer, 9*, 415–428.
59. Fidler, I. J. (2003). The pathogenesis of cancer metastasis: The 'seed and soil' hypothesis revisited. *Nature Reviews. Cancer, 3*, 453–458.
60. Micalizzi, D. S., Maheswaran, S., & Haber, D. A. (2017). A conduit to metastasis: Circulating tumor cell biology. *Genes & Development, 31*, 1827–1840.
61. Ramsköld, D., Luo, S., Wang, Y. C., Li, R., Deng, Q., Faridani, O. R., et al. (2012). Full-length mRNA-Seq from single-cell levels of RNA and individual circulating tumor cells. *Nature Biotechnology, 30*, 777–782.
62. Ting, D. T., Wittner, B. S., Ligorio, M., Vincent Jordan, N., Shah, A. M., Miyamoto, D. T., et al. (2014). Single-cell RNA sequencing identifies extracellular matrix gene expression by pancreatic circulating tumor cells. *Cell Reports, 8*, 1905–1918.
63. Théry, C., Ostrowski, M., & Segura, E. (2009). Membrane vesicles as conveyors of immune responses. *Nature Reviews. Immunology, 9*, 581–593.
64. Liu, Y., Gu, Y., & Cao, X. (2015). The exosomes in tumor immunity. *Oncoimmunology, 4*, e1027472.
65. James Hurley, L. H., Brock, G., Sinclair, I., Sullivan, R. J., O'Neill, V. J., Skog, J., & Flaherty, K. (2015). Profiling exosomal mRNAs in patients undergoing immunotherapy for malignant melanoma. *Journal of Clinical Oncology, 33*, e22159.
66. Shih, A. J., Menzin, A., Whyte, J., Lovecchio, J., Liew, A., Khalili, H., et al. (2018). Identification of grade and origin specific cell populations in serous epithelial ovarian cancer by single cell RNA-seq. *PLoS One, 13*, e0206785.
67. Izar, B., Tirosh, I., Stover, E. H., Wakiro, I., Cuoco, M. S., et al. (2020). A single-cell landscape of high-grade serous ovarian cancer. *Nature Medicine, 26*(8), 1271–1279.
68. HERCULES Project: Comprehensive characterization and effective combinatorial targeting of high-grade serous ovarian cancer via single-cell analysis. Retrieved from http://www.project-hercules.eu.

Enforced Expression of METCAM/ MUC18 Decreases In Vitro Motility and Invasiveness and Tumorigenesis and In Vivo Tumorigenesis of Human Ovarian Cancer BG-1 Cells

8

Guang-Jer Wu

Abbreviations

CAM	Cell adhesion molecule
huMETCAM/MUC18	Human METCAM/ MUC18
IHC	Immunohistochemistry
IP	Intraperitoneal
METCAM	Metastasis-regulating cell adhesion molecule
SC	Subcutaneous

8.1 Introduction

Epithelial ovarian cancer (EOC) is the fifth leading cause of female cancers in the USA with a high fatality rate of about 65% [1]. The reason for its high lethality is due to that early disease is mostly asymptomatic thus cancer remains undiagnosed until it has disseminated throughout the peritoneal cavity [2]. While early-stage ovarian cancer can be effectively treated; however, the lack of a good biomarker to detect the disease at an early stage is the major problem. This is because currently the only validated marker for ovarian cancer, CD125, is not a diagnostic or prognostic marker in spite of its presence in the serum of more than 80% of women with ovarian carcinoma [3]. Furthermore, effective therapy for advanced-stage disease is lacking. Major challenges for dealing with ovarian cancer are: (a) ovarian cancer is histologically and molecularly heterogeneous with at least four major subtypes (serous adenocarcinoma (40%), endometrioid adenocarcinoma (20%), mucinous adenocarcinoma (10%), and clear cell carcinomas (5%)) [4, 5], (b) there is a lack of reliable specific diagnostic markers for an effective early diagnosis of each subtype, though molecular signatures of the major subtypes are available [6], and (c) how ovarian cancer emerges and progresses to malignant form remain elusive ([7] for a review). Thus, there is an urgent need for a new diagnostic marker to detect cancer at an early stage and new therapeutic targets designed from a better understanding of the mechanisms in the progression of cancer. Since altered expression of cell adhesion molecules (CAMs) affects the motility and invasiveness of many tumor cells in vitro and metastasis in vivo [8], we have focused on the

G.-J. Wu (✉)
Department of Microbiology & Immunology and Winship Cancer Institute, Emory University School of Medicine, Atlanta, GA, USA

Department of Bioscience Technology and Center for Biomedical Technology, Chung Yuan Christian University, Taoyuan City, Taiwan
e-mail: gwu@emory.edu

H. Schatten (ed.), *Ovarian Cancer: Molecular & Diagnostic Imaging and Treatment Strategies*, Advances in Experimental Medicine and Biology 1330, https://doi.org/10.1007/978-3-030-73359-9_8

expression of METCAM/MUC18 in ovarian tissues and its effects on the development of cancer [9, 10].

METCAM/MUC18, a cell adhesion molecule (CAM) in the immunoglobulin gene superfamily, contains an N-terminal extracellular domain of 558 amino acids, a transmembrane domain, and a short intracellular cytoplasmic domain (64 amino acids) at the C-terminus [11–13], as shown in Fig. 8.1. The extracellular domain of the protein comprises a signal peptide sequence and five immunoglobulin-like domains and one X domain. The protein is highly glycosylated by N-glycan in the extracellular domains. The cytoplasmic domain contains five consensus sequences potentially to be phosphorylated by PKA, PKC, and CK2 [11–13]. Thus, human METCAM/MUC18 is capable of performing typical functions of CAMs, such as governing the social behaviors by affecting the adhesion status of cells and modulating cell signaling. Therefore, an altered expression of METCAM/MUC18 may affect motility and invasiveness of many tumor cells in vitro and tumorigenesis and metastasis in vivo [14].

We have found that METCAM/MUC18 is expressed in normal ovarian epithelial cells [9]. The METCAM/MUC18 levels in normal ovarian tissues, cancerous ovarian tissues, and metastatic lesions in ovaries were 3.6, 9.1, and 4 times that in benign cystadenomas in ovaries, respectively. From the results of immunohistochemistry (IHC) by using our unique chicken anti-human METCAM/MUC18 antibody, we showed that the expression of METCAM/MUC18 in normal ovarian tissues and benign cystadenomas was mostly not located in epithelial cells; whereas the expression of METCAM/MUC18 in cancer and metastatic lesions was mostly located in the epithelial cells. The percentage of cells stained in IHC was increased in ovarian cancers and metastatic lesions when compared to that in normal ovaries and benign cystadenomas. The percentage of cells stained in IHC was also directly proportional to the pathologically higher grades of ovarian cancer and metastatic lesions. Further mechanical studies indicated that the apoptotic index (Bax) was not much different in different ovarian cancer tissues, but the levels of Bcl2, PCNA, phospho-AKT, and VEGF were all elevated in cancerous tissues in comparison with those in normal ovarian tissues and benign cystadenomas in ovaries. Taken together, we concluded that METCAM/MUC18 may be a biomarker for the early detection of the malignant potential of ovarian carcinomas, implicating its central role in promoting the progression of

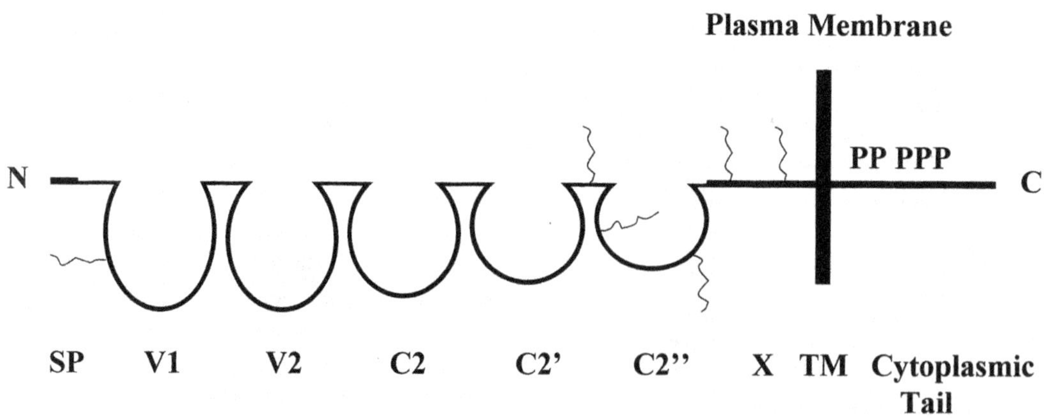

Fig. 8.1 The structure of human METCAM/MUC18 protein. SP stands for signal peptide sequence, V1, V2, C2, C2′, C2″ for five Ig-like domains (each held by a disulfide bond) and X for one domain (without any disulfide bond) in the extracellular region, and TM for transmembrane domain. P stands for five potential phosphorylation sites in the cytoplasmic tail. The six conserved N-glycosylation sites are shown (as wiggled lines) in the V1, between C2′ and C2″, C2″, and X domains

clinical ovarian cancers. Furthermore, elevated anti-apoptosis, proliferation ability, signals in survival pathways, and angiogenesis in ovarian cancer tissues may contribute to the process [9].

However, from these results, we were not clear if the increased expression of METCAM/MUC18 with the malignant progression of epithelial ovarian carcinomas might implicate that it plays an important positive role in initiating tumorigenesis and metastasis of ovarian cancer cells, or the correlation is simply fortuitous and it may play a negative role in the progression of ovarian cancer cells. To test the hypothesis, we used the SK-OV-3 cell line in the previous studies and successfully showed that enforced expression of METCAM/MUC18 decreases tumorigenesis and suppresses the progression of human ovarian cells [15]. Thus, we concluded that METCAM/MUC18 plays a negative role in the progression of ovarian cancer cells. It is not clear if this is only true for the SK-OV-3 cell line or this is generally true for all ovarian cancer cells. In this chapter, we extended our studies to further test the hypothesis by using a different human ovarian cancer cell line, BG-1. The ovarian cancer BG-1 cells were transfected with the wild type METCAM/MUC18 cDNA. G418 resistant ($G418^R$) BG-1 clones expressing different levels of METCAM/MUC18 were isolated, their expression level of the protein was characterized by Western blot analysis, and used for testing in vitro motility, invasiveness, and anchorage-independent colony formation (in vitro tumorigenesis) and in vivo tumorigenesis and metastasis in athymic nude mice. We found that overexpression of METCAM/MUC18 decreased in vitro motility, invasiveness, and in vitro tumorigenesis and in vivo tumorigenesis of nonmetastatic human ovarian cancer cells, BG-1; thus, similar to its effects on SK-OV-3 cells. We conclude that METCAM/MUC18 also plays a suppressor role in the tumorigenesis of BG-1 cells in a xenograft mouse model. Thus, it is generally true that METCAM/MUC18 plays a suppressor role in the progression of human ovarian cancer.

8.2 Materials and Methods

Growth of ovarian cancer cell line BG-1 The human ovarian cancer cell line, BG-1 [16], from Dr. Erin Dickerson at Georgia Institute of Technology, was maintained in DMEM/F12 medium supplemented with 10% fetal bovine serum and penicillin/streptomycin/amphotericin. The media were obtained from Invitrogen/Life Technology/GIBCO/BRL or Cellgro/MediaTech. Fetal bovine serum was from Cellgro/MediaTech. The cell line was maintained in a humidified 37 °C incubator with 5% CO_2. All the other human ovarian cancer cell lines were maintained as previously described [10, 15, 17].

Lipofection of BG-1 cells with the wild type METCAM/MUC18 cDNA gene and selection for human METCAM/MUC18-expressing clones 1×10^6 BG-1 cells were seeded to each 60 mm petri dish (about 60% confluence) one day before lipofection. A standard transfection procedure from the manufacturer (Life Technology) was followed [18, 19]. In brief, 5μg DNA (the wild type METCAM/MUC18 cDNA or the empty vector pcDNA3.1+ DNA) was mixed with 30μg of DMRIE-C (Cat. # 10459-014, Life Technology) in Opti-MEM to form the DNA-lipofection reagent complexes and added to each petri dish. Six hours later, the lipofection solution was removed and replaced with 5 ml of regular growth medium (DMEM/F12 medium with 10% fetal bovine serum). 0.25 mg/ml of G418 (active component 0.2 mg/ml, LD50) was then added and after 6 days it was increased to 0.75 mg/ml to enrich the G418-resistant ($G418^R$) clones, which appeared after about 14 days. Each clone was expanded and huMETCAM/MUC18 expression was determined by Western blot analysis. We found that DMRIE-C was 100 times more efficient than Lipofectamine in transfection of the BG-1 cells, and more than 75% of the $G418^R$-clones obtained by the DMRIE-C transfection expressed a high level of huMETCAM/MUC18.

Determination of in vitro motility and invasiveness A published procedure [20] with slight modifications [11, 18, 19] was followed. To test cell motility, 2×10^5 cells in 0.4 ml of the growth medium containing 0.2% BSA was seeded to each top-well of a 12-well Boyden Transwell system (8.0μm pore size, Falcon #35-3182). 1.2 ml of growth medium containing 10% FBS was added into each bottom well (Fisher #08-771-22 or Falcon 35-3503). After 16 h, cells migrating to the bottom wells were treated with trypsin, concentrated by centrifugation, and counted with a hemocytometer. The mean value and the standard deviation of four measurements of cell numbers migrated to bottom wells were calculated and presented. The data were also analyzed by the student's t-tests.

To test invasiveness, 2×10^5 cells of each clone in 0.4 ml of the growth medium containing 0.2% BSA were initially seeded to each top-well of a 12-well Boyden Transwell system, of which the polycarbonate membrane of 12μm pore size (Fisher Cat # 07200158 or Costar #3403) or sometimes 8μm pore size was coated with 150μg of Matrigel (65μl of 2.3 mg/ml of Matrigel, Becton Dickinson Matrigel Basement membrane Matrix, phenol-red free, Collaborative Research Cat. # 40234C). 1.2 ml of the growth medium containing 10% FBS was added to the bottom well. After 20 h, the cells migrating to the bottom well were treated with trypsin and counted with a hemocytometer as described in the cell motility assay. The mean value and the standard deviation of triplicate values were indicated. The data were also analyzed by the student's t-tests.

Anchorage-independent colony formation assay The published procedures were followed [21] with slight modifications [22, 23]. The 3% noble agar was liquefied by boiling in a water bath and kept in liquid form in a 48 °C water bath until use. An agar plug was formed in each well (with a diameter of 2.2 cm and a surface area of 3.8 cm^2) of a 12-well plate by addition of about 0.95 ml of 0.7% Noble agar, which was prepared by mixing 3% Noble agar with 3.3 volumes of the medium. A monolayer culture of BG-1 cells was treated with trypsin and by pipetting up and down for at least 20–30 times to form single-cell suspension, and the cell numbers were determined by counting in a hemocytometer after staining with 0.5% trypan blue in PBS. 1×10^4 cells in 0.9 ml of medium were mixed with 0.1 ml of 3% Noble agar and seeded into on top of the agar plug in each well of a 12-well plate. The plates were kept in a humidified 37 °C incubator for at least 14 days. The number of colony (20–50 cells per colony) was counted after 14–30 days.

Determination of in vivo tumorigenesis and possible ascites formation in athymic nude mice Ten 40–41 days old female athymic NU/NU (Crl: Foxn <nu>) nude mice from Charles River were used for each clone in the experiment. The approved IACUC protocol according to the NIH animal health care guidelines was followed. 2 or 4×10^6 cells per ml of clone 2-1 (high expression level of WT METCAM/MUC18), clone 2-2 (low expression level of WT METCAM/MUC18), or the empty-vector-transfected pooled clone 3 (as a negative control) were mixed with an equal volume of 10 mg/ml Matrigel [24], and 0.25 ml of the mixture (0.25 or 0.5×10^6 cells) were injected subcutaneously (*SC*) into the dorsal right flank or intraperitoneally (*IP*) into the abdominal cavity of each nude mouse [10, 15, 25, 26]. Each week after *SC* injection, tumor size was determined by a caliper until the end of the experiments (19 weeks) and tumor volume determined by the formula $V = \pi/6(d1 \times d2)^{3/2}$ $(mm)^3$ [20]. Each week after *IP* injection, the size of the abdominal was observed for possible ascites formation and recorded until at the end of the experiment (15 weeks). At the endpoint of the experiment, all mice were euthanatized and tumors at the *SC* sites or tumors in the abdominal cavity were excised. The final tumor weight of each mouse was determined by a balance. Tumors were homogenized in 4 volumes of RB by using a Polytron homogenizer for 15–20 s and mixed with the Western blot buffer for making lysates, as described [9, 15, 18, 19].

Western blot analysis Lysates of cell lines and tumor tissues were prepared as previously described [9, 15, 18, 19]. Protein concentration of each lysate was determined and verified by gel electrophoresis and staining as described [9, 15, 18, 19]. The standard procedure of Western blot analysis with minor modifications was used [9, 15, 18, 19]. 1/200 dilution of the primary antibody (the chicken anti-huMETCAM/MUC18 IgY) [11] and 1/2000 dilution of a secondary antibody (AP-conjugated rabbit anti-chicken IgY antibody (AP162A from Chemicon)) were used for the Western blot analyses. The protein bands on the nitrocellulose membrane were stained with NBT/BCIP. The image of the huMUC18 band from each specimen was scanned with an Epson Photo Scanner model 1260 and its intensity was quantitatively determined by a NIH Image J program version 1.31.

Histology and Immunohistochemistry of ovarian tissue sections Paraffin-embedded tissue sections (5μm) were de-paraffinized, rehydrated with graded alcohol and PBS, and used for histological staining (H&E) and IHC analyses [9, 15, 18, 19]. A tumor section of *SC* tumors derived from a huMETCAM/MUC18-expressing LNCaP clone (LNS239) [18, 19] was used as a positive external control for IHC staining. 1/200 to 1/300 dilution of the chicken anti-huMETCAM/MUC18 IGY antibody was used as the primary antibody and 1/250 dilution of the biotinylated rabbit anti-chicken IGY antibodies (G2891, Promega) as the secondary antibody [9, 15, 18, 19, 22, 23]. A streptavidin-conjugated horseradish peroxidase complex (Dako LSAAB-2 system) and diaminobenzidine were used for color development. Hematoxylin was used as the counterstaining. Negative controls had the primary antibody replaced by nonfat milk or control chicken IGY.

Statistical analysis of data All the data were statistically analyzed by the Student's *t*-test by using the 1 tailed distribution type 1 or 2 methods. Two corresponding sets of data were considered significantly different if the P value was <0.05.

8.3 Results

Expression of huMETCAM/MUC18 in various human ovarian cancer cell lines We determined the expression levels of METCAM/MUC18 in various ovarian cancer cell lines and showed that IOSE, HEY, CAOV3, and OVCAR3 expressed a significant level of the protein when compared to the expression level in a human melanoma cell line, SK-Mel-28, as shown in Fig. 8.2a. We also found that BG-1 cell line expressed no detectable METCAM/MUC18 and SK-OV-3 expressed a very low level of the protein (Fig. 8.2a). Since BG-1 and SK-OV3 cell lines expressed a minimal level of METCAM/MUC18, they should be the two most suitable cell lines to test the effects of enforced expression of METCAM/MUC18 on the tumorigenesis and malignant progression of ovarian cancer cells in vitro and in animal models with anticipation of obtaining the most clear-cut and dramatic effect. Previously we used the SK-OV-3 cell line [15] and in this chapter the BG-1 cell line for the test.

Expression of huMETCAM/MUC18 in various G418^R-BG-1 clones For the studies, we have transfected the huMETCAM/MUC18 cDNA gene (in a mammalian cell expressible vector, pcDNA3.1+) into the human ovarian cell line, BG-1, and used G418 to kill most of the cells that were not successfully transfected with the DNA. The transfection was very efficient by using the DMRIE-C lipofection reagent, similar to that in LNCaP cells [18, 19]. We found that more than 90% of the clones expressed huMETCAM/MUC18. Thus, the pooled clone #2 significantly expressed a high level of METCAM/MUC18 (66%), and the pooled vector control clone #3 did not express any of this protein, as shown in Fig. 8.2b. Figure 8.2b also shows typical examples of several G418-resistant (G418^R)-clones which expressed different levels of METCAM/MUC18: clones 2-1, 2-6, 2-18, and

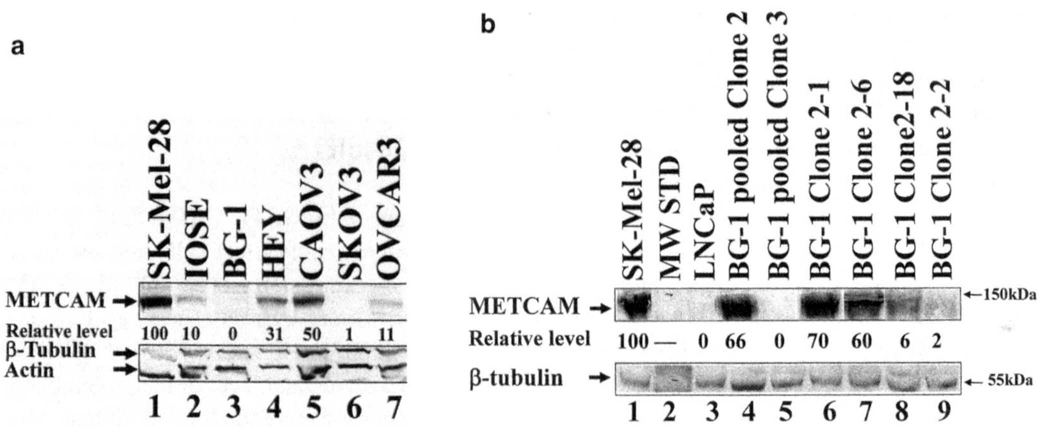

Fig. 8.2 Expression of METCAM/MUC18 in various ovarian cancer cell lines (**a**) and in various BG-1 clones/cells (**b**). (**a**) Shows expression of METCAM/MUC18 in various ovarian cancer cell lines. Lane 1 shows SK-Mel-28 (as a positive control and assumed as 100%), lane 2 IOSE (10%), lane 3 BG-1 (0%), lane 4 HEY (31%), lane 5 CAOV3 (50%), lane 6 SK-OV-3 (1%), and lane 7 NIH OVCAR3 (11%). β-tubulin and actin were used as the loading controls. (**b**) Lane 1 shows SK-Mel-28, a human melanoma cell line (100%). Lane 2 MW Std. Lane 3 shows a human prostate cancer cell line LNCaP (0%). Lanes 4 and 5 show pooled clones of #2 (66%) and #3 (0%). Lanes 6–9 show clones with various expression levels: clone 2-1 (70%), clone 2-6 (60%), clone 2-18 (6%), and clone 2-2 (2%). β-tubulin was used as the loading control

2-2 expressed 70%, 60%, 6%, and 2% of the protein, respectively. Some of these clones were used for testing the effect of METCAM/MUC18 expression on in vitro motility and invasiveness, in vitro tumorigenesis (*anchorage*-independent colony formation in soft agar), and on in vivo tumorigenesis at the SC sites and in the intraperitoneal cavity, as described next.

The effect of enforced expression of METCAM/MUC18 on the in vitro motility and invasiveness of BG-1 clones Figure 8.3a shows that the 2-1 clone, which expressed 70% of huMETCAM/MUC18, had a 1.6-fold less motility than the vector control clone #3, which did not express any of the protein. Figure 8.3b shows that the 2-1 clone, which expressed a high level of huMETCAM/MUC18, only had a 66% invasiveness that of the vector control clone #3, and the 2-2 clone, which expressed a very low level of the protein, had invasiveness slightly less than that of the vector control clone #3. Taken together, we concluded that enforced expression of METCAM/MUC18 reduced the in vitro motility and invasive ability of the BG-1 cells.

The reduced motility and invasiveness of the BG-1 cells appeared to be directly due to the expression of METCAM/MUC18 since both cellular behaviors were significantly decreased in proportion to the dosage of the protein expressed in the cells (comparing the invasiveness of clones 2-1 and 2-2 to clone #3).

The effect of enforced expression of METCAM/MUC18 on in vitro tumorigenesis (anchorage-independent colony formation) Anchorage-independent colony formation in soft agar has been shown to be an in vitro method to determine the tumorigenicity of most cancer cells as an alternative to the determination of the tumorigenicity in model animals (in vivo tumorigenicity) [21]. As shown in Fig. 8.4, in vitro tumor formation of the 2-1 clone, which expressed a high level of METCAM/MUC18, was reduced 5.5-fold compared to that of the empty vector control clone pooled #3, suggesting that enforced expression of METCAM/MUC18 in BG-1 cells repressed the formation of in vitro tumor in the soft agar colony formation assay. However, this notion was further scrutinized by the in vivo tumorigenesis test, as described next.

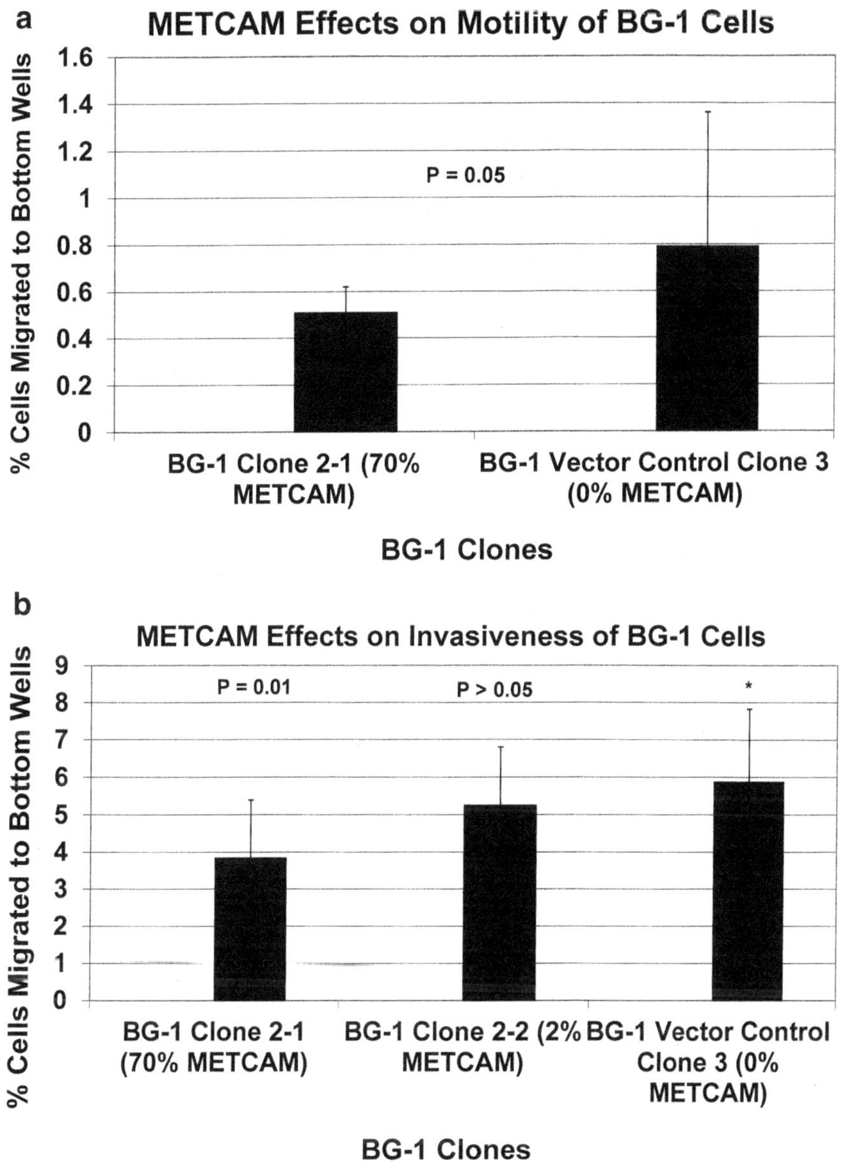

Fig. 8.3 Effects of METCAM/MUC18 expression on the in vitro cellular motility (**a**) and the in vitro cellular invasiveness (**b**) in BG-1 cells. (**a**) Shows the motility of one METCAM/MUC18-expressing clone (clones 2-1) and a vector control clone 3. Cells migrating to the bottom wells were determined as described in Sect. 8.2. (**b**) Shows the invasiveness of two METCAM/MUC18-expressing clones (clones 2-1 and 2-2) and one vector control clone 3. Cells migrating to the bottom wells were determined as described in Sect. 8.2

The effect of enforced expression of METCAM/MUC18 on in vivo tumorigenesis of BG-1 clones To test the effect of METCAM/MUC18 expression on in vivo tumorigenesis and metastasis, the METCAM/MUC18-expressing clones were used for testing their ability to form tumor and ascites in female nude mice [25, 26]. As shown in Fig. 8.5 after *SC* injection (a non-orthotopic route) of the clones/cells the expression of METCAM/MUC18 had a minimal effect on the final tumor weights since the final tumor weights of the METCAM/MUC18-expressing clones and the empty vector control clone were not statistically different. We concluded that

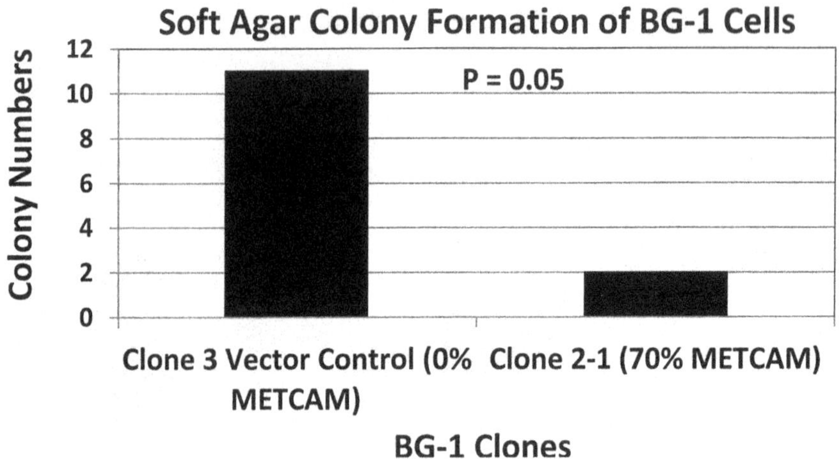

Fig. 8.4 Effects of METCAM/MUC18 expression on the in vitro anchorage-independent colony formation. The in vitro anchorage-independent colony formation assay was used as described in Sect. 8.2

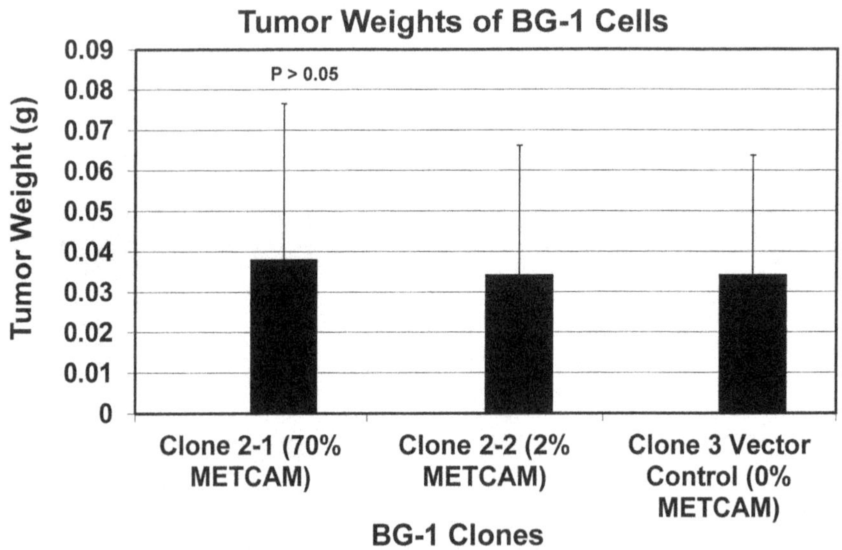

Fig. 8.5 Effects of enforced expression of METCAM in BG-1 cells on in vivo tumorigenicity. The tumorigenicity of the two clones expressed high (clone 2-1) and low (clone 2-2) levels of the huMETCAM/MUC18 and the vector control clone #3 was determined after *SC* injection of the cells as described in Sect. 8.2. Final tumor weights of the three clones were determined and indicated. The tumorigenicity at the (non-orthotopic) *SC* sites were not prominent; thus, it was not surprising that the *P* values (>0.05) were not significantly different

overexpression of METCAM/MUC18 did not significantly affect the in vivo tumorigenesis when non-orthotopic site was injected with the cells.

Figure 8.6 shows that METCAM/MUC18 expression significantly decreased the final tumor weight of BG-1 cells when they were injected intraperitoneally (at the orthotopic site) in nude mice. Taken together, expression of METCAM/MUC18 reduced the in vivo tumorigenesis. However, no ascites were found in the abdominal cavity either from injecting the huMETCAM/MUC18-expressing clones or the empty vector control clone.

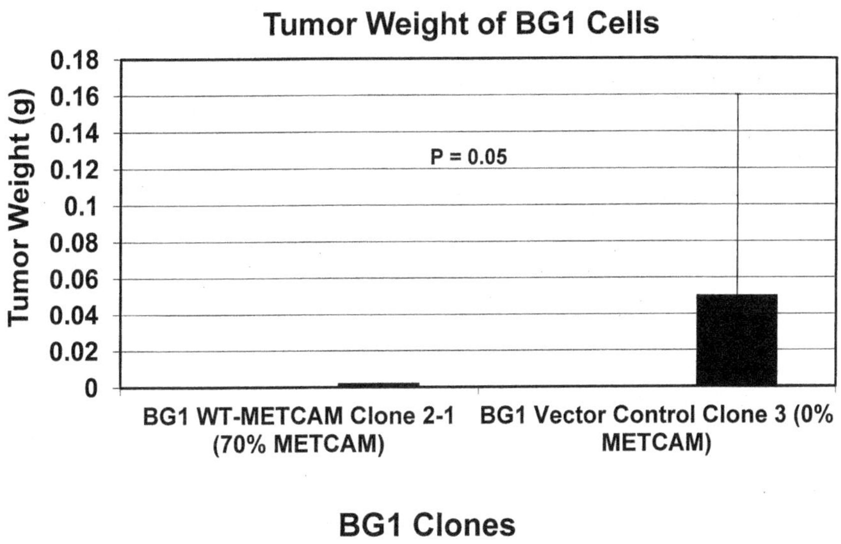

Fig. 8.6 METCAM effects on the tumor proliferation of BG-1 cells in intraperitoneal cavity of nude mice. The tumorigenicity of two clones (2-1 and 3) was determined at the orthotopic site (*IP* injection) as described in Sect. 8.2. The final tumor weights are indicated

Figure 8.7 shows the histology and IHC of tumor sections from *IP* injection of the BG-1 clone/cells which was transfected with the empty vector control (clone #3). As expected, the tumor sections should not be stained with the anti-huMETCAM/MUC18 antibody since the vector control clone did not express any of the protein, suggesting that the tumor was from the injected cells. No histology and IHC results are shown for the tumor sections from *IP* injected huMETCAM/MUC18-expressing BG-1 clone/cells since they did not induce any tumor formation.

8.4 Discussion

In this chapter, we extended our study of the effects of enforced expression of METCAM/MUC18 on the tumorigenesis and progression of human ovarian cancer cell line to BG-1 cell line. We provided evidence to show that a higher level of METCAM/MUC18 expression significantly reduced the in vitro motility and invasiveness and in vitro tumorigenesis (anchorage-independent colony formation) of ovarian cancer BG-1 cells. We also found that a higher level of METCAM/MUC18 expression significantly reduced the tumor proliferation at the orthotopic route (intraperitoneal cavity) in female athymic nude mouse; however, a higher level of METCAM/MUC18 expression did not significantly affect the tumor proliferation at the non-orthotopic route (subcutaneous sites) in the same mouse model. Taken together, we concluded that in addition to SK-OV-3 cells [15], overexpression of METCAM/MUC18 also suppresses the progression of human ovarian cancer BG-1 cells, suggesting this conclusion is generally applicable to human ovarian cancer cells [14]. This conclusion also contradicts the results of a positive correlation of clinical prognosis with the increased expression of METCAM/MUC18 in malignant ovarian cancer specimens [9, 27, 28], suggesting that the positive correlation in this case is fortuitous; thus, we should not assume a positive role of METCAM/MUC18 in the progression of ovarian cancer without the support of tests in an animal model [9, 10, 15].

Surprisingly, we found that BG-1 clones/cells were not as tumorigenic as SK-OV-3 cells either at a non-orthotopic (*SC*) site or at the orthotopic (*IP*) site. We also found that BG-1 clones/cells

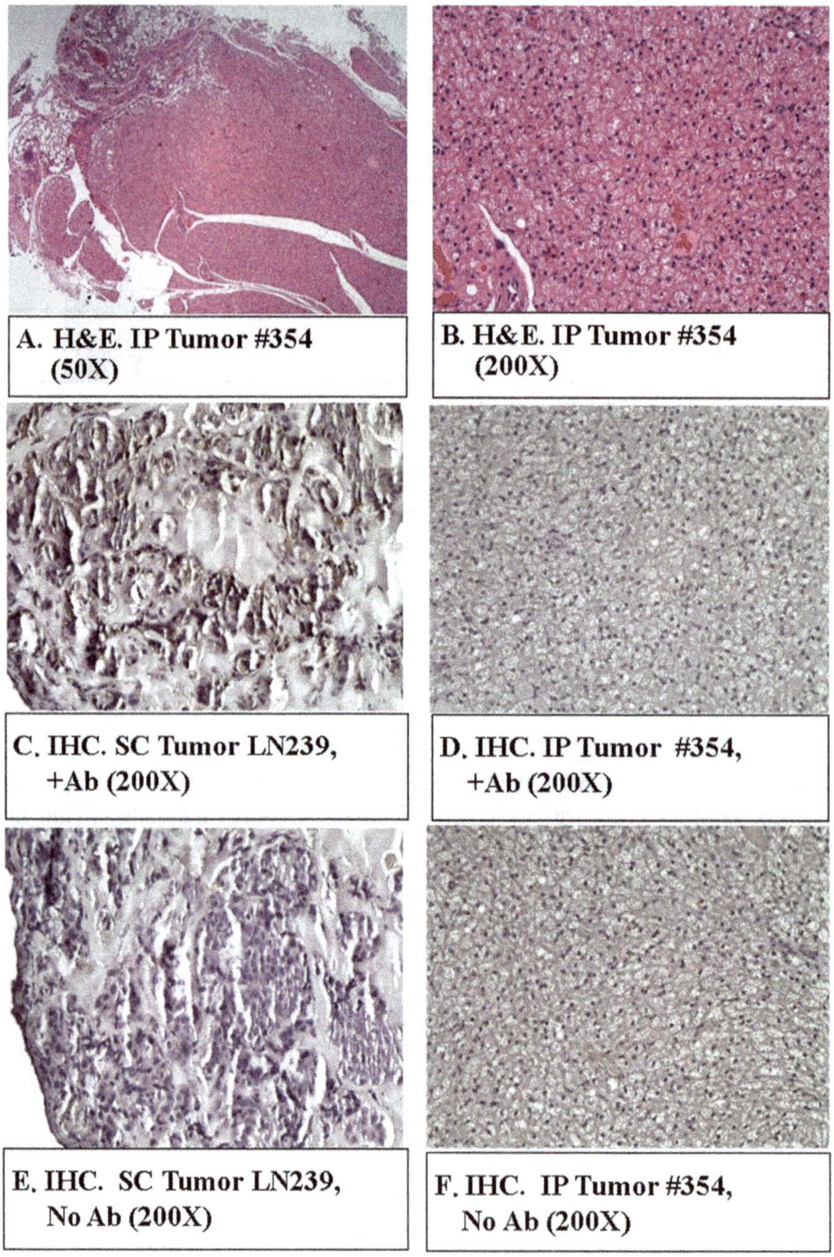

Fig. 8.7 Histology and IHC of tumor sections. The H&E stain and immunohistochemistry (IHC) of tumor sections were determined as described in Sect. 8.2. (**a** and **b**) Show the histology of a tumor from IP injection of the BG-1 clone/cells transfected with the empty vector control (clone #3). (**c** and **d**) Show the IHC stained with anti-METCAM/MUC18 antibody of a tumor section from SC injection of the clone LNCaP 239, as an IHC positive control, and of a tumor section from *IP* injection of the BG-1 clone/cells transfected with the empty vector control (clone #3). (**e** and **f**) Show the IHC with no antibody controls (or control IgY) of the adjacent tumor sections of (**c** and **d**)

did not form ascites in the abdominal cavity when METCAM/MUC18-expressing or control clones/cells were *IP* injected. We did not know the reason for this difference between BG-1 and SK-OV-3 cell lines. One possibility is that the BG-1 cell line, which was established from a poorly differentiated adenocarcinoma [16], may not be as advanced in malignancy as the SK-OV-3 cell line, which was established from an adenocarcinoma metastasis as malignant ascites [17]; thus, requires additionally altered physiological conditions to manifest the effect of huMETCAM/MUC18. One possible condition is that the BG-1 cell line may require estrogen for augmentation of the tumorigenicity since it contains estrogen and progesterone receptors [16]. Other altered physiological factors are not ruled out and may require systematic investigation [25, 26].

The suppressor role played by the overexpression of METCAM/MUC18 in human ovarian cancer cells now has been extended from a mouse melanoma cell line, K1735-9 [29], one nasopharyngeal carcinoma cell line, NPC-TW01 ([14, 30, 31] and Wu, unpublished results), and possibly hemangioma [32]. In contrast, METCAM/MUC18 has also been shown previously to serve as a tumor promoter in both prostate cancer cells [19], breast cancer cells [22, 23], and another NPC cell line, NPC-TW04 ([14, 30, 31] and Wu, unpublished results), and as a metastasis promoter in most melanoma cell lines [14, 33], prostate cancer [18], and breast cancer [34]. This obviously strengthens the notion that METCAM/MUC18 has an intriguing, unique biological function in tumorigenesis and metastasis in that it plays a dual role in the progression of several tumor cell lines [35]. It is not clear why METCAM/MUC18 plays a dual role in tumorigenicity and metastasis. One point is clear in that METCAM/MUC18 plays an opposite role in different cancer types or in different clones/sublines of the same cancer type [14, 35]. Thus, it is logical to propose that the effect of METCAM/MUC18 on the progression of epithelial cancers is modulated by different intrinsic factors in different tumor cells/types. The dual role of METCAM/MUC18 is very likely due to the presence of different interacting partners intrinsic to each cancer cell type and different clone, or perhaps due to different heterophilic ligands, which unfortunately have not been identified [12, 14, 35]. Interactions of METCAM/MUC18 with different sets of intrinsic partners may result in the promotion or suppression of tumorigenicity and metastasis via increasing or decreasing aerobic glycolysis, proliferation, angiogenesis, other growth-promoting pathways, as well as altering tumor cell motility, invasiveness, and vascular metastasis. In the future, the identification of these partners and/or heterophilic ligands is essential to understand further detailed mechanisms [12, 14, 35].

The dual function of METCAM/MUC18 in the progression of human cancers is not an unusual surprise since many biological molecules have recently also been revealed to play a dual role in the progression of cancer. The most well-known examples are TGF-β, which is context-dependent and acts as a tumor suppressor in the early stage of tumorigenesis, but as a progression promoter in the late stage [8], VEGF, which plays a dual role in tumor progression dependent upon the levels of its expression and the context and timing of its modulation [36], and c-myc, which is modulated by different partners to play a dual role in tumor progression [37].

8.5 Perspectives and Clinical Applications

The suppressor role of METCANM/MUC18 in the progression of human ovarian cancer may be intimately associated with tumor dormancy [38]; thus, this may serve as a model to study the mechanisms of tumor dormancy, which may be due to intrinsic growth inhibition, immunological suppression, and/or angiogenic suppression [38]. In light of this, the tumor suppressor role of METCANM/MUC18 in the progression of human ovarian cancer may be useful for clinical treatment of clinical ovarian cancer by using various strategies, as previously described [10], to keep ovarian cancer cells in a dormant state or arresting the cancer cells at the stage of micro-metastases.

Acknowledgments I thank Eugene Lee Son, Jonathan Geekai Chang, and Guofang Zeng for technical supports. This project was supported by the research funding of Emory University School of Medicine, Atlanta, GA, USA and funds of National Science Council, Taiwan.

Conflicts of Interest The author declares no conflicts of interest.

References

1. Siegel, R., Ma, J., Zou, Z., & Jemal, A. (2014). Cancer statistics 2014. *CA: A Cancer Journal for Clinicians, 64*(1), 9–29.
2. Clarke-Pearson, D. L. (2009). Screening for ovarian cancer. *New England Journal of Medicine, 361*, 170–177.
3. Zhao, J., Guo, N., Zhang, L., & Wang, L. (2018). Serum CA125 in combination with ferritin improves diagnostic accuracy for epithelial ovarian cancers. *British Journal of Biomedical Science, 75*(2), 66–70. https://doi.org/10.1080/09674845.2017.1394051.
4. McCluggage, W. G., & Wilkinson, N. (2005). Metastatic neoplasms involving the ovary: A review with an emphasis on morphological and immunohistochemical features. *Histopathology, 47*, 231–247.
5. Wei, W., Dizon, D., Vathipadiekal, V., & Birrer, M. J. (2013). Ovarian cancer: Genomic analysis. *Annals Oncology, 24*(Suppl 10), x7–x15.
6. Menon, U., Griffin, M., & Gentry-Maharaj, A. (2014). Ovarian cancer screening-current status, future directions. *Gynecologic Oncology, 132*, 490–495.
7. Jacob, F., Nixdorf, S., Hacker, N. F., & Heinzelmann-Schwarz, V. A. (2014). Reliable in vitro studies require appropriate ovarian cancer cell lines. *Journal of Ovarian Research, 7*, 60.
8. Harahan, D., & Weinberg, R. A. (2011). Hallmarks of cancer: The next generation. *Cell, 144*, 646–674.
9. Wu, G. J., & Dickerson, E. B. (2014). Frequent and increased expression of human METCAM/MUC18 in cancer tissues and metastatic lesions associates with the clinical progression of human ovarian carcinoma. *Taiwanese Journal of Obstetrics and Gynecology, 53*, 509–517.
10. Wu, G. J. (2017). METCAM/MUC18 plays a novel tumor and metastasis suppressor role in the progression of human ovarian cancer cells. *Obstetrics & Gynecology International Journal, 6*(4), 00210.
11. Wu, G. J., Wu, M. W. H., Wang, S. W., Liu, Z., Peng, Q., Qu, P., Yang, H., Varma, V. A., Sun, Q., Petros, J. A., Lim, S., & Amin, M. B. (2001). Isolation and characterization of the major form of human MUC18 cDNA gene and correlation of MUC18 overexpression in prostate cancer cells and tissues with malignant progression. *Gene, 279*, 17–31.
12. Wu, G. J. (2005). METCAM/MUC18 expression and cancer metastasis. *Current Genomics, 6*, 333–349.
13. Lehmann, J. M., Reithmuller, G., & Johnson, J. P. (1989). MUC18, a marker of tumor progression in human melanoma. *Proceedings of National Academy of Sciences of the United States of America, 86*, 9891–9895.
14. Wu, G. J. (2018). Chapter 13: Dual role of METCAM/MUC18 in the progression of human cancer cells. In F. Uchiumi (Ed.), *Gene expression and regulation in mammalian cells* (pp. 257–289). Rijeka, Croatia. ISBN 978-953-51-3856-3, Print ISBN 978-953-51-3855-6: InTech-Open Access Publisher.
15. Wu, G. J., & Zeng, G. F. (2016). METCAM/MUC18 is a novel tumor and metastasis suppressor for the human ovarian cancer SKOV3 cells. *BMC Cancer, 16*, 136. https://doi.org/10.1186/S12885-016-2181-9.
16. Geisinger, K. R., Kute, T. E., Pettenati, M. J., Welander, C. E., Dennard, Y., Collins, L. A., et al. (1989). Characterization of a human ovarian carcinoma cell line, BG-1, with estrogen and progesterone receptors. *Cancer, 63*, 280–288.
17. Buick, R. N., Pullano, R., & Trent, J. M. (1985). Comparative properties of five human ovarian adenocarcinoma cell lines. *Cancer Research, 45*, 3668–3676.
18. Wu, G. J., Qiong, P., Fu, P., Wang, S.-W., Chiang, C. F., Dillehay, D. L., & Wu, M. W. H. (2004). Ectopical expression of human MUC18 increases metastasis of human prostate cancer cells. *Gene, 327*, 201–213.
19. Wu, G. J., Wu, M. W. H., & Liu, Y. (2011). Enforced expression of human METCAM/MUC18 increases the tumorigenesis of human prostate cancer cells in nude mice. *Journal of Urology, 185*, 1504–1512.
20. Passaniti, A., Isaacs, J. T., Haney, J. A., Adler, S. W., Cujdik, T. J., Long, P. V., & Kleinman, H. K. (1992). Stimulation of human prostatic carcinoma tumor growth in athymic mice and control of migration in culture by extracellular matrix. *International Journal of Cancer, 51*, 318–324.
21. Leone, A., Flatow, U., King, C. R., Sandeen, M. A., Margulies, I. M., Liotta, L. A., & Steeg, P. S. (1991). Reduced tumor incidence, metastatic potential and cytokine responsiveness of nm-23 transfected melanoma cells. *Cell, 65*, 25–35.
22. Zeng, G., Cai, S., & Wu, G. J. (2011). Up-regulation of METCAM/MUC18 promotes motility, invasiveness and tumorigenesis of human breast cancer cells. *BMC Cancer, 11*, 113. https://doi.org/10.1186/1471-2407-11-113.
23. Zeng, G., Cai, S., Liu, Y., & Wu, G. J. (2012). METCAM/MUC18 augments promotes migration, invasion and tumorigenicity of human breast cancer SK-BR-3 cells. *Gene, 492*, 229–238.
24. Pretlow, T. G., Delmoror, A. M., Dilley, G. G., Spadafora, E. G., & Pretlow, T. P. (1991). Transplantation of human prostatic carcinoma into nude mice in Matrigel. *Cancer Research, 51*, 3814–3817.
25. Shaw, T. J., Senterman, M. K., Dawson, K., Crane, C. A., & Vanderhyden, B. C. (2004). Characterization

of intraperitoneal, orthotopic, and metastatic xenograft models of human ovarian cancer. *Molecular Therapy, 10*, 1032–1042.

26. Yi, X. F., Yuan, S. T., Lu, L. J., Ding, J., & Feng, Y. J. (2005). A clinically relevant orthotopic implantation nude mouse model of human epithelial ovarian cancer—Based on consecutive observation. *International Journal of Gynecological Cancer, 15*, 850–855.
27. Aldovini, D., Demichelis, F., Doglioni, C., Di Vizio, D., Galligioni, E., Brugnara, S., et al. (2006). M-CAM expression as marker of poor prognosis in epithelial ovarian cancer. *International Journal of Cancer, 119*(8), 1920–1926.
28. Wu, Z., Wu, Z. Y., Li, J., Yang, X., Wang, Y., Yu, Y., et al. (2012). MCAM is a novel metastasis marker and regulates spreading, apoptosis and invasion of ovarian cancer cells. *Tumor Biology, 33*, 1619–1628.
29. Wu, G. J. (2016). Ectopic expression of MCAM/MUC18 increases in vitro motility and invasiveness, but decreases in vivo tumorigenesis and metastasis of a mouse melanoma K1735-9 subline in a syngeneic mouse model. *Clinical & Experimental Metastasis, 33*(8), 817–828. https://doi.org/10.1007/s10585-016-9812-z.
30. Lin, J. C., Chiang, C. F., Wang, S. W., Wang, W. Y., Kwan, P. C., & Wu, G. J. (2014). Significance and expression of human METCAM/MUC18 in nasopharyngeal carcinoma (NPC) and metastatic lesions. *Asian Pacific Journal of Cancer Prevention, 15*(1), 245–252.
31. Liu, Y. C. (2014). *The putative role of human METCAM/MUC18 in modulating the development and progression of nasopharyngeal carcinoma*. M.S. thesis under the supervision of Dr. Guang-Jer Wu, Department of Bioscience Technology, Chung Yuan Christian University, Chung Li, Taiwan. http://www.lib.cycu.edu.tw/thesis.
32. Li, Q., Yu, Y., Bischoff, J., Milliken, J. B., & Olsen, B. R. (2003). Differential expression of CD146 in tissues and endothelial cells derived from infantile haemangioma and normal human skin. *The Journal of Pathology, 201*, 296–302.
33. Wu, G. J., Fu, P., Wang, S.-W., & Wu, M. W. H. (2008). Enforced expression of MCAM/MUC18 increased *in vitro* motility and invasiveness and *in vivo* metastasis of two mouse melanoma K1735 Sublines in a syngeneic mouse model. *Molecular Cancer Research, 6*(11), 1666–1677.
34. Zeng, Q., Li, W., Lu, D., Wu, Z., Duan, H., Luo, Y., et al. (2012). CD146, an epithelial-mesenchymal transition inducer, is associated with triple-negative breast cancer. *Proceedings of the National Academy of Sciences of the United States of America, 109*(4), 1127–1132.
35. Wu, G. J. (2012). Dual roles of METCAM in the progression of different cancers. *Journal of Oncology, 2012*, 853797. https://doi.org/10.1155/2012/853797.
36. Vecchiarelli-Federico, L. M., Cervi, D., Haeri, M., Li, Y., Nagy, A., & Ben-David, Y. (2010). Vascular endothelial growth factor—A positive and negative regulator of tumor growth. *Cancer Research, 70*, 863–867.
37. Adhikary, S., & Eilers, M. (2005). Transcriptional regulation and transformation by Myc protein. *Nature Review. Molecular Cell Biology, 6*, 635–645.
38. Aguirre-Chiso, J. A. (2007). Models, mechanisms and clinical evidence for cancer dormancy. *Nature Review. Cancer, 7*, 834–846.

3D Models for Ovarian Cancer

9

Verena Kast and Daniela Loessner

Abbreviations

2D	Two-dimensional
3D	Three-dimensional
ECM	Extracellular matrix
GelMA	Gelatin methacryloyl
HGSC	High-grade serous carcinoma
MAL	Maleimide
PARP	Poly-ADP ribose polymerase
PEG	Polyethylene glycol
TME	Tumour microenvironment

V. Kast
Leibniz Institute of Polymer Research Dresden, Max Bergmann Center of Biomaterials Dresden, Hohe Straβe, Dresden, Germany

D. Loessner (✉)
Leibniz Institute of Polymer Research Dresden, Max Bergmann Center of Biomaterials Dresden, Hohe Straβe, Dresden, Germany

Department of Chemical Engineering, Faculty of Engineering, Monash University, Melbourne, VIC, Australia

Department of Materials Science and Engineering, Faculty of Engineering, Monash University, Melbourne, VIC, Australia

Department of Anatomy and Developmental Biology, Faculty of Medicine, Nursing and Health Sciences, Monash University, Melbourne, VIC, Australia
e-mail: daniela.loessner@monash.edu

9.1 Introduction

Ovarian cancer is the deadliest gynaecological malignancy in the Western world [1]. The majority (over 75%) of patients have metastatic disease at the time of diagnosis, and survival rates have not changed over the past four decades. Two reasons for the slow progress in improving survival outcomes are (1) the 'one-size-fits-all' therapeutic approach and (2) the lack of clinically relevant experimental models that mimic the advanced stages of the human disease to find better therapeutic options [2]. In 2015, the first whole-genome study of chemoresistant ovarian cancer was reported [3]. This worldwide, largest DNA analysis revealed key regulatory mechanisms involving interactions with the tumour microenvironment (TME) that pointed to how cancer cells hijack chemotherapy. Thus, there is an urgent need for more targeted strategies and cell models more representative of the TME, specifically new 3D models [4], for treating and studying the most aggressive form—high-grade serous carcinoma (HGSC).

Since then, advances have been made in our understanding of ovarian cancer and the role of the TME in cancer progression and treatment resistance [5, 6]. Scientists need to consider the cellular and extracellular TME when developing experimental cancer models [7]. The majority of our current knowledge about cellular processes

H. Schatten (ed.), *Ovarian Cancer: Molecular & Diagnostic Imaging and Treatment Strategies*, Advances in Experimental Medicine and Biology 1330, https://doi.org/10.1007/978-3-030-73359-9_9

and mechanism has been derived from cancer cells grown attached to flat plastic culture dishes as monolayers, namely two-dimensional (2D) cell cultures [8]. However, 2D cell cultures are limited in terms of their complexity and cellular interactions that govern cell behaviour and drug responses [9]. Dynamic processes, such as epithelial-to-mesenchymal transition, cell invasion, treatment resistance or angiogenesis, cannot be adequately explained [10]. Shifting from 2D to three-dimensional (3D) cell cultures allows for the reconstruction of cell–cell and cell–matrix interactions and other critical TME components and enables studies of physiologically relevant cell behaviours, for example, the dynamics in spatial-temporal oxygen, growth factor gradients and shear stress [11].

Other experimental cancer models include xenograft approaches and murine models of ovarian cancer. These models replicate the human disease in terms of disease development and progression, metastasis and partial immune responses and are useful tools to study responses to treatment [12, 13]. However, murine cell infiltration may lead to mouse-specific tumour evolution, and these models are costly and labour-intensive [14]. In the following sections, we will briefly describe the ovarian TME and discuss some of the clinically relevant experimental approaches that have been used to model selected elements of the TME in 3D.

9.2 Ovarian Cancer

Ovarian cancer is a heterogeneous disease, which makes the design of experimental cancer models even harder. There are different subtypes with distinct biological characteristics and molecular aberrations and treatment strategies are stereotypically applied, in particular aggressive surgical debulking and platinum-based chemotherapy [15].

Epithelial ovarian cancer occurs in over 90% of ovarian malignancies, whereas non-epithelial forms, including germ cell and sex cord-stromal tumours, account for about 5% [16]. Epithelial malignancies develop from the fallopian tubes or other epithelial sites. They are categorised into distinct histo-morphological subtypes: low-grade serous, endometrioid, mucinous and clear cell carcinoma ('type I ovarian tumours') and HGSC, mixed Mullerian malignancies and high-grade endometrioid ovarian carcinomas ('type II ovarian tumours'). Both type I and type II ovarian tumours differ in their point-of-origin, gene mutations, disease progression and clinical outcomes [15]. HGSC is the most frequently diagnosed sub-type, accounting for most ovarian cancer deaths (70–80%), and its point-of-origin is still under debate [6].

At an early stage, ovarian cancer is asymptomatic and specific biomarkers do not exist. Consequently, patients are diagnosed at an advanced stage, after metastatic spread has already occurred. Dissemination of ovarian cancer is a major clinical problem and results in even further reduced survival rates for patients. Metastatic spread occurs through intraperitoneal dissemination within the tumour fluid (ascites) to secondary sites. Ascites is associated with chemoresistance and disease recurrence and contains single cancer cells as well as multicellular cancer spheroids. These cells and spheroids adhere to mesothelium-lined organs, such as the peritoneum, the small and large bowel serosa, or the omentum and invade into the underlying extracellular matrix (ECM) to form macro-metastases [17, 18]. The cellular composition and cellular states of ascites vary significantly between patients, and within a patient's primary tumour and metastatic lesions, and need to be considered when developing 3D models using ascites-derived cells or recreating the ascites-specific TME. In a single-cell analysis of ascites-derived cells from patients with advanced HGSC, diverse subpopulations of immunomodulatory fibroblasts, cancer-associated fibroblasts and macrophages were identified [19].

9.3 3D Models That Recreate The Ovarian TME

Signals between malignant and non-malignant cells give rise to a tissue-specific TME that promotes cancer progression, metastasis and chemoresistance. The ovarian TME is highly complex and consists of various non-malignant cell types, including cells of the tumour vasculature, fibroblasts, adipocytes, mesothelial and inflammatory cells, and extracellular components, for example, the ECM, growth factors, cytokines and proteolytic factors [20]. The intrinsically heterogeneous tumour-immune microenvironment contains co-existing regions with immune-cell-excluded and inflammatory areas within the same patient or tumour site. Cytotoxic chemotherapy has an immunogenic effect and induces immune cell infiltration in HGSC as demonstrated, for example, by an enrichment of natural killer cells or an oligoclonal expansion of T cell subsets [21].

Cancer cells recruit and reprogram these non-malignant cells to create tumourigenic niches. The reciprocal crosstalk between cancer cells and the different TME components supports tumour growth and spread. For example, a common matrix response is associated with metastasis and poor survival [20]. Cancer cell dynamics are intricate and challenging to model experimentally [6]. However, 3D cancer models that accurately replicate the different components and the diversity of the TME are urgently needed to study cancer biology and physiology and to identify more effective treatments to improve the clinical outcomes for this disease.

In recent years, great progress in the fields of 3D in vitro cancer models and tumour tissue engineering have been made. A variety of 3D cancer models and bioengineered microenvironments that integrate key elements of the TME in a spatially and biomechanical relevant manner have been developed. These models enable the study of cancer cell behaviour and drug responses under physiological cell culture conditions [7]. Among others, multicellular cancer spheroids, organoid and organotypic models, as well as hydrogel-based and scaffold-based systems are well established. And new technologies, such as cancer-on-a-chip devices, are gaining more attention within the cancer research community.

9.3.1 Cancer Spheroids

Cancer spheroids are an adequate and versatile tool to culture cells in 3D. They are widely used as 3D in vitro cancer model because they closely resemble in vivo tumours in terms of the 3D structure and organisation of tissues or organs [22, 23]. Spheroids are either self-assembling or are forced to grow as cell clusters or aggregates from a single-cell suspension in the absence or presence of exogenous ECM components [24]. Cancer spheroids range in their diameter from 30 to 750 μm and are referred to as bona fide metastatic units with an outermost layer of proliferating cells and a central area of quiescent cells [14]. Larger spheroids (>200 μm) exhibit gradients of oxygen, nutrient, catabolic and soluble factors, including cytokines and growth factors as seen in physiological micro-metastases and avascular tumours, which make them a great 3D model for cancer research [4, 14].

Ovarian cancer spheroids can be isolated from patient's tumour fluid (ascites) or fabricated by using scaffold-free approaches by which cells produce their own ECM, for example, non-adherent cell culture dishes, hanging drop methods and spinner flask cultures [4]. By using scaffold-based approaches, cells attach and grow on or within polymeric scaffolds, such as matrices and hydrogels [25]. Ovarian cancer spheroids can either be monotypic (malignant cells only) or heterotypic (a mixture of malignant and non-malignant cells).

9.3.2 Hydrogel-Based Models

The ECM provides structural support and triggers biomechanical and biochemical signals that are essential for cancer cell behaviour. ECM properties, including the stiffness, permeability and spatial arrangements, are critical features that influence cancer progression and therapy success and thus need to be considered when developing bioengineered microenvironments. Hydrogels are water-absorbing and water-swollen 3D scaffolds generated from crosslinked biomaterials for the study of cell–cell and cell–matrix interactions. They have physiological properties comparable to native tissues, and their biomechanical and biochemical properties can be widely tailored, which makes them an exciting 3D tool [7].

There are different types of hydrogels, depending on the origin of their main component. Natural hydrogels are generated from naturally derived biomaterials, such as alginate, collagen, hyaluronic acid and Matrigel. Their low cytotoxicity and the presence of cell binding sites make them compatible for 3D cell cultures. However, naturally derived biomaterials have several drawbacks, including a high batch-to-batch variation, undefined and mixed compositions, uncontrolled degradation and poor or low mechanical properties [7]. Synthetic biomaterials, such as polyethylene glycol (PEG) [26] or self-assembling peptide amphiphiles [25], overcome some of these limitations and are applied to increase the experimental reproducibility and to precisely control the biomechanical and biochemical properties. As synthetic biomaterials lack cell adhesion and proteolytic degradation sites, bioactive peptides and molecules are integrated to facilitate cell-stimulatory processes. To combine the biological characteristics of the native ECM with the stable and well-defined properties of synthetic matrices, semi-synthetic hydrogels have been produced [7]. For example, gelatin methacryloyl (GelMA) hydrogels are an alternative 3D model that supports the formation and growth of ovarian cancer spheroids [27]. GelMA hydrogels retain their cell binding and proteolytic degradation sites and have tuneable physical properties, allowing a high degree of experimental control and reproducibility.

The omentum is the primary metastatic site for ovarian cancer. The development of 3D in vitro cancer models that mimic the omental microenvironment may improve the prediction of drug responses. To capture critical omentum-specific ECM protein characteristics [28], omentum-inspired polyethylene glycol-maleimide (PEG-MAL) hydrogels have been developed (Fig. 9.1) [9]. Ovarian cancer cells, allowed to aggregate in microwells, or patient-derived ascites spheroids, were encapsulated into PEG-MAL hydrogels. To support cell viability and proliferation, omentum-specific integrin-binding and ECM-related peptides were added into the hydrogel network. Subsequently, the effects of several anticancer drugs, including the clinically used chemotherapeutic paclitaxel, on cancer spheroids were analysed. While cancer spheroids did not respond to paclitaxel, cell monolayer controls grown on plastic culture dishes were sensitive to paclitaxel and had a reduced cell viability. Moreover, ascites-derived spheroids from patients that had been already treated with paclitaxel did not respond either when grown in PEG-MAL hydrogels. Additionally, cancer spheroids produced their own ECM when cultured in PEG-MAL hydrogels compared to cell monolayers. These findings indicate that the omentum-inspired 3D model may be used as a clinically relevant drug screening platform for ovarian cancer and to identify ECM-related factors that are involved in drug resistance [9].

9.3.3 Organoids

Organoid cultures have been used to study human development and diseases, as well as clinically relevant drug screening platforms and as models of the TME. Patient-derived tumour organoids (or tumouroids) are a great 3D tool for cancer research as they reconstruct the tumour profile in terms of the morphology and gene expression from which they originate [6]. They are established from primary human tumour cells, or murine oviductal (fallopian tube in humans) and ovarian surface epithelium cells that harbour mutagenic modifications [29], and form 3D structures to recapitulate and study tumour heterogeneity and the origin of HGSC [30]. In the presence of growth factors, small molecules and a supporting matrix, mostly Matrigel and collagen gels, organoid cultures are maintained over several months. Organoids grow within days and allow untransformed and precancerous cells to expand [31]. Organoid cultures are cheaper and easier to establish compared to patient-derived xenografts and murine models of ovarian cancer [30].

About 50% of HGSC harbour DNA repair defects, which are targeted by inhibition of a nuclear enzyme, poly-ADP ribose polymerase (PARP). PARP is an important protein that repairs

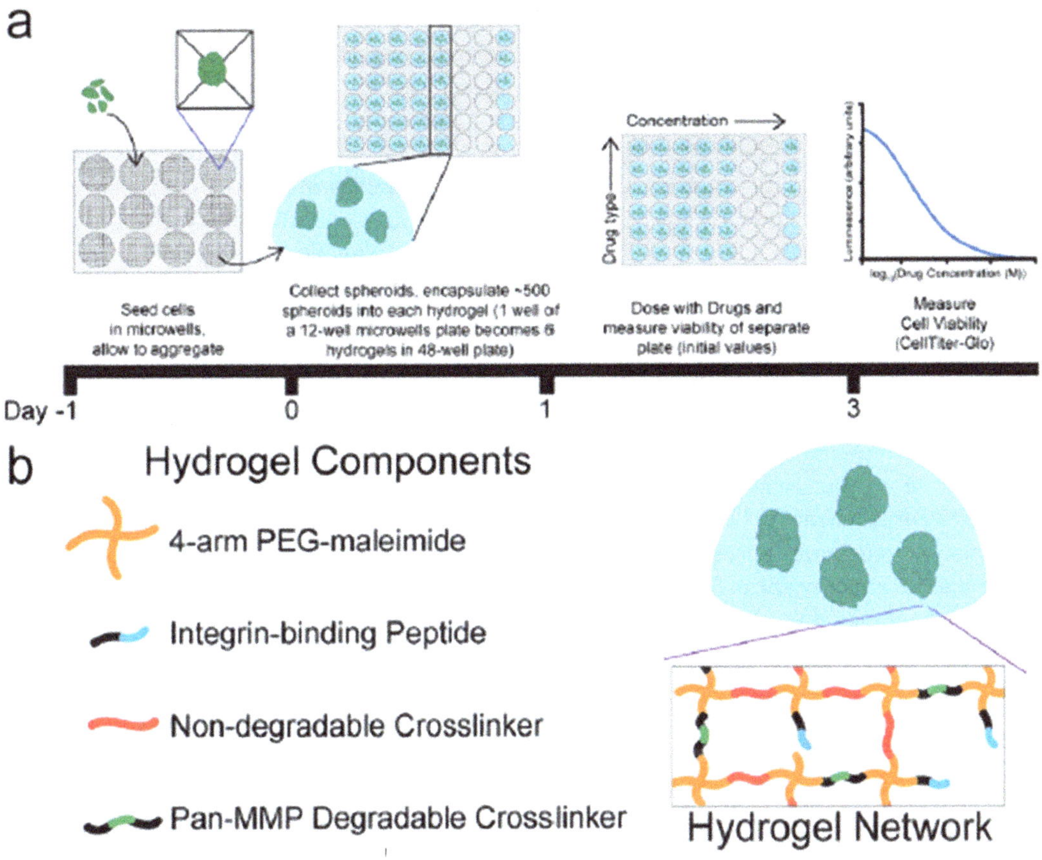

c

Entrez Gene Symbol	Peptide Sequence	Mole % in Cocktail	Amount in Cocktail (mM)
COL1A1, COL1A2	GFOGER	85.5	12.3
FN1	PHSRN-RGD	2	0.655
FBN1, FGA, FGB, FGG, NID1, NID2, VTN, VWF, COL6A1, COL6A2, COL6A3, COL6A5	RGD	12	8.04
COL4A1, COL4A2	DGEA	0.5	0.376

Fig. 9.1 3D culture of ovarian cancer cells using polyethylene glycol-maleimide (PEG-MAL) hydrogels. Cancer cells seeded in microwells aggregate and form cancer spheroids, which are then collected and grown encapsulated in PEG-MAL hydrogels. The components of the hydrogel network include PEG-MAL, integrin-binding peptides and two different crosslinkers. To recreate the omental microenvironment, different peptide sequences that represent extracellular matrix proteins (e.g. collagen, fibronectin) of the human omentum are crosslinked into the hydrogel network [9]

damaged DNA. One of the DNA damage repair mechanism is homologous recombination involving the breast cancer susceptibility genes, *BRCA1* and *BRCA2*. Deficiency, such as mutations, within either gene results in defective homologous recombination, loss of efficient DNA repair and responsiveness to PARP inhibition. For functional profiling of DNA repair and defects in homologous recombination, HGSC patient-derived organoids were established. Regardless of the mutational profile of the DNA repair genes, a functional defect in homologous recombination in the organoids positively correlated with PARP inhibitor sensitivity. In combination with genomic screenings, the functional testing of ovarian tumour organoids is a valid 3D tool for the identification of targetable defects in the repair of DNA damage [30].

In another study, 56 organoids from 32 ovarian cancer patients were established, with a success rate of 65%. For the first time, the generated organoid lines covered all major ovarian cancer subtypes. Moreover, patient-specific genomic features were maintained. These organoid cultures allowed long-term expansion and manipulation, thus offering a platform for drug screening approaches for the different ovarian cancer subtypes [6].

9.3.4 Organotypic Cultures

Organotypic models are composed of multiple cell types found in the cellular TME and an organ-specific ECM to mimic tumour tissues as seen in patients [32]. Ovarian cancer cells preferentially metastasise to the omentum, which is lined by a layer of mesothelial cells. To capture this omental microenvironment, a 3D organotypic model was developed (Fig. 9.2) [18]. Primary omental fibroblasts were mixed with a collagen matrix, followed by addition of mesothelial cells and subsequently co-cultured with fluorescently labelled ovarian cancer cells. This model was used to screen a compound library for their potential to inhibit cell functions in an automated and quantitative high-throughput screen using different HGSC cell lines. Over 44,000 compounds and pharmacologically active small molecules were tested, and only 3 compounds were found to inhibit ovarian cancer cell adhesion, invasion and metastasis, to prolong survival and to reduce omental tumour growth [5].

Ovarian cancer initiation and progression, including changes in gene expression during early cancer cell dissemination, are poorly understood. To analyse early events in ovarian cancer spread, the aforementioned 3D organotypic model was used. A comprehensive RNA sequencing analysis of healthy fallopian tubes, primary tumours and metastatic lesions was compared with the profiles of their cultures using the 3D organotypic model to identify changes in gene expression. Significant changes in gene expression and key pathways during ovarian cancer initiation, metastasis and early colonisation were identified, which includes the deregulation of ECM proteins and ECM-related factors [33]. The results may help to improve our understanding of critical pathways and their role in ovarian cancer progression in order to develop new and more effective treatment options.

9.3.5 Scaffold-Based Organotypic Cultures

The organ-specific characteristics of the cellular and extracellular TME in disease progression are important, in particular, when designing hydrogel- and scaffold-based 3D models of the omental microenvironment. Tissue engineering approaches that have been used for regenerative medicine can be repurposed and applied to cancer research. Using this interdisciplinary strategy, controllable and reproducible 3D organotypic models have been developed [7]. For example, a 3D TME model was designed to mimic the integral steps in the dissemination of ovarian cancer and its spread to the omentum. Hereby, ovarian cancer spheroids, which formed within PEG hydrogels, were assembled with medical-grade polycaprolactone fibrous scaffolds that were seeded with a layer of mesothelial cells to create a 3D co-culture model (Fig. 9.3) [34]. These 3D co-cultures were carried out for 2 weeks and then, both cell types were mechanically separated for subsequent molecular profiling. Proliferation assays and a high-throughput gene expression and signalling analysis indicated that cancer spheroid growth was enhanced upon 3D co-culture compared to the corresponding 3D monoculture controls and that genes linked to cell growth (e.g. *IGFBP7*,

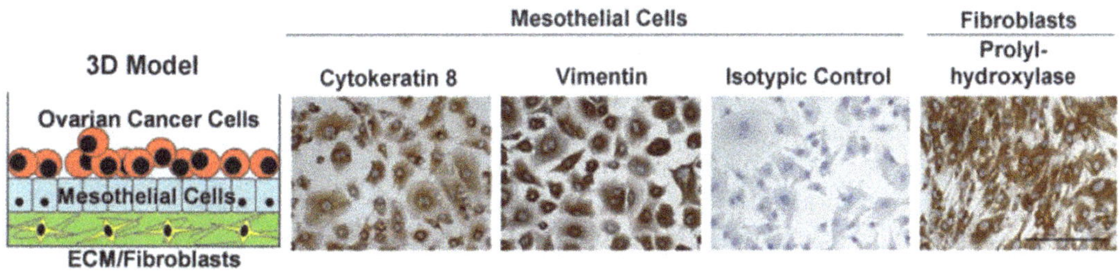

Fig. 9.2 3D co-culture of ovarian cancer cells with non-malignant cells using an organotypic model. The 3D organotypic model mimics the human omentum through the 3D co-culture of ovarian cancer cells with mesothelial cells and fibroblasts in a collagen matrix. Human primary mesothelial cells and fibroblasts are extracted from human omentum and assessed for the presence of cell type-specific markers (e.g. cytokeratin 8, vimentin and proline-hydroxylase). To mimic the omental basement membrane, patient-derived omental fibroblasts are mixed with a collagen matrix. After cell adhesion, mesothelial cells are seeded on top to reconstruct the mesothelial lining. Subsequently, fluorescently labelled ovarian cancer cells are added and subjected to fluorescence-based assays [18]

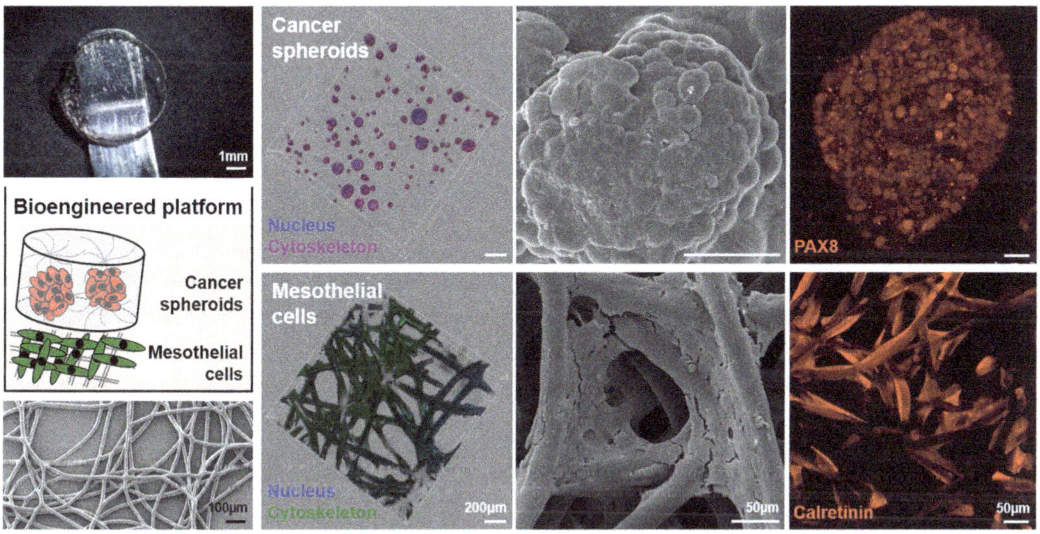

Fig. 9.3 3D co-culture of ovarian cancer cells with mesothelial cells using a combination of polyethylene glycol (PEG) hydrogels and fibrous scaffolds. To recreate the omental microenvironment, fibrous scaffolds are seeded with mesothelial cells and assembled with ovarian cancer cell-containing PEG hydrogels to form tumour constructs. This combined hydrogel/scaffold model enables the separation of the individual cell types and analysis of low cell yields from 3D co-cultures. Upon cell separation, the effect of 3D co-culture is assessed by confocal and scanning electron microscopy. The presence of cell type-specific markers (e.g. PAX8 and calretinin) confirms the complete separation of the different cell populations. (Modified from [34])

FGF2, *VEGFC*, *COX2*) and proteolytic factors (e.g. *KLK5*, *KLK6*, *KLK7*) were increased in 3D co-cultured cancer spheroids. Upon implantation of 3D co-culture constructs intra-peritoneally into NOD/SCID mice, tumour growth and spread were significantly increased compared to 3D monoculture implants [34]. This tailored and clinically relevant experimental model recreates the organ-specific pattern of early ovarian cancer dissemination within the peritoneal cavity. It represents a quantitative 3D approach to identify the regulatory cellular and molecular mechanisms involved and may help to identify targeted therapies to increase survival rates or to molecular stratify the design of clinical trials for a subset of patients with HGSC.

9.3.6 Cancer-on-a-Chip Devices

A variety of experimental models have been developed to study the tumour biology and drug responses of ovarian cancer. Some of the early 3D in vitro cancer models are often oversimplified and unsuited to accurately mimic the complexity of the TME. Consequently, more complex 3D in vitro cancer models, such as multicellular cancer spheroids, organoids and organotypic systems evolved. To overcome their drawbacks, for example, the lack of tissue–tissue interfaces, fluid flow and biochemical cues, new tools, namely organ-on-a-chip devices, have evolved [35]. Organ-on-a-chip models provide cells with fluidic stimuli by perfusing medium in a laminar flow through a porous membrane that separates the individual compartments. This allows the fluidic 3D co-culture of different cell populations to recapitulate complex tissue–tissue interactions. A unique advantage of this technology is its inherent ability to integrate multiple organ functions into a closed microfluidic system, which represents the physiology and metabolism as seen in native tissues, allowing disease modelling and preclinical drug studies [36].

Early and specific biomarkers that allow for HGSC detection at a less-advanced stage do not exist. However, exosomes or extracellular vesicles that are essential for cell–cell communication may be used as a promising biomarker. A cancer-on-chip device has been used for the isolation of intact exosomes from the culture medium of HGSC cells. The proteasome profile was characterised and compared to healthy patient-derived donor cells of the ovarian surface epithelium and the fallopian tube secretory epithelium. Notably, 25 exosomal proteins were differentially expressed in HGSC compared to the controls [37]. These findings may potentially help to detect the disease early and to design targeted therapies.

9.4 Conclusion

Despite intensive research, ovarian cancer remains the leading cause of mortality among gynaecological malignancies for which the treatment options are limited. Given its complexity, experimental models that faithfully mimic the complex microenvironmental stimuli during disease development and progression are urgently needed. In the last decade, it became apparent that traditional cell monolayer models and animal studies are not entirely suited for the modelling of the human disease and treatment response. Hence, new TME models that reconstruct critical elements of the TME in a spatially, physically and chemically relevant manner have been designed. These new 3D platforms have proven more efficient for drug testing and drug discovery and hold enormous potential to improve the treatment options and to screen personalised medicines for patients suffering from ovarian cancer. Personalised medicines aim to move away from the 'one-size-fits-all' therapy for patients in order to stratify the variation between individual patients or specific subgroups of patients or subsets of tumours. Personalising cancer treatment includes the molecular profiling of patients to identify biomarkers or genetic profiles that help to select patients for targeted therapies.

Promising new therapies for patients with HGSC, or for patients that have developed resistance to platinum-based chemotherapy, include PARP inhibitors, antiangiogenic therapies and immunotherapies. While PARP inhibitors have shown excellent activity in ovarian cancer, immunotherapies exhibit only modest activity. However, combined PARP and immune checkpoint inhibition has yielded encouraging results for the treatment of ovarian cancer and immunogenomic profiling may identify predictive biomarkers of treatment response [38]. Resistance to PARP inhibitors and platinum-based chemotherapy may be overcome by combining PARP inhibitors with inhibitors of alternative DNA repair pathways, depending on the genetic profile of the individual patient [39, 40]. For ovarian cancer, 3D models that recapitulate physiological aspects and matrix composition of tumour tissues and integrate patient-derived cell populations can be used as patient surrogates to directly test responses to targeted therapies or personalised medicines.

References

1. Siegel, R. L., Miller, K. D., Fuchs, H. E. & Jemal, A. (2021). Cancer statistics, 2021. CA: A Cancer Journal for Clinicians, 71(1), 7–33.
2. Bowtell, D. D., Bohm, S., Ahmed, A. A., Aspuria, P. J., Bast, R. C., Jr., Beral, V., Berek, J. S., Birrer, M. J., Blagden, S., Bookman, M. A., Brenton, J. D., Chiappinelli, K. B., Martins, F. C., Coukos, G., Drapkin, R., Edmondson, R., Fotopoulou, C., Gabra, H., Galon, J., Gourley, C., Heong, V., Huntsman, D. G., Iwanicki, M., Karlan, B. Y., Kaye, A., Lengyel, E., Levine, D. A., Lu, K. H., McNeish, I. A., Menon, U., Narod, S. A., Nelson, B. H., Nephew, K. P., Pharoah, P., Powell, D. J., Jr., Ramos, P., Romero, I. L., Scott, C. L., Sood, A. K., Stronach, E. A., & Balkwill, F. R. (2015). Rethinking ovarian cancer II: Reducing mortality from high-grade serous ovarian cancer. *Nature Reviews. Cancer, 15*(11), 668–679.
3. Patch, A. M., Christie, E. L., Etemadmoghadam, D., Garsed, D. W., George, J., Fereday, S., Nones, K., Cowin, P., Alsop, K., Bailey, P. J., Kassahn, K. S., Newell, F., Quinn, M. C., Kazakoff, S., Quek, K., Wilhelm-Benartzi, C., Curry, E., Leong, H. S., Australian Ovarian Cancer Study Group, Hamilton, A., Mileshkin, L., Au-Yeung, G., Kennedy, C., Hung, J., Chiew, Y. E., Harnett, P., Friedlander, M., Quinn, M., Pyman, J., Cordner, S., O'Brien, P., Leditschke, J., Young, G., Strachan, K., Waring, P., Azar, W., Mitchell, C., Traficante, N., Hendley, J., Thorne, H., Shackleton, M., Miller, D. K., Arnau, G. M., Tothill, R. W., Holloway, T. P., Semple, T., Harliwong, I., Nourse, C., Nourbakhsh, E., Manning, S., Idrisoglu, S., Bruxner, T. J., Christ, A. N., Poudel, B., Holmes, O., Anderson, M., Leonard, C., Lonie, A., Hall, N., Wood, S., Taylor, D. F., Xu, Q., Fink, J. L., Waddell, N., Drapkin, R., Stronach, E., Gabra, H., Brown, R., Jewell, A., Nagaraj, S. H., Markham, E., Wilson, P. J., Ellul, J., McNally, O., Doyle, M. A., Vedururu, R., Stewart, C., Lengyel, E., Pearson, J. V., Waddell, N., de Fazio, A., Grimmond, S. M., & Bowtell, D. D. (2015). Whole-genome characterization of chemoresistant ovarian cancer. *Nature, 521*(7553), 489–494.
4. Lengyel, E., Burdette, J. E., Kenny, H. A., Matei, D., Pilrose, J., Haluska, P., Nephew, K. P., Hales, D. B., & Stack, M. S. (2014). Epithelial ovarian cancer experimental models. *Oncogene, 33*(28), 3619–3633.
5. Kenny, H. A., Lal-Nag, M., Shen, M., Kara, B., Nahotko, D. A., Wroblewski, K., Fazal, S., Chen, S., Chiang, C. Y., Chen, Y. J., Brimacombe, K. R., Marugan, J., Ferrer, M., & Lengyel, E. (2020). Quantitative high-throughput screening using an organotypic model identifies compounds that inhibit ovarian cancer metastasis. *Molecular Cancer Therapeutics, 19*(1), 52–62.
6. Kopper, O., de Witte, C. J., Lohmussaar, K., Valle-Inclan, J. E., Hami, N., Kester, L., Balgobind, A. V., Korving, J., Proost, N., Begthel, H., van Wijk, L. M., Revilla, S. A., Theeuwsen, R., van de Ven, M., van Roosmalen, M. J., Ponsioen, B., Ho, V. W. H., Neel, B. G., Bosse, T., Gaarenstroom, K. N., Vrieling, H., Vreeswijk, M. P. G., van Diest, P. J., Witteveen, P. O., Jonges, T., Bos, J. L., van Oudenaarden, A., Zweemer, R. P., Snippert, H. J. G., Kloosterman, W. P., & Clevers, H. (2019). An organoid platform for ovarian cancer captures intra- and interpatient heterogeneity. *Nature Medicine, 25*(5), 838–849.
7. Loessner, D., Holzapfel, B. M., & Clements, J. A. (2014). Engineered microenvironments provide new insights into ovarian and prostate cancer progression and drug responses. *Advanced Drug Delivery Reviews, 79–80*, 193–213.
8. Baker, B. M., & Chen, C. S. (2012). Deconstructing the third dimension: How 3D culture microenvironments alter cellular cues. *Journal of Cell Science, 125*(Pt 13), 3015–3024.
9. Brooks, E. A., Gencoglu, M. F., Corbett, D. C., Stevens, K. R., & Peyton, S. R. (2019). An omentum-inspired 3D PEG hydrogel for identifying ECM-drivers of drug resistant ovarian cancer. *APL Bioengineering, 3*(2), 026106.
10. Al Ameri, W., Ahmed, I., Al-Dasim, F. M., Ali Mohamoud, Y., Al-Azwani, I. K., Malek, J. A., & Karedath, T. (2019). Cell type-specific TGF-beta mediated EMT in 3D and 2D models and its reversal by TGF-beta receptor kinase inhibitor in ovarian cancer cell lines. *International Journal of Molecular Sciences, 20*(14), 3568.
11. Masiello, T., Dhall, A., Hemachandra, L. P. M., Tokranova, N., Melendez, J. A., & Castracane, J. (2018). A dynamic culture method to produce ovarian cancer spheroids under physiologically-relevant shear stress. *Cell, 7*(12), 277.
12. Maniati, E., Berlato, C., Gopinathan, G., Heath, O., Kotantaki, P., Lakhani, A., McDermott, J., Pegrum, C., Delaine-Smith, R. M., Pearce, O. M. T., Hirani, P., Joy, J. D., Szabova, L., Perets, R., Sansom, O. J., Drapkin, R., Bailey, P., & Balkwill, F. R. (2020). Mouse ovarian cancer models recapitulate the human tumor microenvironment and patient response to treatment. *Cell Reports, 30*(2), 525–540.e7.
13. Kim, O., Park, E. Y., Klinkebiel, D. L., Pack, S. D., Shin, Y. H., Abdullaev, Z., Emerson, R. E., Coffey, D. M., Kwon, S. Y., Creighton, C. J., Kwon, S., Chang, E. C., Chiang, T., Yatsenko, A. N., Chien, J., Cheon, D. J., Yang-Hartwich, Y., Nakshatri, H., Nephew, K. P., Behringer, R. R., Fernandez, F. M., Cho, C. H., Vanderhyden, B., Drapkin, R., Bast, R. C., Jr., Miller, K. D., Karpf, A. R., & Kim, J. (2020). In vivo modeling of metastatic human high-grade serous ovarian cancer in mice. *PLoS Genetics, 16*(6), e1008808.
14. Loessner, D., Little, J. P., Pettet, G. J., & Hutmacher, D. W. (2013). A multiscale road map of cancer spheroids—Incorporating experimental and mathematical modelling to understand cancer progression. *Journal of Cell Science, 126*(Pt 13), 2761–2771.
15. Loessner, D., Goettig, P., Preis, S., Felber, J., Bronger, H., Clements, J. A., Dorn, J., & Magdolen, V. (2018). Kallikrein-related peptidases represent attractive ther-

apeutic targets for ovarian cancer. *Expert Opinion on Therapeutic Targets, 22*(9), 745–763.
16. Torre, L. A., Trabert, B., DeSantis, C. E., Miller, K. D., Samimi, G., Runowicz, C. D., Gaudet, M. M., Jemal, A., & Siegel, R. L. (2018). Ovarian cancer statistics, 2018. *CA: A Cancer Journal for Clinicians, 68*(4), 284–296.
17. Al Habyan, S., Kalos, C., Szymborski, J., & McCaffrey, L. (2018). Multicellular detachment generates metastatic spheroids during intra-abdominal dissemination in epithelial ovarian cancer. *Oncogene, 37*(37), 5127–5135.
18. Kenny, H. A., Kaur, S., Coussens, L. M., & Lengyel, E. (2008). The initial steps of ovarian cancer cell metastasis are mediated by MMP-2 cleavage of vitronectin and fibronectin. *The Journal of Clinical Investigation, 118*(4), 1367–1379.
19. Izar, B., Tirosh, I., Stover, E. H., Wakiro, I., Cuoco, M. S., Alter, I., Rodman, C., Leeson, R., Su, M. J., Shah, P., Iwanicki, M., Walker, S. R., Kanodia, A., Melms, J. C., Mei, S., Lin, J. R., Porter, C. B. M., Slyper, M., Waldman, J., Jerby-Arnon, L., Ashenberg, O., Brinker, T. J., Mills, C., Rogava, M., Vigneau, S., Sorger, P. K., Garraway, L. A., Konstantinopoulos, P. A., Liu, J. F., Matulonis, U., Johnson, B. E., Rozenblatt-Rosen, O., Rotem, A., & Regev, A. (2020). A single-cell landscape of high-grade serous ovarian cancer. *Nature Medicine, 26*(8), 1271–1279.
20. Pearce, O. M. T., Delaine-Smith, R. M., Maniati, E., Nichols, S., Wang, J., Bohm, S., Rajeeve, V., Ullah, D., Chakravarty, P., Jones, R. R., Montfort, A., Dowe, T., Gribben, J., Jones, J. L., Kocher, H. M., Serody, J. S., Vincent, B. G., Connelly, J., Brenton, J. D., Chelala, C., Cutillas, P. R., Lockley, M., Bessant, C., Knight, M. M., & Balkwill, F. R. (2018). Deconstruction of a metastatic tumor microenvironment reveals a common matrix response in human cancers. *Cancer Discovery, 8*(3), 304–319.
21. Jimenez-Sanchez, A., Cybulska, P., Mager, K. L., Koplev, S., Cast, O., Couturier, D. L., Memon, D., Selenica, P., Nikolovski, I., Mazaheri, Y., Bykov, Y., Geyer, F. C., Macintyre, G., Gavarro, L. M., Drews, R. M., Gill, M. B., Papanastasiou, A. D., Sosa, R. E., Soslow, R. A., Walther, T., Shen, R., Chi, D. S., Park, K. J., Hollmann, T., Reis-Filho, J. S., Markowetz, F., Beltrao, P., Vargas, H. A., Zamarin, D., Brenton, J. D., Snyder, A., Weigelt, B., Sala, E., & Miller, M. L. (2020). Unraveling tumor-immune heterogeneity in advanced ovarian cancer uncovers immunogenic effect of chemotherapy. *Nature Genetics, 52*(6), 582–593.
22. Dean, M., Jin, V., Bergsten, T. M., Austin, J. R., Lantvit, D. D., Russo, A., & Burdette, J. E. (2019). Loss of PTEN in fallopian tube epithelium results in multicellular tumor spheroid formation and metastasis to the ovary. *Cancers (Basel), 11*(6), 884.
23. Singh, M. S., Goldsmith, M., Thakur, K., Chatterjee, S., Landesman-Milo, D., Levy, T., Kunz-Schughart, L. A., Barenholz, Y., & Peer, D. (2020). An ovarian spheroid based tumor model that represents vascularized tumors and enables the investigation of nanomedicine therapeutics. *Nanoscale, 12*(3), 1894–1903.
24. Tomas-Bort, E., Kieler, M., Sharma, S., Candido, J. B., & Loessner, D. (2020). 3D approaches to model the tumor microenvironment of pancreatic cancer. *Theranostics, 10*(11), 5074–5089.
25. Yang, Z., Xu, H., & Zhao, X. (2020). Designer self-assembling peptide hydrogels to engineer 3D cell microenvironments for cell constructs formation and precise oncology remodeling in ovarian cancer. *Advanced Science (Weinh), 7*(9), 1903718.
26. Loessner, D., Stok, K. S., Lutolf, M. P., Hutmacher, D. W., Clements, J. A., & Rizzi, S. C. (2010). Bioengineered 3D platform to explore cell-ECM interactions and drug resistance of epithelial ovarian cancer cells. *Biomaterials, 31*(32), 8494–8506.
27. Kaemmerer, E., Melchels, F. P., Holzapfel, B. M., Meckel, T., Hutmacher, D. W., & Loessner, D. (2014). Gelatine methacrylamide-based hydrogels: An alternative three-dimensional cancer cell culture system. *Acta Biomaterialia, 10*(6), 2551–2562.
28. Naba, A., Pearce, O. M. T., Del Rosario, A., Ma, D., Ding, H., Rajeeve, V., Cutillas, P. R., Balkwill, F. R., & Hynes, R. O. (2017). Characterization of the extracellular matrix of normal and diseased tissues using proteomics. *Journal of Proteome Research, 16*(8), 3083–3091.
29. Lohmussaar, K., Kopper, O., Korving, J., Begthel, H., Vreuls, C. P. H., van Es, J. H., & Clevers, H. (2020). Assessing the origin of high-grade serous ovarian cancer using CRISPR-modification of mouse organoids. *Nature Communications, 11*(1), 2660.
30. Hill, S. J., Decker, B., Roberts, E. A., Horowitz, N. S., Muto, M. G., Worley, M. J., Jr., Feltmate, C. M., Nucci, M. R., Swisher, E. M., Nguyen, H., Yang, C., Morizane, R., Kochupurakkal, B. S., Do, K. T., Konstantinopoulos, P. A., Liu, J. F., Bonventre, J. V., Matulonis, U. A., Shapiro, G. I., Berkowitz, R. S., Crum, C. P., & D'Andrea, A. D. (2018). Prediction of DNA repair inhibitor response in short-term patient-derived ovarian cancer organoids. *Cancer Discovery, 8*(11), 1404–1421.
31. Maru, Y., & Hippo, Y. (2019). Current status of patient-derived ovarian cancer models. *Cell, 8*(5), 505.
32. Lu, M., Henry, C. E., Lai, H., Khine, Y. Y., Ford, C. E., & Stenzel, M. H. (2019). A new 3D organotypic model of ovarian cancer to help evaluate the anti-metastatic activity of RAPTA-C conjugated micelles. *Biomaterials Science, 7*(4), 1652–1660.
33. Mitra, S., Tiwari, K., Podicheti, R., Pandhiri, T., Rusch, D. B., Bonetto, A., Zhang, C., & Mitra, A. K. (2019). Transcriptome profiling reveals matrisome alteration as a key feature of ovarian cancer progression. *Cancers (Basel), 11*(10), 1513.
34. Loessner, D., Rockstroh, A., Shokoohmand, A., Holzapfel, B. M., Wagner, F., Baldwin, J., Boxberg, M., Schmalfeldt, B., Lengyel, E., Clements, J. A., & Hutmacher, D. W. (2019). A 3D tumor microenvironment regulates cell proliferation, peritoneal growth

and expression patterns. *Biomaterials, 190–191*, 63–75.
35. Sontheimer-Phelps, A., Hassell, B. A., & Ingber, D. E. (2019). Modelling cancer in microfluidic human organs-on-chips. *Nature Reviews. Cancer, 19*(2), 65–81.
36. Ronaldson-Bouchard, K., & Vunjak-Novakovic, G. (2018). Organs-on-a-chip: A fast track for engineered human tissues in drug development. *Cell Stem Cell, 22*(3), 310–324.
37. Dorayappan, K. D. P., Gardner, M. L., Hisey, C. L., Zingarelli, R. A., Smith, B. Q., Lightfoot, M. D. S., Gogna, R., Flannery, M. M., Hays, J., Hansford, D. J., Freitas, M. A., Yu, L., Cohn, D. E., & Selvendiran, K. (2019). A microfluidic chip enables isolation of exosomes and establishment of their protein profiles and associated signaling pathways in ovarian cancer. *Cancer Research, 79*(13), 3503–3513.
38. Farkkila, A., Gulhan, D. C., Casado, J., Jacobson, C. A., Nguyen, H., Kochupurakkal, B., Maliga, Z., Yapp, C., Chen, Y. A., Schapiro, D., Zhou, Y., Graham, J. R., Dezube, B. J., Munster, P., Santagata, S., Garcia, E., Rodig, S., Lako, A., Chowdhury, D., Shapiro, G. I., Matulonis, U. A., Park, P. J., Hautaniemi, S., Sorger, P. K., Swisher, E. M., D'Andrea, A. D., & Konstantinopoulos, P. A. (2020). Immunogenomic profiling determines responses to combined PARP and PD-1 inhibition in ovarian cancer. *Nature Communications, 11*(1), 1459.
39. Kim, H., Xu, H., George, E., Hallberg, D., Kumar, S., Jagannathan, V., Medvedev, S., Kinose, Y., Devins, K., Verma, P., Ly, K., Wang, Y., Greenberg, R. A., Schwartz, L., Johnson, N., Scharpf, R. B., Mills, G. B., Zhang, R., Velculescu, V. E., Brown, E. J., & Simpkins, F. (2020). Combining PARP with ATR inhibition overcomes PARP inhibitor and platinum resistance in ovarian cancer models. *Nature Communications, 11*(1), 3726.
40. Sanij, E., Hannan, K. M., Xuan, J., Yan, S., Ahern, J. E., Trigos, A. S., Brajanovski, N., Son, J., Chan, K. T., Kondrashova, O., Lieschke, E., Wakefield, M. J., Frank, D., Ellis, S., Cullinane, C., Kang, J., Poortinga, G., Nag, P., Deans, A. J., Khanna, K. K., Mileshkin, L., McArthur, G. A., Soong, J., Berns, E., Hannan, R. D., Scott, C. L., Sheppard, K. E., & Pearson, R. B. (2020). CX-5461 activates the DNA damage response and demonstrates therapeutic efficacy in high-grade serous ovarian cancer. *Nature Communications, 11*(1), 2641.

Ovarian Cancer Stem Cells: Characterization and Role in Tumorigenesis

10

Sarama Saha, Seema Parte, Partha Roy, and Sham S. Kakar

10.1 Introduction

Ovarian cancer originates from the female organ responsible for producing eggs. This cancer most often remains undetected until it has spread locally within the pelvis. In the early stage, patient usually remains asymptomatic. Although in later stages, patients develop symptoms such as anorexia, loss of weight which are nonspecific, causing more confusion. Moreover, there is no better way to detect cancer at an early stage. Late detection is primarily the most important factor contributing to difficult-to-treat cases and increased fatality [1].

Ovarian cancer is the most lethal gynecological cancer in women worldwide. According to World Ovarian Cancer Coalition, ovarian cancer is the fifth most common cause of death from cancer. Approximately 295,000 women are diagnosed with ovarian cancer every year worldwide and account for more than 184,000 deaths per year. The 5-year survival rate is approximately 30% compared to 80% in case of breast cancer. By the year 2035, the diagnosis of new cases of ovarian cancer is expected to increase by 55% and the number of deaths would increase by 70% [2]. In the United States, out of 21,750 new cases estimated in 2020, 13,940 women will die with it within 5 years of diagnosis [3]. Despite substantial improvement in technology, lack of early diagnosis remains a clinical problem and contributes to the highest mortality among female gynecological cancers.

Ovary comprises of different cell types such as surface epithelial cells, germ cells, and sex cord-stromal cells. All these different cell types can give rise to different tumors. If we consider the 5-year survival rate of different stages of the disease, it has been observed that if it can be diagnosed at an early stage, it is highly curable. Moreover, if the disease is confined to the ovary, the survival rate is approximately 90% while it drops significantly as it proceeds further higher stages. Unfortunately, ovarian cancer patients frequently present with advanced disease and hence survival rate drops to about 40% after 5 years of diagnosis [2].

S. Saha
Department of Biochemistry, All India Institute of Medical Sciences, Rishikesh, India

S. Parte
Department of Biochemistry and Molecular Biology, University of Nebraska, Omaha, NE, USA

P. Roy
Department of Biotechnology, Indian Institute of Technology Roorkee, Roorkee, Uttarakhand, India

S. S. Kakar (✉)
Department of Physiology and James Graham Brown Cancer Center, University of Louisville, Louisville, KY, USA
e-mail: sham.kakar@louisville.edu

H. Schatten (ed.), *Ovarian Cancer: Molecular & Diagnostic Imaging and Treatment Strategies*, Advances in Experimental Medicine and Biology 1330, https://doi.org/10.1007/978-3-030-73359-9_10

Ovarian cancer is a heterogeneous disease with variable clinicopathological and molecular mechanisms of carcinogenesis, progression, metastasis, response to oncotherapy, and manifests as different histotypes when examined under microscope [4]. The most frequently presented subgroup is epithelial ovarian cancer which originates from the surface of the ovary and among this group there are different histotypes described briefly as follows: Germ cell tumor is seen in younger age group and stromal cell tumors appear in midlife. Not every ovarian cancer is the same. Different subtypes have different prognoses. Median survival rate varies with the type of ovarian cancer [5]. Hence, it is pertinent to identify and stratify the tumor type/subtype so as to administer appropriate treatment to the patient depending upon their differential response to the chemotherapy regimen.

Primary cytoreductive surgery combined with chemotherapy is initially effective treatment in the annihilation of bulk of tumor thus retaining the cells with stemness properties (self-renewal and quiescence) also termed as cancer stem cells (CSCs) which may get enriched leading to therapy recalcitrance and disease relapse [6, 7]. Chemoresistance is a crucial hindrance to achieve success in ovarian cancer therapy and is a major factor for stage-wise (I, II, III) progression of tumorigenesis [8]. Histopathologically epithelial ovarian cancer is classified as high-grade serous carcinoma, low-grade serous carcinoma, endometrioid clear cell carcinoma, and mucinous carcinoma. Modulation and cross-talk of various signaling pathways (Wnt, Shh, Notch1, etc.) might be implicated in this CSC-mediated therapeutic resistance [7, 8]. Recent report (utilizing genetically engineered mouse models and organoids) implicated both fallopian tube and ovarian surface epithelium (OSE) as the origin of high-grade serous ovarian carcinoma [9]. In addition, intratumoral heterogeneity of CSCs may be responsible for chemoresistance which could be probed to study at single-cell RNA transcriptome level with respect to tumor stage, patient-specific treatment regimen, and clinical outcome to establish corelation and thus effective therapies.

Tumor represents a complex ecosystem comprising of varied subclones differing in their genetic and epigenetic constitution (i.e., mutational burden and promoter hypermethylation) [intrinsic factors] and those from its surrounding microenvironment [extrinsic factors] constituting the spatiotemporal variations within the stromal cells, extracellular matrix components, immune, and endothelial cells [10]. This heterogeneity is a by-product of either or both of the situations defined by the clonal selection and stochastic model of CSCs, respectively. CSCs represent heterogeneity/plasticity in terms of the spectrum provided by their variability from stemness towards differentiation pathway thus providing a hierarchy of cells within a single tumor. Oncogenic transformation of cells culminating into CSCs rendering them with self-renewal ability and hence tumor aggressiveness and therapy resistance represents their genetic/epigenetic and also functional plasticity [11, 12].

Oncogenic transformation of OSE cells have been reported and implicated in epithelial ovarian cancer however there exists a gap in the knowledge about novel oncogenes and their molecular mechanisms [13]. Recently, Securin also known as pituitary transforming gene (PTTG) has been reported to be responsible for the transformation of normal cells to cancer cells. PTTG1, first cloned from ovary and testis is basically involved as a regulator of sister chromatid separation during cell cycle during normal physiology. It is a multi-domain proto-oncogene with pleiotropic functional significance due to its overexpression in different tumor types such as pituitary, thyroid, and breast besides ovarian cancer [14–16]. Recently PTTG1 was represented as a novel candidate which could demarcate the normal stem cells and ovarian CSC compartments (within the OSE layer and cortex region) in benign, borderline, and high-grade human ovarian tumors and ascites derived CSCs in comparison to normal ovaries by its co-expression with CSC-specific markers. Silencing of PTTG1 by gene-specific siRNA, or adenovirus vector expressing PTTG1 siRNA in ovarian cancer cells abrogated and enhanced the expression of PTTG1 leading to suppression of tumor progression and metastasis.

In addition, self-renewal, Wnt/B-Catenin, Notch1, and EMT pathway-specific markers were differentially regulated signifying a definitive role of PTTG1 in CSC self-renewal and EMT [17]. Hence, some fundamental concepts are described below.

10.2 Normal Cell Versus Cancer Cell

NCCP guideline for ovarian cancer 2019

There are three characteristically different features between normal and cancer cells.

1. Unlike normal cells, cancer cells grow in an uncontrolled manner without contact inhibition resulting in the development of tumor.
2. Cancer cell can invade other tissues which are known as invasion. Normal cells lack such property.
3. Unlike normal cells, cancer cells can propagate to a different part of the body by implanting or seeding, or via blood or lymphatic vessels.

10.3 Stem Cells Versus Cancer Stem Cells

10.3.1 Stem Cell

It is a special type of cell that possesses the ability to renew itself through cell division and differentiate into cells of multiple lineages. This cell may be of three types: adult stem cells, embryonic stem cell (ESC), and induced pluripotent stem cells (iPSCs).

(a) *Mesenchymal stem cells (MSCs)* are non-hematopoietic, multipotent adult stem cells and possess a varying degree of propensity to differentiate into Mesodermal lineage. Moreover, it can transdifferentiate into ectodermal and endodermal lineages. Later on, it was observed that these cells are present in almost all tissues such as adipose tissue, amniotic fluid and membrane, dental tissue, endometrium, limb bud, peripheral blood, placenta and fetal membrane, salivary gland, skin and foreskin, sub amniotic umbilical cord lining membrane, synovial fluid, Wharton's jelly, and menstrual blood [18].

(b) *Embryonic stem cells (ESCs)* are pluripotent in nature and have ability to differentiate into any type of somatic cells derived from an embryo. As a result of this, ESCs can be used as a promising biological tool for exploring the complex mechanism of development of multiple organ structures. First time, the embryonic stem cell line was derived successfully from mouse embryo in 1981 [19, 20]. Later on, Thompson and coworkers in 1998 generated the first stable ESC line from human embryos produced by in vitro fertilization [21].

(c) *Induced pluripotent stem cells (iPSCs)*: Despite the highest therapeutic potentiality of human ESC (hESC) in translational medicine, their use was limited to be so popular because of its controversy related to ethical issue. In order to overcome this problem, scientists have developed induced pluripotent stem cells (iPSCs) by introducing specific gene into already specialized mouse adult cells and thus overexpressing the transcription factors such as Oct3/4 (octamer-binding transcription factor 3/4), Sox2 (sex-determining region Y)-box 2, Klf4 (Kruppel-like factor 4), and c-Myc (Avian Myelocytomatosis virus oncogene cellular homolog) [22].

10.3.2 Cancer Stem Cells (CSCs)

CSCs represent a specialized group of cells but a minuscule fraction (~0.1–0.8%, maximum of 30%) within most of the tumors (solid and liquid) capable of initiation of tumor. These cells possess cellular and molecular heterogeneity and capable of self-renewal and reflect pluripotency. Recent experimental and clinical evidence both implicate CSCs in cancer initiation, progression, metastasis, and recurrence, as well as radio- and chemotherapeutic resistance [23, 24]. Expression of several surface/non-surface markers, transcription factors related to stemness and self-renewal, other cellular properties such as autofluorescence [25] and dye-efflux mechanisms for Side population cells [26, 27] exhibiting stemness potential have been utilized to identify them (summarized in Table 10.1).

Although CSCs initially referred as tumor stem cells express distinct cell surface markers, their origin per se is still debatable. Lapidot and group first reported CD34+/CD38− subpopulation in primary acute myeloid leukemia with tumorigenic potential upon transplantation in SCID mice. A recent review by Nimmakayala and group [54] summarized the dynamics of CSCs from origin to metastasis whereby they explained cell fusion, horizontal gene transfer, and mutations driving cellular transformation and reprogramming into CSCs and metabolic shifts from glycolytic to oxidative phosphorylation or vice versa implicated in cancer stemness. Upon extensive reconciliation of the literature, it was interestingly proposed in this review that CSC populations with specific phenotypes, metabolic profiles, and clonogenic potential may metastasize to specific organs.

According to one theory, the tissue stem cells undergo mutation and behave as a cancer stem cell. Another theory is that the cancer cells acquire stemness following oncogenic hit [54–56] (Fig. 10.1). This detailed hierarchy could not be defined in solid tumor since there are subpopulations of cells residing within the same tumor such as "resident cancer stem cells" which can initiate the tumor and "migrating stem cells" which are responsible for propagation of tumor growth and metastasis [59]. Thus, the alternative model of carcinogenesis has been originated and states that tumor is composed of heterogeneous clones of cells resulting from different types of mutation and accounts for different phases of tumor development [60].

CSCs should be considered as one of the main targets of novel experimental and therapeutic strategy such as tissue repair in various clinical fields such as cardiac, orthopedic, plastic, and breast surgery [61–63]. CSC targeted therapies include drugs targeting cell surface, signaling pathways, chief components of tumor microenvironment, those aimed at reversing drug resistance, those focussed upon differentiation of CSCs, and other miscellaneous cellular features of CSCs [64]. However, several therapeutic interventions targeting CSCs per se are still immature and clinical trial outcomes are yet in pipeline to conclude constructively.

Similar to CSCs targeting, the origin of CSCs is enigmatic because as per the Clonal or Stochastic model all the cells may possess tumor-initiating properties, whereas the Hierarchical or CSC model suggests the persistence of a small fraction of CSCs [65, 66]. A tumor as an entity as such may reflect complex hierarchy in terms of the CSC profile because the normal tissue resident stem cells may acquire mutations and exhibit transformed phenotype and subsequently altered key cellular properties; or the progenitor cells, a progeny of stem cells may acquire mutations or their terminally differentiated progeny, in turn, may exhibit mutated version of cancer/tumor cells thus implicating self-renewal, differentiation, and proliferation as key mechanisms guiding the putative origin of CSCs [67].

10.4 Genetics of Ovarian Cancer

Ovarian cancer contributes to nearly 3% of all cancers among women. In 2035, it will account for more than 200,000 deaths all over the world [2]. Considering the facts, it is of crucial importance to identify women who are at enhanced risk of developing ovarian cancer so that preventive measures can be ensured. Early onset of menarche, late menopause, and being nulliparous are considered as well-known risk factors for ovarian cancer [68, 69]. However, presence of family history of ovarian cancer especially in first-degree relative has been found to elevate the lifetime risk of developing ovarian cancer. Hereditary ovarian cancer contributes to 20% of all ovarian cancer and results from mutation in BRCA1 and BRCA2 genes [70, 71]. This mutation varies with ethnicity giving rise to the higher prevalence of ovarian cancer in certain ethnic populations such as Polish, French, Canadian, and Ashkenazi Jews

Table 10.1 Cancer stem cell markers in ovarian cancer

Marker		Experimental design	Outcome	Authors [References]
CD24	Transmembrane glycoprotein	Ovarian serous tumor from patients	Presence of CD24 in the cytoplasm independently predicts poor survival	Choi et al. [28]
CD24	Transmembrane glycoprotein	Human ovarian cancer cell line Caov3	CD 24 is responsible for metastasis and chemoresistance through the induction of epithelial to mesenchymal transition via Akt -ERK signaling mechanism	Nakamura et al. [29]
CD44+/CD24−	CD44: Hyaluronate receptor	Ovarian cancer cell line (SKOV3 and OV90), Cancer cell isolated from ascites of ovarian cancer patients	These markers are predictor of chemoresistance, relapse, and poor prognosis	Meng et al. [30]
CD117/c-kit	Receptor/Oncoprotein having tyrosine kinase activity	Paraffin-embedded specimens of human serous ovarian carcinoma	Indicative of chemoresistance	Raspollini et al. [31]
CD133 (Prominin-1)	Transmembrane glycoprotein	Flow cytometric analysis of various Cancer cell lines and cells isolated from ascitic fluid of ovarian cancer patients	CD133 is an indicator of tumorigenicity and its expression is modulated by epigenetics	Baba et al. [32]
ALDH1A1	Intracellular enzyme, one of 17 isoforms of ALDH	Several cancer cell lines and primary xenograft developed from omental tissue of metastatic ovarian cancer patients	Predictor of tumor initiation, identification of chemoresistant cells	Landen et al. [33]
CD44+/CD117+	CD117: Stem cell factor receptor	Xenograft experiment	Indicative of greater tumorigenicity	Zhang et al. [34]
SP cells	Having dye exclusion property	H2B-GFP transgenic mice models	Identification and characterization of ovarian cancer stem cells	Szotek et al. [26]
CD133	Transmembrane glycoprotein	Primary tumor and cells isolated from ascitic fluid of ovarian cancer	Contributes to angiogenesis for driving metastasis	Kusumbe and Bapat [35]
CD44+, MyD88+	MYD88: Innate immune signal transduction adaptor	Ascites sample from advanced ovarian cancer patients	Maintenance of cell survival and chemoresistance via TLR4-MyD88 and NF kappa B pathway	Alvero et al. [36]
CD34+	Transmembrane phosphoglycoprotein	Xenograft tumor	Role in angiogenesis	Alvero et al. [37]

(continued)

Table 10.1 (continued)

Marker		Experimental design	Outcome	Authors [References]
CD105, CD44, CD106	CD105: Type I membrane glycoprotein; CD106: vascular cell adhesion molecule-1 (VCAM-1)	Human ovarian cancer cell line OVCAR3	Significant association with progression of disease, relapse, and chemoresistance	Zhang et al. 2019 [38]
Epithelial cell adhesion molecule (EpCAM)	Type I transmembrane glycoprotein	In vitro study in human ovarian cancer cell line and in vivo study in C57BL/6 mice and finally clinical correlation	Role in chemoresistance and prognostication	Tayama et al. [39]
CD44-EpCAM		Ovarian cancer cell line OVCAR8, SKOV3, OCC1, ES2, and HEK293	Effects on tumor growth	Zheng et al. [40]
SOX2 (SRY-box transcription factor 2)	Transcription factor	Human epithelial ovarian cancer line SKOV3 and HO8910	Required for maintenance of ovarian cancer stem cells	Wen et al. [41]
ROR1 (receptor-tyrosine-kinase-like orphan receptor 1)	Transmembrane protein from receptor tyrosine kinase family	ROR1 expression investigated in ovarian cancer patients	Independently prognosticated with disease-free survival	Zhang et al. [42]
NANOG	Homeobox transcription factor	Expression checked in ovarian cancer cell line e.g., SKOV3 and ovarian cancer patients	Prognostic factor of ovarian cancer and chemoresistance	Lee et al. [44]
OCT4	Transcription factor	Side population cells isolation from ovarian cancer cell lines (SKOV3 and A2780) by Hoechst dye exclusion method	Responsible for tumor progression via JAK/STAT pathway	Ruan et al. [45]
MYC	Oncogenic transcription factor	Ovarian cancer cell line SKOV3 and OVCAR and tissues from ovarian cancer patients	Expression of c-Myc has significant association with disease progression	Ning et al. [46]
ABCG2	Member of the ATP-binding cassette (ABC) transporter	Ovarian cancer cell line A2780	Related to drug resistance	Duo et al. [47]
PTTG1 (Securin)	Oncogene	Ovarian cancer cell line A2780 and CSCs isolated from ascitic fluid of ovarian cancer patients	Relation with disease progression via regulation of EMT	Parte et al. [17]

(continued)

Table 10.1 (continued)

Marker		Experimental design	Outcome	Authors [References]
LGR5 (leucine-rich repeat-containing G-protein coupled receptors)	Transmembrane receptors	Mice model	Identification of stem cells/ progenitor cells	Schindler et al. [48]; Ng et al. [49]
VASA	ATP-dependent RNA helicases	Ovarian cancer cell line and tissues from epithelial ovarian cancer patients	Impact in disease progression by abrogation of DNA damage-induced G2 checkpoint	Hashimoto et al. [50]
NANOG, SOX2, SSEA4 (stage-specific embryonic antigen-4)	SSE4: glycosphingolipid	Paraffin-embedded ovarian tissue from high-grade serous ovarian carcinoma	Involved with tumorigenesis	Virant-Klun et al. [51]

Note: Information were collected from Zuber et al. [7], Bapat [52], Padilla et al. [53]

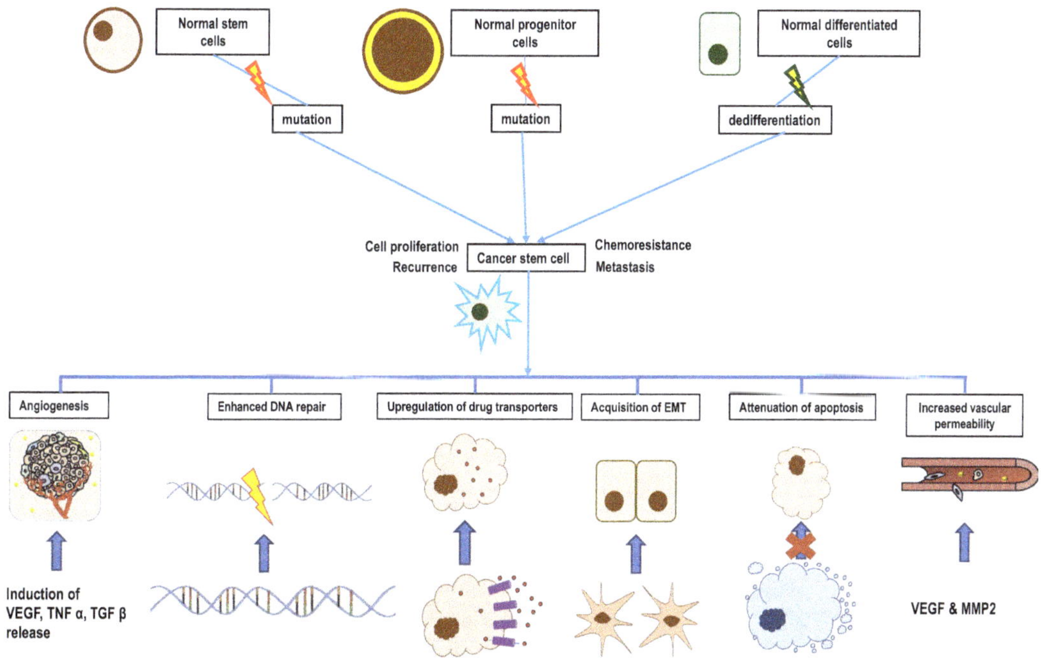

Fig. 10.1 Potential mechanisms of generation of cancer stem cells and their functions in tumor initiation, progression, and metsatsis. (Adopted from [6, 56–58])

[72, 73]. Since ovarian cancer is strongly associated with BRCA1/2 mutation, it is quintessential to carry out genetic testing and counseling of all ovarian cancer cases irrespective of age at onset and presence or absence of family history. BRCA1/2 gene mutations are responsible for earlier age of onset of ovarian cancer compared to women without mutation [70, 74]. It has been reported that BRCA1 gene mutation has a higher contribution nearly 48% compared to BRCA2 gene (nearly 29%) in elevating the lifetime risk of ovarian cancer [75]. BRCA1/2 associated OC

primarily originates from surface epithelium. They belong to high-grade serous histotype and are invasive in nature [76]. However, a recent large pooled cohort analysis revealed that OC with BRCA mutation carriers showed a better prognosis in terms of 5-year survival rate compared to that with noncarriers [77]. This could be explained by the better response of BRCA mutation to treatment with conventional platinum-based chemotherapy [78, 79].

Other contributing genes include BARD1, BRIP1, C14EK2, MRE11, MSH6, NBN, PALB2, RAD50, RAD51c, and TP53. RAD51C and RAD51D (paralog of RAD51) originate from the same ancestral gene of RAD51 and intrinsic component of homologous recombination-mediated double-strand break repair pathway. As opposed to BRCA1/2 linked OC, the average age at onset is older in RAD51C carrier [80–82]. Ovarian cancer related to RAD51C and RAD51D are found to originate from surface epithelial cells. Given the fact of their predisposition to develop OC, the role of RAD51 needs to be validated in mutation positive larger cohorts in order to determine whether germline mutation has any clinical implications as well as to establish screening and therapeutic strategies [82, 83].

Recent advanced gene testing technologies would help us in understanding the pathogenesis of ovarian cancer in a better way and thus would help in identifying women who are at risk of developing OC before the development of the disease and in turn help in implementing preventive strategies in time.

10.5 Identification of Ovarian CSCs

CSCs are identified by their tissue-specific expression of proteins known as biomarkers [84]. The stemness and tumorigenicity of isolated CSCs are validated by spheroid forming assay and limiting dilution assay, respectively on experimental models [85–87]. Ovarian CSC (OCSC) was detected first time from ascites of an ovarian cancer having the capacity of tumorigenicity in mice for several generations [88]. Ovarian CSCs express two types of markers: cell surface and nonsurface markers which are used either alone or in combination to identify and isolate the CSCs from the primary tumor and metastasized colony. The cell surface marker which was identified first time on ovarian CSC was CD117, a tyrosine kinase receptor [89–91] while more commonly documented ovarian CSC surface marker is CD133, transmembrane glycoprotein [32, 92–94]. Other reported surface markers are CD44, epithelial cell adhesion molecule (EpCAM), ROR1, and CD24 [28, 29, 95, 96]. These surface markers are significantly associated with tumor initiation, cancer propagation, prognosis, drug resistance, and recurrence of the disease [28, 32, 89, 90]. The nonsurface marker detected in ovarian CSC is aldehyde dehydrogenase family 1A2 (ALDH1A2). In an animal model, attenuation of ALDH1 improved the sensitivity of the cells to therapy [33]. Several studies have documented the association of the enzyme with the promotion of cell proliferation, facilitation of propagation, chemoresistance, unfavorable prognosis, and survival [97–99]. Some transcription factors, such as NANOG, OCT4, and SOX2, which are crucial for maintaining the stemness of embryonic stem cells [100], are also identified in ovarian CSCs [41]. Ovarian CSC biomarkers and their significance in various preclinical and clinical experiments have been presented in Table 10.1.

A special type of CSCs present in ovary having capacity to efflux the DNA binding dye are known as SP "side population" cells [101]. These cells are heterogeneous in nature because of differential expression of surface markers such as higher expression of Oct4, CD117, and CD44 compared to others. This heterogeneity in one tumor as well as difference between individual patients makes some ovarian cancer more chemoresistant and difficult to treat by universal treatment [102]. Hence, the concept of personalized therapy should be adopted to obtain effective outcome. Functional significance of variability and heterogeneity in expression of various CSC markers have been reported for example ALDH+/CD133+ possess higher tumorigenic potential than ALDH+/CD133− population, whereas

CD133+ CSCs induced CD133− cells to undergo EMT and metastatic potential via CCL5 and NFKB signaling. Similarly, ALDH1 and CD44 coexpressing cells may elicit a chemotherapeutic response and poor clinical outcome in patients rather than ALDH1+ only populations [23]. Despite encouraging results, clinical trials reveal only limited success attributed to intra- and inter-tumor heterogeneity among CSC and non-CSC compartments. In another study, Parte et al. [103, 104] extensively studied the expression of CSC surface markers in Benign, Borderline, and High-Grade ovarian tumors compared to normal ovaries which co-expressed with germline stem cell-specific markers initially implicated in ovarian stem cells. Similarly, they also explored and characterized actively dividing cell marker Ki67 with CSC populations distributed across the OSE and cortex regions of the tumorous ovaries. Novel insights about the tumor stage- and cellular compartment-specific distribution of ovarian CSCs co-expressing germline stem cell and proliferation markers were first demonstrated in these studies. Knowledge about the marker expression of tumor-initiating populations coupled with molecular mechanisms could improve their effective targeting in future and prevention of metastatic dissemination into peritoneum. Against this background several researchers and oncologists are striving hard to achieve success by addressing the bottlenecks and much remains to be researched. Recently Udoh et al. [105] and Carter et al. [106] during independent studies explored the potential of a fungal metabolite from *Myrothecium verrucaria* known as Verrucarin J (VJ) to target lung and ovarian CSCs via inhibition of the Wnt/β-Catenin and Notch1 stemness related signaling pathways. Similarly, Kakar and his group in the last couple of years have pronounced the effect of a herbal supplement Withaferin A (WFA) from *Withania somnifera* aka Ashwagandha plant extract and delineated its significant role in inhibiting ovarian ALDH1+ CSC populations while combining with standard chemotherapeutic drug Cisplatin [107]. ALDH1+ CSCs are differentially distributed within Benign, Borderline, and High-Grade ovarian tumors compared to normal ovaries within the OSE and cortex compartments [108, 109]. Further WFA while acting in synergy with Doxil exhibited a phenomenal inhibition of tumorigenic potential of ovarian CSCs thus attenuating the side effects of higher dose of Doxil and thus revealing potential against curbing recurrence by destroying the CSC population. Even in in vitro (A2780 cells) and in vivo models (mice treated with Cisplatin and WFA either alone or in combination), revealed significant reduction of CSC markers and thus proving the efficacy of WFA with conventional chemotherapy.

10.6 Tumor Microenvironment and OCSCs

One of the crucial contributing factors for cancer progression is interaction between ovarian cancer stem cells and tumor microenvironment (TME) which involves the following:

(a) *Extracellular matrix* which includes cytokines, chemokines, matrix metalloproteinases (MMP), and integrins [110].
(b) *Cancer-associated fibroblasts* derived from mesenchymal cells or as a result of trans-differentiation of pericytes and epithelial cells following exposure to various growth factors such as vascular endothelial growth factor (VEGF) [111]. They promote tumor growth via increased expression of CXCL14, IL-6, STAT3, and promote dissemination via neovascularisation, and immune suppression through intrusion of regulatory T lymphocytes. They modulate chemosensitivity and induce recurrence through higher expression of fibroblast activation protein alpha. They can remain in a quiescent state as well as in the active state. Cancer cells through the release of cytokines can activate CAFs which in turn can control the activity of immune cells [112]. This interaction between CSCs and other components of TME through various signaling pathways (TGF-β, Hedgehog, JAK) promotes tumorigenesis. Hence, targeting CAF might be promising therapeutic

approach for the prevention of disease progression [113].

(c) *Endothelial cells* comprise of the blood vessels and are primarily linked to neo-angiogenesis which are activated by VEGF, TGF-β, TNF-α, prostaglandin E2 and inhibited by angiopoietin, thrombospondin1 [114]. In presence of hypoxia, VEGF acts through its receptors present on endothelial cells and co-receptors neuropilins while angiopoietin attenuated signal transduction via modulation of tyrosine phosphorylation [115].

(d) *Immune cells* comprise of macrophages, dendritic cells, myeloid-derived suppressor cells (MDSC), and lymphocytes. Macrophages present within the tumor microenvironment known as tumor-associated macrophages (TAM). In the presence of interferon gamma and lipopolysaccharide, these macrophages suppress tumor growth through its cytotoxic effect by the release of IL-1, IL-12, and TNF-α. However, in presence of IL-4 and IL-10, TAM can promote tumorigenesis via immune suppression through inhibition of T cell multiplication [116]. MDSCs are GR1 and CD11b expressing myeloid cells, that facilitate tumorigenesis through exhaustion of nutrients required for the survival of T lymphocytes such as L-arginine and L-cysteine and induction of T cell apoptosis [117]. MDSCs further promote neo-angiogenesis in the presence of ischemia through the release of VEGF and EGF2 as well as activation of STAT3. They can also promote dissemination via the release of MMP 9. Therefore, targeting the molecular pathways involved in the activation of various components of TME might be promising for inhibition of tumorigenesis, invasion, metastasis, and recurrence of ovarian cancer [118].

(e) *Extracellular vesicles* released from ovarian cancer stem cells form a part of "premetastatic niche" and helps in communication with other components of the microenvironment for example stromal cells and extracellular matrix via its biological content such as lipids, proteins, ds-DNA, mRNA, and micro RNA [119]. This exosome model explains the metastatic role of ovarian cancer stem cells in transportation of biological material including CD44 along the bloodstream to recipient cells of distant organs. Exosomes act as vehicles for transferring the miR222-3p to the macrophages giving rise to Tumor-associated macrophages (TAM). Moreover, exosomes containing miR-21, miR-103, miR-205, miR-200 are linked to adverse outcomes in OC patients. Having its immense diagnostic and therapeutic potential, an extracellular vesicle is considered as one of the recently investigated target for the treatment of refractory ovarian cancers [119].

10.7 Epithelial to Mesenchymal Transition (EMT)

Following downregulation of factors responsible for intercellular adhesion such as E-cadherin, Epcam, occludin, claudin and upregulation of vimentin, fibronectin, and MMPs, epithelial cells reversibly convert to mesenchymal cells having spindle shape. This process is known as EMT which passes through a more aggressive intermediate dynamic stage having both the properties of epithelial and mesenchymal cells [120]. EMT is a characteristic feature for embryogenesis, wound healing, and tumor progression. This process makes the cancer cell mobile and acts as a driving force for maintaining the stemness of the cancer cells thus resulting in the migration, invasion, and resistance to chemotherapy and immunotherapy. The controlling transcription factors involved in EMT include zinc-finger E-box-binding homeobox factors Zeb1 and Zeb2, Snail (SNAI1), Slug (SNAI2), Twist1, and Twist2 [121]. Various experimental studies documented that chemoresistant cancer cells acquire the features of mesenchymal cells, indicating the implication of EMT in refractoriness to therapy [122]. EMT could induce chemoresistance by altered expression of class III beta tubulin, increase in drug efflux via overexpression of ATP-binding cassette transporters [123], accentuating DNA repair mechanism by Sirtuin6-mediated activation of poly

ADP ribose polymerase (PARP) enzyme [124] attenuation of p-53 regulated apoptosis and modulation of the expression of various microRNA (miR200b represses [125] and miR20a induces EMT [126]). EMT causes immune-resistance through upregulation of programmed cell death-ligand 1(PD-L1) [127]. Hence, EMT could be a potential target to regain the sensitivity to chemo and immunotherapy. Atezolizumab and Bevacizumab (anti-PDL1 and Anti-VEGF) therapy might improve sensitivity symbiotically to cisplatin targeting EMT via attenuation of STAT3 phosphorylation [128]. Recent studies on administration of TWIST-targeted siRNA and miR15a and miR-16 in the form of nanoparticles could alleviate the drug resistance in an experimental animal model [129].

Since it is challenging to target the effector molecules of EMT, inhibitors targeting various metabolic pathways involved in EMT would be promising. Hence, recently various clinical trials are on-going to repurpose various metabolic inhibitors such as phosphodiesterase 4 (PDE4) inhibitor—Rolipram, 3-hydroxy-3-methylglutaryl-coenzyme A (HMG-CoA) reductase inhibitor—Simvastatin, Heparinase-inhibitor—Suramin for prevention of cancer progression in combination with already established standard therapy [130].

10.8 Targeted Therapy for OCSC

Even after satisfactory response to traditional chemotherapy, almost 70% of ovarian cancer patients return within 5 years with features suggestive of recurrence and chemoresistance. Hence, in order to improve the survival rate of advanced ovarian cancers, there is a critical need to find out novel therapeutic approaches to target specific molecular pathways and their complex interplay responsible for carcinogenesis. This is known as targeted therapy having lesser toxic effects compared to conventional chemotherapy which can also affect normal dividing cells because of its DNA damaging effects. Potential therapeutic targets have been depicted in Fig. 10.2. Very recent preclinical study including patient-derived xenograft model documented the significant role of Axitinib, tyrosine kinase inhibitor, in inhibition of tumor growth via modulation of VEGFR signaling pathway indicating the crucial role of angiogenesis along with overexpression of VEGF in ovarian cancer progression [133]. Hence, targeting angiogenesis could be one effective mode of therapy for the prevention of disease recurrence.

(a) *Angiogenesis Inhibitor*

Four double-blind, placebo-controlled phase III trials were conducted on chemotherapy with or without monoclonal VEGF antibody, Bevacizumab. In (GOG-0218) [134] and International Standard Randomised Controlled Trial, (ISRCTN91273375) [135] treatment was given to newly diagnosed advanced ovarian cancer patients, while in OCEANS [136] and AURELIA [137], treatment was given to platinum-sensitive and resistant recurrent epithelial ovarian cancer cases, respectively. Another, VEGF-independent angiogenesis pathway targeted, double-blind phase III trial (TRINOVA-1) conducted on trebananib which inhibits binding of angiopoietin to its receptor Tie2 for recurrent ovarian cancer reported a significant improvement in progression-free survival [138]. Various phase II/III clinical trials on multiple tyrosine kinase inhibitors such as Pazopanib, Nintedanib (*BIBF 1120*), Cediranib, Sunitinib targeting VEGF receptors, PDGF receptors, C-Tyrosine kinase, and FMS-like tyrosine kinase-3 (c-KIT) in combination showed promising results [139].

(b) *Poly-adenosine-diphosphate-ribose-polymerase (PARP) Inhibitor*

PARP inhibitors convert single-strand DNA damage to double-strand break in ovarian cancer patients with a mutation in BRCA 1/2 tumor suppressor protein which contributes to double-strand DNA repair and thus produces synthetic lethality to the cells. Various phase III clinical trials as maintenance therapy of PARP inhibitors either alone or in combination such as SOLO1 (Olaparib), PRIMA (Niraparib), PAOLA

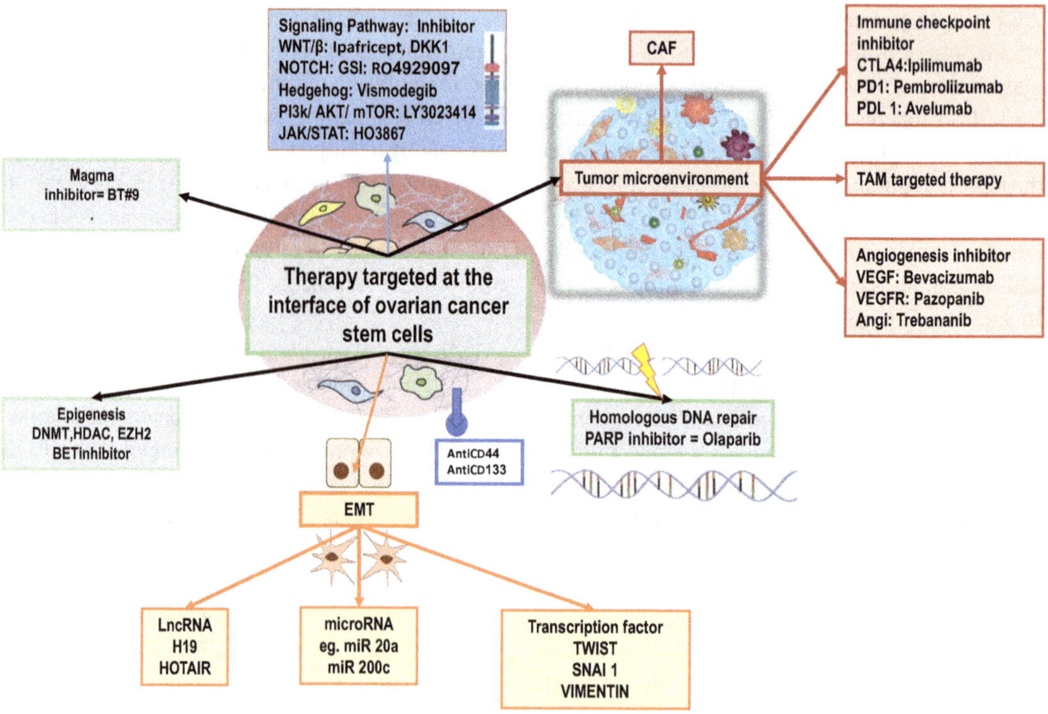

Fig. 10.2 Potential targets for the elimination of ovarian cancer stem cells. *CAF* cancer-associated fibroblast, *VEGF* vascular endothelial growth factor, *VEGFR* VEGF receptor, *EMT* epithelial to mesenchymal transition, *Magma* mitochondrial associated granulocyte macrophage colony stimulating factor, *LncRNA* long noncoding RNA, *EZH2* enhancer of zeste homolog 2, *TAM* tumor-associated macrophage, *PD1* programmed death 1, *PDL1* programmed death-1 (PD-1) ligand1, *PARP* poly (ADP-ribose) polymerase 1, *DKK1* Dickopf 1, *GSI* gamma secretase inhibitor. (Sources: [6, 52, 57, 110, 131, 132])

(Olaparib + bevacizumab) for newly diagnosed advanced ovarian cancer [140] and SOLO2/ENGOT-Ov21 on Olaparib as maintenance therapy for BRCA mutant relapsed ovarian cancer patients showed significant improvement in progression-free survival (PFS) [141].

(c) *Targeting Underlying Signaling Mechanisms for OCSC*

Cumulating evidence suggested that various signaling pathways such as Wnt/β, Hedgehog, Notch, JAK-STAT play an important role in the proliferation of ovarian CSCs and initiation of metastasis. Hence, various clinical studies on signaling pathway blockers are in process to validate their clinical implications in the elimination of the OCSCs and in turn prevention of recurrence.

A phase I, dose-escalation trial (NCT01608867) on Ipafricept a recombinant fusion protein targeting Wnt signaling pathway, was found to have better tolerance in combination with conventional chemotherapy for advanced ovarian cancer patients [142]. The interplay between Notch and Wnt-β catenin signaling pathways is attributed to tumor growth through the survival of CSCs. Enoticumab (REGN421), a Delta-like Ligand 4 (Dll4) monoclonal antibody was documented to be a safe drug in phase I, human study (NCT00871559) conducted on advanced ovarian cancer patients [143]. Various phase II clinical trials on mTOR inhibitor Temsirolimus either alone or in combination with Bevacizumab were found to be reasonably tolerated in advanced ovarian cancer cases (https://clinicaltrials.gov/ct2/home).

(d) *Immune Checkpoint Inhibitors for OCSC*

Considering the sensible role of immune checkpoint pathways in maintaining stem related features of ovarian CSCs, various phase I/II clinical trials were conducted on immune checkpoint inhibitors such as anti-PDL1 [Avelumab (NCT01772004) and Atezolizumab (NCT01375842)], anti-PD-1 Pembrolizumab (NCT02674061) for recurrent, advanced ovarian cancer cases. However, preliminary results obtained were far from being satisfactory. In order to improve the outcome, various phase III clinical trials (NCT03598270, NCT02891824, NCT02659384) are on-going as a combination of immune checkpoint inhibitor and PARP inhibitor and/or anti-VEGF drugs [144]. Preliminary results are awaited.

(e) *Epigenetic Therapy for Annihilation of OCSCs*

Dysregulation of epigenesis contributes to the survival and gain in plasticity resulting in the development of metastatic features of ovarian CSCs. Guadecitabine (SGI-110) effectively helps in the differentiation of ALDH+ OCSCs and thus regained the chemosensitivity in ovarian cancer cell lines [145]. Moreover, Histone deacetylase inhibitors could restore the differentiation of epithelial cells via attenuation of gene expression of HIF-1α, Notch-1, and STAT3 [146]. Similarly, Bromodomain and Extra-terminal inhibitor JQ1 could inhibit tumorigenesis by suppression of ALDH activity [147]. These preclinical studies emphasize the role of epigenetic reprogramming in OCSC aggressiveness and implicate as OCSCs differentiation strategy.

Gening et al. [148] documented that circulating long noncoding RNA (LncRNA), MALAT and HOTAIR, were significantly associated with recurrence-free periods in ovarian cancer patients indicating their role in the prognosis of the disease. Silencing of LncRNA, LINC00152, could regress tumor growth via modulation of miR-125b-mediated mitochondrial apoptosis [149] as well as regain the sensitivity towards Cisplatin in ovarian cancer cell line. Moreover, LncRNA HotairM1 attributes to a critical role in maintaining the stemness properties of CSCs through downstream effector HOXA1-Nanog [150]. These studies provide the evidence of a cross-talk between epigenetic regulation via LncRNA and ovarian cancer progression indicating the role of LncRNA as a potential therapeutic target along with conventional therapy to prevent recurrence in ovarian cancer through the elimination of CSCs.

10.9 Concluding Remarks/Foresights

It is surmised hence that a comprehensive understanding of various cellular events, types, and dynamics of cellular functions and activities of the tumor bulk cells as well as CSCs within a tumor in context of patient tumor grade, oncotherapy administrated to the patient, and the interaction of cancer cells and CSCs with each other, and that of a tumor as an entity with its surrounding microenvironmental components are very pertinent aspects that require to be understood in greater details to further develop effective therapies targeting CSCs. Collectively, the most recent developments of clinical trials (vividly covered in this chapter) reflect remarkable improvements on this front. Nevertheless worth noting are the dismal setbacks which rather serve as newer milestones/targets to be achieved by cancer researchers and oncologists alike, to improvise on the translational front thus providing greater hopes for better clinical management of patients which remains the ultimate goal.

References

1. Lheureux, S., Gourley, C., Vergote, I., & Oza, A. M. (2019). Epithelial ovarian cancer. *The Lancet, 393*(10177), 1240–1253.
2. World ovarian cancer coalition. (2020). Retrieved from https://worldovariancancercoalition.org/.
3. Key Statistics for Ovarian Cancer. American Cancer Society. (2020). Retrieved from https://www.cancer.org/cancer/ovarian-cancer/about/key-statistics.html.

4. Motohara, T., & Katabuchi, H. (2019). Ovarian cancer stemness: Biological and clinical implications for metastasis and chemotherapy resistance. *Cancers, 11*(7), 907.
5. Stewart, C., Ralyea, C., & Lockwood, S. (2019). Ovarian cancer: An integrated review. *Seminars in Oncology Nursing, 35*(2), 151–156.
6. Ayob, A. Z., & Ramasamy, T. S. (2018). Cancer stem cells as key drivers of tumour progression. *Journal of Biomedical Science, 25*(1), 1–18.
7. Zuber, E., Schweitzer, D., Allen, D., Parte, S., & Kakar, S. S. (2020). Stem cells in ovarian cancer and potential therapies. *Proceedings of Stem Cell Research and Oncogenesis, 8*, e1001.
8. Pieterse, Z., Amaya-Padilla, M. A., Singomat, T., Binju, M., Madjid, B. D., Yu, Y., et al. (2019). Ovarian cancer stem cells and their role in drug resistance. *The International Journal of Biochemistry & Cell Biology, 106*, 117–126.
9. Zhang, Y., Roos, M., Himburg, H., Termini, C. M., Quarmyne, M., Li, M., et al. (2019). PTPσ inhibitors promote hematopoietic stem cell regeneration. *Nature Communications, 10*(1), 1–15.
10. Prasetyanti, P. R., & Medema, J. P. (2017). Intra-tumor heterogeneity from a cancer stem cell perspective. *Molecular Cancer, 16*(1), 1–9.
11. Meacham, C. E., & Morrison, S. J. (2013). Tumour heterogeneity and cancer cell plasticity. *Nature, 501*(7467), 328–337.
12. Melo, F. D. S. E., Vermeulen, L., Fessler, E., & Medema, J. P. (2013). Cancer heterogeneity—A multifaceted view. *EMBO Reports, 14*(8), 686–695.
13. Sasaki, R., Narisawa-Saito, M., Yugawa, T., Fujita, M., Tashiro, H., Katabuchi, H., et al. (2009). Oncogenic transformation of human ovarian surface epithelial cells with defined cellular oncogenes. *Carcinogenesis, 30*(3), 423–431.
14. Kakar, S., & Jennes, L. (1999). Molecular cloning and characterization of the tumor transforming gene (TUTR1): A novel gene in human tumorigenesis. *Cytogenetic and Genome Research, 84*(3–4), 211–216.
15. Puri, R., Tousson, A., Chen, L., & Kakar, S. S. (2001). Molecular cloning of pituitary tumor transforming gene 1 from ovarian tumors and its expression in tumors. *Cancer Letters, 163*(1), 131–139.
16. Hamid, T., Malik, M. T., & Kakar, S. S. (2005). Ectopic expression of PTTG1/securin promotes tumorigenesis in human embryonic kidney cells. *Molecular Cancer, 4*(1), 3.
17. Parte, S., Virant-Klun, I., Patankar, M., Batra, S. K., Straughn, A., & Kakar, S. S. (2019). PTTG1: A unique regulator of stem/cancer stem cells in the ovary and ovarian cancer. *Stem Cell Reviews and Reports, 15*(6), 866–879.
18. Berebichez-Fridman, R., & Montero-Olvera, P. R. (2018). Sources and clinical applications of mesenchymal stem cells: State-of-the-art review. *Sultan Qaboos University Medical Journal, 18*(3), e264.
19. Evans, M. J., & Kaufman, M. H. (1981). Establishment in culture of pluripotential cells from mouse embryos. *Nature, 292*(5819), 154–156.
20. Martin, G. R. (1981). Isolation of a pluripotent cell line from early mouse embryos cultured in medium conditioned by teratocarcinoma stem cells. *Proceedings of the National Academy of Sciences, 78*(12), 7634–7638.
21. Thomson, J. A., Itskovitz-Eldor, J., Shapiro, S. S., Waknitz, M. A., Swiergiel, J. J., Marshall, V. S., et al. (1998). Embryonic stem cell lines derived from human blastocysts. *Science, 282*(5391), 1145–1147.
22. Takahashi, K., & Yamanaka, S. (2006). Induction of pluripotent stem cells from mouse embryonic and adult fibroblast cultures by defined factors. *Cell, 126*(4), 663–676.
23. Shibata, M., & Hoque, M. O. (2019). Targeting cancer stem cells: A strategy for effective eradication of cancer. *Cancers, 11*(5), 732.
24. Konrad, C. V., Murali, R., Varghese, B. A., & Nair, R. (2017). The role of cancer stem cells in tumor heterogeneity and resistance to therapy. *Canadian Journal of Physiology and Pharmacology, 95*(1), 1–15.
25. Miranda-Lorenzo, I., Dorado, J., Lonardo, E., Alcala, S., Serrano, A. G., Clausell-Tormos, J., et al. (2014). Intracellular autofluorescence: A biomarker for epithelial cancer stem cells. *Nature Methods, 11*(11), 1161.
26. Szotek, P. P., Chang, H. L., Brennand, K., Fujino, A., Pieretti-Vanmarcke, R., Celso, C. L., et al. (2008). Normal ovarian surface epithelial label-retaining cells exhibit stem/progenitor cell characteristics. *Proceedings of the National Academy of Sciences, 105*(34), 12469–12473.
27. Hu, L., McArthur, C., & Jaffe, R. (2010). Ovarian cancer stem-like side-population cells are tumourigenic and chemoresistant. *British Journal of Cancer, 102*(8), 1276–1283.
28. Choi, Y.-L., Kim, S.-H., Shin, Y. K., Hong, Y.-C., Lee, S.-J., Kang, S. Y., et al. (2005). Cytoplasmic CD24 expression in advanced ovarian serous borderline tumors. *Gynecologic Oncology, 97*(2), 379–386.
29. Nakamura, K., Terai, Y., Tanabe, A., Ono, Y. J., Hayashi, M., Maeda, K., et al. (2017). CD24 expression is a marker for predicting clinical outcome and regulates the epithelial-mesenchymal transition in ovarian cancer via both the Akt and ERK pathways. *Oncology Reports, 37*(6), 3189–3200.
30. Meng, E., Long, B., Sullivan, P., McClellan, S., Finan, M. A., Reed, E., et al. (2012). CD44+/CD24− ovarian cancer cells demonstrate cancer stem cell properties and correlate to survival. *Clinical & Experimental Metastasis, 29*(8), 939–948.
31. Raspollini, M., Amunni, G., Villanucci, A., Baroni, G., Taddei, A., & Taddei, G. (2004). c-KIT expression and correlation with chemotherapy resistance in ovarian carcinoma: An immunocytochemical study. *Annals of Oncology, 15*(4), 594–597.

32. Baba, T., Convery, P., Matsumura, N., Whitaker, R., Kondoh, E., Perry, T., et al. (2009). Epigenetic regulation of CD133 and tumorigenicity of CD133+ ovarian cancer cells. *Oncogene, 28*(2), 209–218.
33. Landen, C. N., Goodman, B., Katre, A. A., Steg, A. D., Nick, A. M., Stone, R. L., et al. (2010). Targeting aldehyde dehydrogenase cancer stem cells in ovarian cancer. *Molecular Cancer Therapeutics, 9*(12), 3186–3199.
34. Zhang, S., Balch, C., Chan, M. W., Lai, H.-C., Matei, D., Schilder, J. M., et al. (2008). Identification and characterization of ovarian cancer-initiating cells from primary human tumors. *Cancer Research, 68*(11), 4311–4320.
35. Kusumbe, A. P., & Bapat, S. A. (2009). Cancer stem cells and aneuploid populations within developing tumors are the major determinants of tumor dormancy. *Cancer Research, 69*(24), 9245–9253.
36. Alvero, A. B., Chen, R., Fu, H.-H., Montagna, M., Schwartz, P. E., Rutherford, T., et al. (2009). Molecular phenotyping of human ovarian cancer stem cells unravels the mechanisms for repair and chemoresistance. *Cell Cycle, 8*(1), 158–166.
37. Alvero, A. B., Fu, H. H., Holmberg, J., Visintin, I., Mor, L., Marquina, C. C., et al. (2009). Stem-like ovarian cancer cells can serve as tumor vascular progenitors. *Stem Cells, 27*(10), 2405–2413.
38. Zhang, J., Yuan, B., Zhang, H., & Li, H. (2019). Human epithelial ovarian cancer cells expressing CD105, CD44 and CD106 surface markers exhibit increased invasive capacity and drug resistance. *Oncology Letters, 17*(6), 5351–5360.
39. Tayama, S., Motohara, T., Narantuya, D., Li, C., Fujimoto, K., Sakaguchi, I., et al. (2017). The impact of EpCAM expression on response to chemotherapy and clinical outcomes in patients with epithelial ovarian cancer. *Oncotarget, 8*(27), 44312.
40. Zheng, J., Zhao, S., Yu, X., Huang, S., & Liu, H. Y. (2017). Simultaneous targeting of CD44 and EpCAM with a bispecific aptamer effectively inhibits intraperitoneal ovarian cancer growth. *Theranostics, 7*(5), 1373.
41. Wen, Y., Hou, Y., Huang, Z., Cai, J., & Wang, Z. (2017). SOX 2 is required to maintain cancer stem cells in ovarian cancer. *Cancer Science, 108*(4), 719–731.
42. Zhang, H., Qiu, J., Ye, C., Yang, D., Gao, L., Su, Y., et al. (2014). ROR1 expression correlated with poor clinical outcome in human ovarian cancer. *Scientific Reports, 4*, 5811.
43. Lee, M., Nam, E. J., Kim, S. W., Kim, S., Kim, J. H., & Kim, Y. T. (2012). Prognostic impact of the cancer stem cell–related marker NANOG in ovarian serous carcinoma. *International Journal of Gynecologic Cancer, 22*(9), 1489–1496.
44. Ruan, Z., Yang, X., & Cheng, W. (2019). OCT4 accelerates tumorigenesis through activating JAK/STAT signaling in ovarian cancer side population cells. *Cancer Management and Research, 11*, 389.
45. Ning, Y.-X., Luo, X., Xu, M., Feng, X., & Wang, J. (2017). Let-7d increases ovarian cancer cell sensitivity to a genistein analog by targeting c-Myc. *Oncotarget, 8*(43), 74836.
46. Dou, J., Jiang, C., Wang, J., Zhang, X., Zhao, F., Hu, W., et al. (2011). Using ABCG2-molecule-expressing side population cells to identify cancer stem-like cells in a human ovarian cell line. *Cell Biology International, 35*(3), 227–234.
47. Schindler, A. J., Watanabe, A., & Howell, S. B. (2018). LGR5 and LGR6 in stem cell biology and ovarian cancer. *Oncotarget, 9*(1), 1346.
48. Ng, A., Tan, S., Singh, G., Rizk, P., Swathi, Y., Tan, T. Z., et al. (2014). Lgr5 marks stem/progenitor cells in ovary and tubal epithelia. *Nature Cell Biology, 16*(8), 745–757.
49. Hashimoto, H., Sudo, T., Mikami, Y., Otani, M., Takano, M., Tsuda, H., et al. (2008). Germ cell specific protein VASA is over-expressed in epithelial ovarian cancer and disrupts DNA damage-induced G2 checkpoint. *Gynecologic Oncology, 111*(2), 312–319.
50. Virant-Klun, I., Kenda-Suster, N., & Smrkolj, S. (2016). Small putative NANOG, SOX2, and SSEA-4-positive stem cells resembling very small embryonic-like stem cells in sections of ovarian tissue in patients with ovarian cancer. *Journal of Ovarian Research, 9*(1), 12.
51. Bapat, S. A. (2010). Human ovarian cancer stem cells. *Reproduction (Cambridge, England), 140*(1), 33.
52. Padilla, M. A. A., Binju, M., Wan, G., Rahmanto, Y. S., Kaur, P., & Yu, Y. (2019). Relationship between ovarian cancer stem cells, epithelial mesenchymal transition and tumour recurrence. *Cancer Drug Resistance, 2*, 1127–1135.
53. Nimmakayala, R. K., Batra, S. K., & Ponnusamy, M. P. (2019). Unraveling the journey of cancer stem cells from origin to metastasis. *Biochimica et Biophysica Acta (BBA)-Reviews on Cancer, 1871*(1), 50–63.
54. Tirino, V., Desiderio, V., Paino, F., De Rosa, A., Papaccio, F., Fazioli, F., et al. (2011). Human primary bone sarcomas contain CD133+ cancer stem cells displaying high tumorigenicity in vivo. *The FASEB Journal, 25*(6), 2022–2030.
55. Hanahan, D., & Weinberg, R. A. (2011). Hallmarks of cancer: THE next generation. *Cell, 144*(5), 646–674.
56. Moharil, R. B., Dive, A., Khandekar, S., & Bodhade, A. (2017). Cancer stem cells: An insight. *Journal of Oral and Maxillofacial Pathology, 21*(3), 463.
57. Abildgaard, C., Do Canto, L. M., Steffensen, K. D., & Rogatto, S. R. (2019). Long non-coding RNAs involved in resistance to chemotherapy in ovarian cancer. *Frontiers in Oncology, 9*, 1549.
58. Papaccio, F., Paino, F., Regad, T., Papaccio, G., Desiderio, V., & Tirino, V. (2017). Concise review: Cancer cells, cancer stem cells, and mesenchymal

stem cells: Influence in cancer development. *Stem Cells Translational Medicine, 6*(12), 2115–2125.
59. Brabletz, T., Jung, A., Spaderna, S., Hlubek, F., & Kirchner, T. (2005). Migrating cancer stem cells—An integrated concept of malignant tumour progression. *Nature Reviews. Cancer, 5*(9), 744–749.
60. Diaz-Cano, S. J. (2012). Tumor heterogeneity: Mechanisms and bases for a reliable application of molecular marker design. *International Journal of Molecular Sciences, 13*(2), 1951–2011.
61. Koźlik, M., & Wójcicki, P. (2014). The use of stem cells in plastic and reconstructive surgery. *Advances in Clinical and Experimental Medicine, 23*(6), 1011–1017.
62. Giuseppe Longo, U., Rizzello, G., Berton, A., Ciuffreda, M., Migliorini, F., Khan, W. S., et al. (2013). Potential of adipose derived stem cells in orthopaedic surgery. *Current Stem Cell Research & Therapy, 8*(6), 418–421.
63. Gimble, J. M., Bunnell, B. A., & Guilak, F. (2012). Human adipose-derived cells: An update on the transition to clinical translation. *Regenerative Medicine, 7*(2), 225–235.
64. Pan, Y., Ma, S., Cao, K., Zhou, S., Zhao, A., Li, M., et al. (2018). Therapeutic approaches targeting cancer stem cells. *Journal of Cancer Research and Therapeutics, 14*(7), 1469.
65. Rahman, M., Deleyrolle, L., Vedam-Mai, V., Azari, H., Abd-El-Barr, M., & Reynolds, B. A. (2011). The cancer stem cell hypothesis: Failures and pitfalls. *Neurosurgery, 68*(2), 531–545.
66. O'Flaherty, J. D., Barr, M., Fennell, D., Richard, D., Reynolds, J., O'Leary, J., et al. (2012). The cancer stem-cell hypothesis: Its emerging role in lung cancer biology and its relevance for future therapy. *Journal of Thoracic Oncology, 7*(12), 1880–1890.
67. Ponnusamy, M. P., & Batra, S. K. (2008). Ovarian cancer: Emerging concept on cancer stem cells. *Journal of Ovarian Research, 1*(1), 4.
68. Nelson, H. D., Westhoff, C., Piepert, J., & Berg, A. (2004). *Screening for ovarian cancer: Brief evidence update*. Agency for Healthcare Research and Quality: Rockville, MD.
69. Roett, M. A., & Evans, P. (2009). Ovarian cancer: An overview. *American Family Physician, 80*(6), 609–616.
70. Pal, T., Permuth-Wey, J., Betts, J. A., Krischer, J. P., Fiorica, J., Arango, H., et al. (2005). BRCA1 and BRCA2 mutations account for a large proportion of ovarian carcinoma cases. *Cancer: Interdisciplinary International Journal of the American Cancer Society, 104*(12), 2807–2816.
71. Wooster, R., Bignell, G., Lancaster, J., Swift, S., Seal, S., Mangion, J., et al. (1995). Identification of the breast cancer susceptibility gene BRCA2. *Nature, 378*(6559), 789–792.
72. Struewing, J. P., Hartge, P., Wacholder, S., Baker, S. M., Berlin, M., McAdams, M., et al. (1997). The risk of cancer associated with specific mutations of BRCA1 and BRCA2 among Ashkenazi Jews. *New England Journal of Medicine, 336*(20), 1401–1408.
73. Moslehi, R., Chu, W., Karlan, B., Fishman, D., Risch, H., Fields, A., et al. (2000). BRCA1 and BRCA2 mutation analysis of 208 Ashkenazi Jewish women with ovarian cancer. *The American Journal of Human Genetics, 66*(4), 1259–1272.
74. Risch, H. A., McLaughlin, J. R., Cole, D. E., Rosen, B., Bradley, L., Kwan, E., et al. (2001). Prevalence and penetrance of germline BRCA1 and BRCA2 mutations in a population series of 649 women with ovarian cancer. *The American Journal of Human Genetics, 68*(3), 700–710.
75. Chen, S., & Parmigiani, G. (2007). Meta-analysis of BRCA1 and BRCA2 penetrance. *Journal of Clinical Oncology, 25*(11), 1329.
76. Evans, D., Young, K., Bulman, M., Shenton, A., Wallace, A., & Lalloo, F. (2008). Probability of BRCA1/2 mutation varies with ovarian histology: Results from screening 442 ovarian cancer families. *Clinical Genetics, 73*(4), 338–345.
77. Bolton, K. L., Chenevix-Trench, G., Goh, C., Sadetzki, S., Ramus, S. J., Karlan, B. Y., et al. (2012). Association between BRCA1 and BRCA2 mutations and survival in women with invasive epithelial ovarian cancer. *Journal of the American Medical Association, 307*(4), 382–389.
78. Ben David, Y., Chetrit, A., Hirsh-Yechezkel, G., Friedman, E., Beck, B., Beller, U., et al. (2002). Effect of BRCA mutations on the length of survival in epithelial ovarian tumors. *Journal of Clinical Oncology, 20*(2), 463–466.
79. Cass, I., Baldwin, R. L., Varkey, T., Moslehi, R., Narod, S. A., & Karlan, B. Y. (2003). Improved survival in women with BRCA-associated ovarian carcinoma. *Cancer: Interdisciplinary International Journal of the American Cancer Society, 97*(9), 2187–2195.
80. Meindl, A., Hellebrand, H., Wiek, C., Erven, V., Wappenschmidt, B., Niederacher, D., et al. (2010). Germline mutations in breast and ovarian cancer pedigrees establish RAD51C as a human cancer susceptibility gene. *Nature Genetics, 42*(5), 410–414.
81. Osorio, A., Endt, D., Fernández, F., Eirich, K., de la Hoya, M., Schmutzler, R., et al. (2012). Predominance of pathogenic missense variants in the RAD51C gene occurring in breast and ovarian cancer families. *Human Molecular Genetics, 21*(13), 2889–2898.
82. Pelttari, L. M., Heikkinen, T., Thompson, D., Kallioniemi, A., Schleutker, J., Holli, K., et al. (2011). RAD51C is a susceptibility gene for ovarian cancer. *Human Molecular Genetics, 20*(16), 3278–3288.
83. Loveday, C., Turnbull, C., Ramsay, E., Hughes, D., Ruark, E., Frankum, J. R., et al. (2011). Germline mutations in RAD51D confer susceptibility to ovarian cancer. *Nature Genetics, 43*(9), 879–882.
84. Ahmed, N., Kadife, E., Raza, A., Short, M., Jubinsky, P. T., & Kannourakis, G. (2020). Ovarian cancer,

Cancer stem cells and current treatment strategies: A potential role of magmas in the current treatment methods. *Cell, 9*(3), 719.
85. Virant-Klun, I., & Stimpfel, M. (2016). Novel population of small tumour-initiating stem cells in the ovaries of women with borderline ovarian cancer. *Scientific Reports, 6*, 34730.
86. Visvader, J. E., & Lindeman, G. J. (2008). Cancer stem cells in solid tumours: Accumulating evidence and unresolved questions. *Nature Reviews. Cancer, 8*(10), 755–768.
87. Agro, L., & O'Brien, C. (2015). In vitro and in vivo limiting dilution assay for colorectal cancer. *Bio-Protocol, 5*(22), 1.
88. Bapat, S. A., Mali, A. M., Koppikar, C. B., & Kurrey, N. K. (2005). Stem and progenitor-like cells contribute to the aggressive behavior of human epithelial ovarian cancer. *Cancer Research, 65*(8), 3025–3029.
89. Luo, L., Zeng, J., Liang, B., Zhao, Z., Sun, L., Cao, D., et al. (2011). Ovarian cancer cells with the CD117 phenotype are highly tumorigenic and are related to chemotherapy outcome. *Experimental and Molecular Pathology, 91*(2), 596–602.
90. Štemberger-Papić, S., Vrdoljak-Mozetič, D., Verša Ostojić, D., Rubeša-Mihaljević, R., Krištofić, I., Brnčić-Fischer, A., et al. (2015). Expression of CD133 and CD117 in 64 serous ovarian cancer cases. *Collegium Antropologicum, 39*(3), 745–753.
91. Conic, I., Stanojevic, Z., Jankovic Velickovic, L., Stojnev, S., Ristic Petrovic, A., Krstic, M., et al. (2015). Epithelial ovarian cancer with CD117 phenotype is highly aggressive and resistant to chemotherapy. *Journal of Obstetrics and Gynaecology Research, 41*(10), 1630–1637.
92. Curley, M. D., Therrien, V. A., Cummings, C. L., Sergent, P. A., Koulouris, C. R., Friel, A. M., et al. (2009). CD133 expression defines a tumor initiating cell population in primary human ovarian cancer. *Stem Cells, 27*(12), 2875–2883.
93. Skubitz, A. P., Taras, E. P., Boylan, K. L., Waldron, N. N., Oh, S., Panoskaltsis-Mortari, A., et al. (2013). Targeting CD133 in an in vivo ovarian cancer model reduces ovarian cancer progression. *Gynecologic Oncology, 130*(3), 579–587.
94. Zhang, J., Guo, X., Chang, D. Y., Rosen, D. G., Mercado-Uribe, I., & Liu, J. (2012). CD133 expression associated with poor prognosis in ovarian cancer. *Modern Pathology, 25*(3), 456–464.
95. Burgos-Ojeda, D., Rueda, B. R., & Buckanovich, R. J. (2012). Ovarian cancer stem cell markers: Prognostic and therapeutic implications. *Cancer Letters, 322*(1), 1–7.
96. Jaggupilli, A., & Elkord, E. (2012). Significance of CD44 and CD24 as cancer stem cell markers: An enduring ambiguity. *Clinical and Developmental Immunology, 2012*, 708036.
97. Wang, Y., Shao, F., & Chen, L. (2018). ALDH1A2 suppresses epithelial ovarian cancer cell proliferation and migration by downregulating STAT3. *OncoTargets and Therapy, 11*, 599.
98. Wang, Y.-C., Yo, Y.-T., Lee, H.-Y., Liao, Y.-P., Chao, T.-K., Su, P.-H., et al. (2012). ALDH1-bright epithelial ovarian cancer cells are associated with CD44 expression, drug resistance, and poor clinical outcome. *The American Journal of Pathology, 180*(3), 1159–1169.
99. Januchowski, R., Wojtowicz, K., Sterzyſska, K., Sosiſska, P., Andrzejewska, M., Zawierucha, P., et al. (2016). Inhibition of ALDH1A1 activity decreases expression of drug transporters and reduces chemotherapy resistance in ovarian cancer cell lines. *The International Journal of Biochemistry & Cell Biology, 78*, 248–259.
100. Ng, P. M.-L., & Lufkin, T. (2011). Embryonic stem cells: Protein interaction networks. *Biomolecular Concepts, 2*(1–2), 13–25.
101. Szotek, P. P., Pieretti-Vanmarcke, R., Masiakos, P. T., Dinulescu, D. M., Connolly, D., Foster, R., et al. (2006). Ovarian cancer side population defines cells with stem cell-like characteristics and Mullerian inhibiting substance responsiveness. *Proceedings of the National Academy of Sciences, 103*(30), 11154–11159.
102. Vathipadiekal, V., Saxena, D., Mok, S. C., Hauschka, P. V., Ozbun, L., & Birrer, M. J. (2012). Identification of a potential ovarian cancer stem cell gene expression profile from advanced stage papillary serous ovarian cancer. *PLoS One, 7*(1), e29079.
103. Parte, S. C., Smolenkov, A., Batra, S. K., Ratajczak, M. Z., & Kakar, S. S. (2017). Ovarian cancer stem cells: Unraveling a germline connection. *Stem Cells and Development, 26*(24), 1781–1803.
104. Parte, S. C., Batra, S. K., & Kakar, S. S. (2018). Characterization of stem cell and cancer stem cell populations in ovary and ovarian tumors. *Journal of Ovarian Research, 11*(1), 69.
105. Udoh, K., Parte, S., Carter, K., Mack, A., & Kakar, S. S. (2019). Targeting of lung cancer stem cell self-renewal pathway by a small molecule verrucarin. *Journal of Stem Cell Reviews and Reports, 15*(4), 601–611.
106. Carter, K., Rameshwar, P., Ratajczak, M. Z., & Kakar, S. S. (2017). Verrucarin J inhibits ovarian cancer and targets cancer stem cells. *Oncotarget, 8*(54), 92743.
107. Kakar, S. S., Parte, S., Kelsey Carter, I. G. J., Worth, C., Rameshwar, P., & Ratajczak, M. Z. (2017). Withaferin A (WFA) inhibits tumor growth and metastasis by targeting ovarian cancer stem cells. *Oncotarget, 8*(43), 74494.
108. Kakar, S. S., Worth, C. A., Wang, Z., Carter, K., Ratajczak, M., & Gunjal, P. (2016). DOXIL when combined with Withaferin A (WFA) targets ALDH1 positive cancer stem cells in ovarian cancer. *Journal of Cancer Stem Cell Research, 4*, e1002.
109. Kakar, S. S., Ratajczak, M. Z., Powell, K. S., Moghadamfalahi, M., Miller, D. M., Batra, S. K., et al. (2014). Withaferin a alone and in combination with cisplatin suppresses growth and metastasis

of ovarian cancer by targeting putative cancer stem cells. *PLoS One, 9*(9), e107596.

110. Yang, Y., Yang, Y., Yang, J., Zhao, X., & Wei, X. (2020). Tumor microenvironment in ovarian cancer: Function and therapeutic strategy. *Frontiers in Cell and Developmental Biology, 8*, 758.
111. Yu, Y., Xiao, C., Tan, L., Wang, Q., Li, X., & Feng, Y. (2014). Cancer-associated fibroblasts induce epithelial–mesenchymal transition of breast cancer cells through paracrine TGF-β signalling. *British Journal of Cancer, 110*(3), 724–732.
112. Barrett, R., & Puré, E. (2020). Cancer-associated fibroblasts: Key determinants of tumor immunity and immunotherapy. *Current Opinion in Immunology, 64*, 80–87.
113. Givel, A.-M., Kieffer, Y., Scholer-Dahirel, A., Sirven, P., Cardon, M., Pelon, F., et al. (2018). miR200-regulated CXCL12β promotes fibroblast heterogeneity and immunosuppression in ovarian cancers. *Nature Communications, 9*(1), 1–20.
114. Ahmed, Z., & Bicknell, R. (2009). Angiogenic signalling pathways. In C. Murray & S. Martin (Eds.), *Angiogenesis protocols* (pp. 3–24). New York: Springer.
115. Apte, R. S., Chen, D. S., & Ferrara, N. (2019). VEGF in signaling and disease: Beyond discovery and development. *Cell, 176*(6), 1248–1264.
116. Ghosn, E. E. B., Cassado, A. A., Govoni, G. R., Fukuhara, T., Yang, Y., Monack, D. M., et al. (2010). Two physically, functionally, and developmentally distinct peritoneal macrophage subsets. *Proceedings of the National Academy of Sciences, 107*(6), 2568–2573.
117. Atretkhany, K.-S., & Drutskaya, M. (2016). Myeloid-derived suppressor cells and proinflammatory cytokines as targets for cancer therapy. *Biochemistry (Moscow), 81*(11), 1274–1283.
118. Bruno, A., Mortara, L., Baci, D., Noonan, D. M., & Albini, A. (2019). Myeloid derived suppressor cells interactions with natural killer cells and pro-angiogenic activities: Roles in tumor progression. *Frontiers in Immunology, 10*, 771.
119. Cheng, L., Wu, S., Zhang, K., & Xu, T. (2017). A comprehensive overview of exosomes in ovarian cancer: Emerging biomarkers and therapeutic strategies. *Journal of Ovarian Research, 10*(1), 1–9.
120. Dongre, A., & Weinberg, R. A. (2019). New insights into the mechanisms of epithelial–mesenchymal transition and implications for cancer. *Nature Reviews. Molecular Cell Biology, 20*(2), 69–84.
121. Goossens, S., Vandamme, N., Van Vlierberghe, P., & Berx, G. (2017). EMT transcription factors in cancer development re-evaluated: Beyond EMT and MET. *Biochimica et Biophysica Acta (BBA)-Reviews on Cancer, 1868*(2), 584–591.
122. Loret, N., Denys, H., Tummers, P., & Berx, G. (2019). The role of epithelial-to-mesenchymal plasticity in ovarian cancer progression and therapy resistance. *Cancers, 11*(6), 838.
123. Saxena, M., Stephens, M. A., Pathak, H., & Rangarajan, A. (2011). Transcription factors that mediate epithelial–mesenchymal transition lead to multidrug resistance by upregulating ABC transporters. *Cell Death & Disease, 2*(7), e179.
124. Bae, J. S., Noh, S. J., Kim, K. M., Park, S.-H., Hussein, U. K., Park, H. S., et al. (2018). SIRT6 is involved in the progression of ovarian carcinomas via β-catenin-mediated epithelial to mesenchymal transition. *Frontiers in Oncology, 8*, 538.
125. Mezencev, R., & Wartell, R. (2018). Cisplatin binds to pre-miR-200b and impairs its processing to mature microRNA. *Neoplasma, 65*(2), 222–227.
126. Liu, Y., Han, S., Li, Y., Liu, Y., Zhang, D., Li, Y., et al. (2017). MicroRNA-20a contributes to cisplatin-resistance and migration of OVCAR3 ovarian cancer cell line. *Oncology Letters, 14*(2), 1780–1786.
127. Alsuliman, A., Colak, D., Al-Harazi, O., Fitwi, H., Tulbah, A., Al-Tweigeri, T., et al. (2015). Bidirectional crosstalk between PD-L1 expression and epithelial to mesenchymal transition: Significance in claudin-low breast cancer cells. *Molecular Cancer, 14*(1), 149.
128. Zhang, L., Chen, Y., Li, F., Bao, L., & Liu, W. (2019). Atezolizumab and bevacizumab attenuate cisplatin resistant ovarian cancer cells progression synergistically via suppressing epithelial-mesenchymal transition. *Frontiers in Immunology, 10*, 867.
129. Shahin, S. A., Wang, R., Simargi, S. I., Contreras, A., Echavarria, L. P., Qu, L., et al. (2018). Hyaluronic acid conjugated nanoparticle delivery of siRNA against TWIST reduces tumor burden and enhances sensitivity to cisplatin in ovarian cancer. *Nanomedicine: Nanotechnology, Biology and Medicine, 14*(4), 1381–1394.
130. Ramesh, V., Brabletz, T., & Ceppi, P. (2020). Targeting EMT in cancer with repurposed metabolic inhibitors. *Trends in Cancer, 6*(11), 942–950.
131. Liang, R., Chen, X., Chen, L., Wan, F., Chen, K., Sun, Y., et al. (2020). STAT3 signaling in ovarian cancer: A potential therapeutic target. *Journal of Cancer, 11*(4), 837.
132. Zong, X., & Nephew, K. P. (2019). Ovarian cancer stem cells: Role in metastasis and opportunity for therapeutic targeting. *Cancers, 11*(7), 934.
133. Paik, E. S., Kim, T.-H., Cho, Y. J., Ryu, J., Choi, J.-J., Lee, Y.-Y., et al. (2020). Preclinical assessment of the VeGfR inhibitor axitinib as a therapeutic agent for epithelial ovarian cancer. *Scientific Reports, 10*(1), 1–9.
134. Burger, R. A., Brady, M. F., Bookman, M. A., Fleming, G. F., Monk, B. J., Huang, H., et al. (2011). Incorporation of bevacizumab in the primary treatment of ovarian cancer. *New England Journal of Medicine, 365*(26), 2473–2483.
135. Oza, A. M., Cook, A. D., Pfisterer, J., Embleton, A., Ledermann, J. A., Pujade-Lauraine, E., et al. (2015). Standard chemotherapy with or without bevacizumab for women with newly diagnosed ovarian

cancer (ICON7): Overall survival results of a phase 3 randomised trial. *The Lancet Oncology, 16*(8), 928–936.

136. Aghajanian, C., Blank, S. V., Goff, B. A., Judson, P. L., Teneriello, M. G., Husain, A., et al. (2012). OCEANS: A randomized, double-blind, placebo-controlled phase III trial of chemotherapy with or without bevacizumab in patients with platinum-sensitive recurrent epithelial ovarian, primary peritoneal, or fallopian tube cancer. *Journal of Clinical Oncology, 30*(17), 2039.
137. Pujade-Lauraine, E., Hilpert, F., Weber, B., Reuss, A., Poveda, A., Kristensen, G., et al. (2014). Bevacizumab combined with chemotherapy for platinum-resistant recurrent ovarian cancer: The AURELIA open-label randomized phase III trial. *Obstetrical & Gynecological Survey, 69*(7), 402–404.
138. Monk, B. J., Poveda, A., Vergote, I., Raspagliesi, F., Fujiwara, K., Bae, D.-S., et al. (2015). Impact of trebananib plus weekly paclitaxel on overall survival (OS) in patients (pts) with recurrent ovarian cancer and ascites: Results from the phase III TRINOVA-1 study. *Journal of Clinical Oncology, 33*(15).
139. Diab, Y., & Muallem, M. Z. (2017). Targeted therapy in ovarian cancer. A comprehensive systematic review of literature. *Anticancer Research, 37*(6), 2809–2815.
140. Franzese, E., Diana, A., Centonze, S., Pignata, S., De Vita, F., Ciardiello, F., et al. (2020). PARP inhibitors in first-line therapy of ovarian cancer: Are there any doubts? *Frontiers in Oncology, 10*. https://doi.org/10.3389/fonc.2020.00782.
141. Pujade-Lauraine, E., Ledermann, J. A., Selle, F., Gebski, V., Penson, R. T., Oza, A. M., et al. (2017). Olaparib tablets as maintenance therapy in patients with platinum-sensitive, relapsed ovarian cancer and a BRCA1/2 mutation (SOLO2/ENGOT-Ov21): A double-blind, randomised, placebo-controlled, phase 3 trial. *The Lancet Oncology, 18*(9), 1274–1284.
142. Jimeno, A., Gordon, M., Chugh, R., Messersmith, W., Mendelson, D., Dupont, J., et al. (2017). A first-in-human phase I study of the anticancer stem cell agent ipafricept (OMP-54F28), a decoy receptor for Wnt ligands, in patients with advanced solid tumors. *Clinical Cancer Research, 23*(24), 7490–7497.
143. Chiorean, E. G., LoRusso, P., Strother, R. M., Diamond, J. R., Younger, A., Messersmith, W. A., et al. (2015). A phase I first-in-human study of enoticumab (REGN421), a fully human delta-like ligand 4 (Dll4) monoclonal antibody in patients with advanced solid tumors. *Clinical Cancer Research, 21*(12), 2695–2703.
144. Borella, F., Ghisoni, E., Giannone, G., Cosma, S., Benedetto, C., Valabrega, G., et al. (2020). Immune checkpoint inhibitors in epithelial ovarian cancer: An overview on efficacy and future perspectives. *Diagnostics, 10*(3), 146.
145. Wang, Y., Cardenas, H., Fang, F., Condello, S., Taverna, P., Segar, M., et al. (2014). Epigenetic targeting of ovarian cancer stem cells. *Cancer Research, 74*(17), 4922–4936.
146. Lin, P.-C., Hsieh, H.-Y., Chu, P.-C., & Chen, C. S. (2018). Therapeutic opportunities of targeting histone deacetylase isoforms to eradicate cancer stem cells. *International Journal of Molecular Sciences, 19*(7), 1939.
147. Stathis, A., & Bertoni, F. (2018). BET proteins as targets for anticancer treatment. *Cancer Discovery, 8*(1), 24–36.
148. Gening, S., Dolgova, D., Abakumova, T., Rizvanov, A., & Antoneeva, I. (2019). 29P Expression profiles of serum long non-coding RNA in ovarian cancer patients receiving platinum-containing chemotherapy. *Annals of Oncology, 30*(Suppl_11), mdz447.027.
149. Chen, P., Fang, X., Xia, B., Zhao, Y., Li, Q., & Wu, X. (2018). Long noncoding RNA LINC00152 promotes cell proliferation through competitively binding endogenous miR-125b with MCL-1 by regulating mitochondrial apoptosis pathways in ovarian cancer. *Cancer Medicine, 7*(9), 4530–4541.
150. Li, F., Xu, Y., Xu, X., Ge, S., Zhang, F., Zhang, H., et al. (2020). lncRNA HotairM1 depletion promotes self-renewal of cancer stem cells through HOXA1-Nanog regulation loop. *Molecular Therapy-Nucleic Acids, 22*, 456–470.

Index

H. Schatten (ed.), *Ovarian Cancer: Molecular & Diagnostic Imaging and Treatment Strategies*, Advances in Experimental Medicine and Biology 1330, https://doi.org/10.1007/978-3-030-73359-9

CPSIA information can be obtained
at www.ICGtesting.com
Printed in the USA
BVHW011552090821
614009BV00003BA/49

9 783030 733582